The Archive
and the
Aural City

Sign, Storage, Transmission
A series edited by JONATHAN STERNE AND LISA GITELMAN

The Archive
and the
Aural City

Sound, Knowledge, and the Politics of Listening

ALEJANDRO L. MADRID

Duke University Press   *Durham and London*   2025

© 2025 DUKE UNIVERSITY PRESS
This work is licensed under a Creative Commons Attribution-
NonCommercial-NoDerivatives 4.0 International License, available
at https://creativecommons.org/licenses/by-nc-nd/4.0/.
Printed in the United States of America on acid-free paper ∞
Project Editor: Livia Tenzer
Designed by Matthew Tauch
Typeset in Arno Pro and Quadraat Sans by
Westchester Publishing Services

Library of Congress Cataloging-in-Publication Data
Names: Madrid, Alejandro L. author
Title: The archive and the aural city : sound, knowledge, and the
politics of listening / Alejandro L. Madrid.
Other titles: Sign, storage, transmission
Description: Durham : Duke University Press, 2025. | Series:
Sign, storage, transmission | Includes bibliographical references
and index.
Identifiers: LCCN 2025004911 (print)
LCCN 2025004912 (ebook)
ISBN 9781478032113 paperback
ISBN 9781478028864 hardcover
ISBN 9781478061083 ebook
ISBN 9781478094425 ebook other
Subjects: LCSH: Music—Mexico—Archival resources | Music—
Latin America—Archival resources | Music—Political aspects—
Mexico | Music—Political aspects—Latin America | Sound
recordings in ethnology—Mexico | Sound recordings in
ethnology—Latin America | Sound archives—Mexico—History |
Sound archives—Latin America—History
Classification: LCC ML210.9 .M337 2025 (print) | LCC ML210.9
(ebook) | DDC 780.74—dc23/eng/20250603
LC record available at https://lccn.loc.gov/2025004911
LC ebook record available at https://lccn.loc.gov/2025004912

Cover art: Claudia Padilla Peña, *Inspiración # 3*. Used by permission
of Claudia Padilla Peña.

To Ekaterina and Marina, my reasons

In loving memory of Jonathan Sterne, to whom this book owes so much

Quic oc tlamati noyollo:
nic caqui in cuicatl,
nic itta in xochitl.
Maca in cuetlahuia in tlalticpac!

Finally, my heart understands:
I hear a chant,
I contemplate a flower.
I hope they won't wither!
    —Nezahualcoyotl

Contents

List of Illustrations xi
List of Abbreviations xv
Acknowledgments xvii

Introduction. Questions About the Circulation of
Knowledge at the Sonic Turn 1

1. Performative Listening, Writing, Reading,
and the Assemblage of Archival Constellations 29

2. Patrimony, Objectification, and Representation
at Mexico's Fonoteca Nacional 57

3. *Critical Constellations of the Audio-Machine in
Mexico* and the Performativity of Archiving/Archival Labor 85

4. Things, Sound Objects, and Legacy at the Berliner
Phonogramm-Archiv's Konrad T. Preuss Collection 117

5. Mexican Rarities, *Disco pirata*, and the Promise
of a Sound Archive of Postnational Memory 161

6. Aurality, Materiality, and the Carrillo Pianos as Archives 191

7. In Search of the Aural City: Collective Action
and the Invisible Sound Archive 227

Epilogue. The Relevance of Archives in Times of Post-Truth:
An Essay Against Nihilism in the Neoliberal Age 270

Notes 285
Bibliography 315
Index 341

Illustrations

1.1	Album cover of *A Program of Mexican Music* (1941). 31
1.2	Drawings showing the measurements of two teponaztlis. 46
1.3	Castañeda and Mendoza's acoustic theory of the teponaztli. 46
1.4	Transduction: inscription of sound through light with the vitaphone. 52
2.1	Official iconography of the project México Suena Así. 65
2.2	Screenshot of the México Suena Así website map. 66
2.3	CD covers for *Paisaje sonoro de Veracruz* (2009) and *Paisaje sonoro de San Luis Potosí* (2009). 67
2.4	Official iconography advertising the Pueblos Mágicos program. 76
2.5	Official map of Pueblos Mágicos. 76
3.1	Program of the *Critical Constellations of the Audio-Machine in Mexico* (CCAMM) exhibit in Berlin. 89
3.2	Flyer for *Constelaciones críticas de la audiomáquina*. 94
3.3	Preparing the CCAMM exhibit in Berlin. 97
3.4	Félix Blume's *Memoria del hierro*. 108
3.5	Map of the CCAMM exhibit in Berlin. 109

3.6	Use of space at the CCAMM exhibit in Cuernavaca. 112
3.7	Picture from Félix Blume's series *Mientras escucho*. 115
4.1	A group of musicians from the Orquesta Experimental de Instrumentos Nativos. 119
4.2	Protesters outside of the Humboldt Forum. 120
4.3	Part of the Berliner Phonogramm-Archiv exhibit featured at the Humboldt Forum. 121
4.4	Galvano copies from the Preuss Collection. 135
4.5	The Berliner Phonogramm-Archiv as Memory of the World. 138
4.6	*Walzenaufnahmen der Cora und Huichol aus Mexiko 1905–1907/Grabaciones en cilindros de cera de los coras y los huicholes de México* (2013). 144
4.7	*La expedición al Nayarit* (2020). 145
5.1	Screenshot of the home page of the Mexican Rarities website. 169
5.2	Partial views of the Mexican Rarities physical archive. 177
5.3	Félix Blume, *Disco pirata* (2016). 183
6.1	Plan of the upright third-tone Carrillo Piano. 198
6.2	Third-tone Carrillo Piano made by Federico Buschmann. 199
6.3	Plan of the sixteenth-tone Carrillo Piano. 201
6.4	The Carrillo Pianos at Expo 58 in Brussels. 202
6.5	Poster for the Palais des Beaux-Arts concert. 204
6.6	Cover of the booklet for the Carrillo Pianos 1959 exhibit in Mexico City. 207
6.7	Julián Carrillo, *Balbuceos* (1958), for sixteenth-tone piano and orchestra, mm. 1–15. 213

6.8	Julián Carrillo, *Estudios* (1959), for fifth-tone piano, mm. 1–14. 215
6.9	Juan Felipe Waller, *Lhorong, 31°N 96°E* (2011–15), mm. 73–90. 218
6.10	Carl Sauter Pianofabrik's information about the sixteenth-tone Carrillo Piano. 222
6.11	"Los Estragos del Sonido 13." Newspaper cartoon of Julián Carrillo destroying a piano. 226
7.1	Image accompanying Híbridas y Quimeras' *Compílame'sta* (2019). 241
7.2	Screenshot of the RAM "Manifestx" as archived on the collective's website. 244
7.3	Screenshot of the home page of MUSEXPLAT's website. 246
7.4	Screenshot of the home page of the *Islas Resonantes* archive website. 248
7.5	Screenshot of the home page of the *Bulla* archive website. 249
7.6	Screenshot of the home page of the Rancho Electrónico website. 252
7.7	Screenshot of the home page of the PoéticaSonoraMX website. 257
7.8 and 7.9	Radial and rhizomatic visualizations of a partial network of the Mexican Aural City as rendered visible in the archive(s) articulated in this chapter. 268–69

Abbreviations

ABRDP	Acta betreffend die Reise des Dr. Preuss nach Amerika vom 17 August 1905 bis 22 August 1913
AMNH	American Museum of Natural History
BPA	Berliner Phonogramm-Archiv
CBS	Columbia Broadcasting System
CCAMM	*Critical Constellations of the Audio-Machine in Mexico*
CCWAHF	Coalition of Cultural Workers Against the Humboldt Forum
CENART	Centro Nacional de las Artes
CEOS	Centro Experimental Oído Salvaje
CNM	Conservatorio Nacional de Música
EDM	electronic dance music
ENAH	Escuela Nacional de Antropología e Historia
ET	equal temperament
EVO	Expo Vinylo Oaxaca
FAFL-UNAM	Facultad de Filosofía y Letras, UNAM
FAM-UNAM	Facultad de Música, UNAM
FONAPAS	Fondo Nacional para Actividades Sociales
GDR	German Democratic Republic

GEXLAT	Generx Experimentación Latinoamérica
INAH	Instituto Nacional de Antropología e Historia
INALI	Instituto Nacional de Lenguas Indígenas
INBA	Instituto Nacional de Bellas Artes
KMV	Königliches Museum für Völkerkunde
MoMA	Museum of Modern Art
MUAC	Museo Universitario de Arte Contemporáneo
MUC	Museo Universitario del Chopo
MUSEXPLAT	Música Experimental Latinoamericana
NPR	National Public Radio
OEIN	Orquesta Experimental de Instrumentos Nativos
OFFAL	Orchestra for Females and Laptops
RAM	Redes Autónomas de Memoria
RDA	Repositorio Digital en Audio
RPM	revolutions per minute
12-TET	twelve-tone equal temperament
UNAM	Universidad Nacional Autónoma de México

Acknowledgments

This book was in the making for longer than ten years. That period has afforded me time to think, question my ideas, and rethink my conclusions many times. It has also allowed me to cross paths and share my work with many scholars whose work I admire and whose opinions I highly respect. Needless to say, the concepts in the final book bear little resemblance to the ideas I had when I started writing it, and much of that is due to the many illuminating conversations I have had the privilege to partake in with these colleagues. In 2013 I was invited to offer one of the first graduate seminars in sound studies taught in Mexico, at the then Escuela Nacional de Música of the Universidad Nacional Autónoma de México. My conversations with the students and colleagues attending that seminar, as well as a class visit to the Fonoteca Nacional, were crucial in establishing the foundation for the development of this project. I am grateful to Roberto Kolb Neuhaus for the invitation and to Gonzalo Camacho, Marcela García López, William Herrera, Rossana Lara Velázquez, Lénica Reyes, Omar Soriano, and Cristina Tamariz for the rich discussions that made this class a memorable experience. Special thanks to Omar Soriano for creating the Sound Studies Mexico Facebook group, which allowed for the conversations started in the seminar to continue in the virtual world. I would also like to thank the staff at the Fonoteca Nacional for welcoming me and my students, and to Francisco "Tito" Rivas and Perla Olivia Rodríguez Reséndiz for candidly talking to me about their work at that institution.

On October 25, 2014, I hosted Mapping Sound and Urban Space in the Americas, a daylong conference at Cornell University that offered an ideal space for a meeting of sound scholars and sound artists. Thanks to Steve Pond and Chris Riley for making the event happen, and thanks to Kemi Adeyemi, Marcelo Armani, Leonardo Cardoso, Nina Eidsheim, José Luis Fernández, Josh Kun, Rossana Lara Velázquez, Tom McEnaney, Ramón

Rivera-Servera, Luz María Sánchez, Jennifer Stoever, and Alexandra Vazquez for the invaluable exchange of ideas, as well as my former Cornell colleagues, María Fernández and Karen Jaime, for moderating the sessions.

The Unit for Criticism and Interpretive Theory at the University of Illinois Urbana-Champaign invited me to present a preliminary lecture on my work about the Fonoteca Nacional's soundscapes project in March 2016. I am thankful to Susan Koshy, Mike Silvers, Gabriel Solis, Jessica Hajek, and Roman Friedman for their encouraging comments, and to Marc Adam Hertzman for his insightful response to my talk. I also presented that lecture at the University of Texas (UT) at Austin's Butler School of Music, the Facultad de Artes at the Pontificia Universidad Católica de Chile in Santiago de Chile, and the Music Department at Brown University. Thanks to Luis Achondo and the ethnomusicology graduate students at Brown University for the invitation to present my research at their musicology colloquium. At UT Austin, I would like to thank Carlos Dávalos and the Association of Graduate Ethnomusicology and Musicology Students for organizing the event, as well as Robin Moore, Veit Erlmann, and their students for their warm reception. I want to express my gratitude to Daniel Party for sponsoring my visit to the Pontificia Universidad Católica de Chile, as well as to the colleagues who contributed to the conversation after I presented this talk; among them, Juan Pablo González, Laura Jordán, Miguel Angel Marín, Juan Francisco Sans, Rodrigo Torres, Mauricio Valdebenito, and Alejandro Vera.

On February 25, 2020, I presented a lecture entitled "The Politics of Distinction and Representation in the Aural Turn: Who Gets to Listen in the 'Sounded' City?" It was my last live talk before the long hiatus forced on us by the COVID-19 pandemic. This was the keynote presentation of the symposium Transnational Literary and Sonic Culture across the Americas and the Iberian Peninsula, organized by Duke University's Romance Studies Department. I am grateful to Elia Romera Figueroa, Silvia Serrano, and Marcelo Noah for the invitation and to Silvia Bermúdez, Alejandra Bronfman, David Garcia, and Tom McEnaney for the illuminating conversation.

I would like to thank Elisabeth Le Guin, Farrah O'Shea, and Pheaross Graham for the invitation to participate in the panel "Methods of Analysis in Music Performance Studies" as part of the conference Music and Performance Studies Today at the University of California, Los Angeles (UCLA). Their invitation was the perfect excuse to return to a research project that had been dormant for several years. Thanks to the panelists, Amy Bauer, Roshanak Kheshti, and Richard Pettengill, for their comments and suggestions, as well as Mitchell Morris for his thought-provoking response to the panel. "Making an Archive and Listening to It: The Performativity of Archiving/Archival Labor," an early version of chapter 3 in this book and a revised version of the UCLA lecture, was subsequently presented at the Conférences de Prestige Series of the Université de Montréal's Faculté de Musique and Cornell University's Department of Romance Studies. At the Université de Montréal, I want to thank Jonathan Goldman for the invitation and Jean-Jacques Nattiez for engaging with my ideas. At Cornell University, I am indebted to Edmundo Paz Soldán for the invitation and to the students and colleagues who participated in the intense and in-depth discussion that ensued after the presentation, especially Irina Troconis, Marianthi Papalexandri-Alexandri, Edmundo Paz Soldán, Re'ee Hagay, Dani Hawkins, Alex Nik Pasqualini, Brian Sengdala, Mark Mahoney, and Thomas Cressy. Another version of chapter 3 was also presented as a research lecture at Harvard University's Department of Music. I am particularly grateful to Anne Shreffler, Kay Kaufman Shelemay, Davindar Singh, and Richard Wolf for their insightful comments, questions, and suggestions.

I owe a debt of gratitude to Miguel Olmos Aguilera for inviting me to present my research about Mexico's Fonoteca Nacional as the keynote lecture for the Coloquio Paisaje Sonoro, Música, Ruidos y Sonidos de las Fronteras at El Colegio de la Frontera Norte-Tijuana. This was a rare, unusual, and very welcome opportunity to share my work with Mexican colleagues. Thanks to Gonzalo Camacho, Igael González, and Xilonen Luna as well as the other conference participants for their questions and comments. I am also deeply thankful to Esteban Buch and Vera Wolkowicz for inviting me to present my research about the Preuss Collection as the keynote lecture for the Conference Musique, Politiques Culturelles et Identités: Rencontres Transatlantiques entre la Musique et les Musiciens d'Europe et d'Amerique Latine at L'École des Hautes Études en Sciences Sociales, Paris, France. I am particularly indebted to Andrés Amado, Jacky Avila, Gisela Cánepa Koch, Heidi Feldman, Daniela Fugellie, Lucía Patiño Mayer, Walther Maradiegue, Violeta Nigro Giunta, Raúl Renato Romero, Belén Vega Pichaco, and Alejandro Vera for the elucidating conversations at this conference.

Thanks to Teresa Cascudo for the invitation to present a paper that would become the core of chapter 1 as the inaugural lecture of the Seminario Permanente de Historiografía of the Doctorado Interdisciplinario hosted by the Universidad de La Rioja, Universidad de Valladolid, and Universidad

Complutense de Madrid. A revised version of this presentation was delivered at the Irna Priore Music and Culture Lecture Series of the University of North Carolina Greensboro. My gratitude to Joan Titus for the invitation and to her colleagues, especially David Aarons, Lorena Guillén, and Elizabeth Keathley, as well as their students, for their questions and comments.

My deep appreciation goes to Sydney Hutchinson and Stefanie Alisch for inviting me to present my research about the Carrillo Pianos at the 54th BEAM Meeting: Berlin Ethnomusicology and the Anthropology of Music Research Group, Humboldt-Universität zu Berlin. Thanks to Barbara Alge and Julio Mendívil for inviting me to present my research about the Preuss Collection at the Goethe-Universität Frankfurt am Main and the Universität Wien respectively. In Frankfurt, I am also indebted to Christina Richter-Ibáñez and Lisa-Maria Brusius, and in Vienna to Juan Bermúdez, Natalia Neira Nieto, and Martin Ringsmut, for their comments and suggestions.

I presented early drafts of chapter 6 at the quinquennial conference of the International Musicological Society (IMS) in Athens in August 2022 and the joint meeting of the American Musicological Society (AMS), the Society for Music Theory (SMT), and the Society for Ethnomusicology (SEM) in New Orleans in November 2022. I truly appreciate the comments and feedback from the participants in those panels, especially Ana Alonso-Minutti (chair), Julin Lee, Ayako Kosaka, and Sebastián Zubieta in Athens; and Theodore Gordon (chair), Laura Tunbridge, and Annie Garlid in New Orleans. I also presented an abbreviated version of chapter 4 at the 2023 meetings of the SEM in Ottawa, Canada, and the AMS in Denver. I truly appreciate the feedback from the panelists, especially Judith Klassen (chair), Beatriz Goubert, and Edwin Porras in Ottawa; and Sergio Ospina-Romero (chair, panel organizer, and participant) and Melissa Camp in Denver.

At the Centro Julián Carrillo in San Luis Potosí, I am particularly thankful to Iván Sánchez Martínez and Alejandra Sánchez for processing and providing images of the design plans for the Carrillo Pianos. I am especially grateful to Eduardo Morones Hernández for granting me access to the unpublished research on these instruments written by his late brother, Mario Morones Hernández. I also wish to take a moment to thank Carmen Carrillo de Viramontes and the Carrillo family for their continuous support of my research about Julián Carrillo.

Special thanks to my graduate students at Cornell University, Rachel Horner, Cibele Moura, and Rafael Torralvo, for bouncing back ideas with me during the sessions of a seminar called Politics, Utopia, and Noise in the Sound Archive, which I taught in spring 2022. I am particularly grateful to María Edurne Zuazu and Valzhyna Mort for visiting the class and providing wonderful opportunities to discuss their work about sound art and poetry, respectively, in relation to our own ideas about archives. Likewise, I am grateful to my graduate students at Harvard University, Alyssa Cottle, Ben Gregson, Luis Pabón Rico, María Alejandra Privado, Eloy Ramirez, Shiva Ramkumar, Lucas Reccitelli, and Sunday Ukaewen, for the insightful and equally delightful conversations during a later version of the same seminar, in which they all read and commented on the complete first draft of this book.

Many other individual friends and colleagues contributed to this book in countless manners, including discussing my ideas, sharing their work with me, or pointing me toward specific scholarly and artistic networks. To all of them I owe a great debt of gratitude. They include Lizette Alegre González, Chelsea Burns, Brigid Cohen, Rodrigo de la Mora Pérez Arce, Pablo Dodero Carrillo, Ana Lidia Domínguez, Jorge David García, Veit Erlmann, Susana González Aktories, Glenda Goodman, Carlos Hernández, Travis Johns, Ana Mora Flores, Marianthi Papalexandri-Alexandri, Benjamin Piekut, Ekaterina Pirozhenko, Pepe Rojo, Silvia Spitta, Gary Tomlinson, Paulina Velázquez, and Daniel Walden. I am especially grateful to friends and colleagues who have provided generous and detailed comments and feedback on specific chapters of the book at various stages; they include Ana Alonso-Minutti, Andrea Ancira, Kjetil Klette Bøhler, Thomas Cressy, Eric Johns, Rossana Lara Velázquez, Tamara Levitz, Julio Mendívil, Sergio Ospina-Romero, Carlos Prieto Acevedo, Jesús Ramos-Kittrell, Emilio Ros-Fábregas, Kay Kaufman Shelemay, Margarita Valdovinos, Juan Fernando Velásquez, and María Edurne Zuazu. I remind the reader that I am solely responsible for any errors and oversights throughout the book.

Writing letters of recommendation is always an onerous task, especially when the requests arrive at particularly busy moments of the academic year. I am extremely grateful to Suzanne Cusick, Josh Kun, Carol Oja, and Jonathan Sterne for writing letters on my behalf through the process of applying for grants and funding to complete this project. Also, many thanks to my former Cornell colleagues Andrew Hicks, Simone Pinet, and David Yearsley for sharing their insights about successfully applying for a variety of sought-after yearlong fellowships.

This book would not have been possible without the generous support of the Simon Guggenheim Memorial Foundation. A Guggenheim

Fellowship granted me the privilege of dedicating a whole year to nothing but doing research in Berlin and writing the first draft of this book. In Berlin, my deep appreciation goes to Maurice Mengel, Sydney Hutchinson, and Albrecht Wiedmann for facilitating my research at the Berliner Phonogramm-Archiv. Thanks to Manuela Fischer at the Ethnologisches Museum, Staatliche Museen zu Berlin, for giving me prompt access to all the documents and materials about Konrad T. Preuss that I requested. I am uniquely indebted to Dörte Schmidt and Norbert Palz for their support in securing a postdoctoral visiting appointment at the Universität der Künste Berlin, which made my yearlong stay in Berlin possible. At the Ibero-Amerikanisches Institut, I am very grateful to Barbara Göbel and the institute's efficient and accommodating library staff. I was very fortunate that early during my stay in Berlin, the Ibero-Amerikanisches Institut sponsored the workshop Medialidades de la Convivialidad: Archivos, Registros Sonoros y Nuevas Dinámicas de Acceso y Circulación. This symposium allowed me to informally participate in enlightening conversations about sound archives with experts such as Miguel A. García, Matthias Lewy, Bernd Brabec de Mori, Gisela Cánepa Koch, Walther Maradiegue, and Rita Eloranta. Toward the end of the writing and publication process, the Alexander von Humboldt Stiftung honored me with a coveted Humboldt-Forschungspreis (Humboldt Research Award) for this book project. I am deeply thankful to Sebastian Klotz and the Humboldt-Universität zu Berlin for putting forward the nomination and supporting it. I would also like to thank Paul Fleming and Elke Siegel for their friendship and for allowing me and my family to stay in their fabulously vintage and cozy Berlin apartment for almost a year.

At Duke University Press, I am grateful to Jonathan Sterne and Lisa Gitelman, coeditors of the Sign, Storage, Transmission series; to Ken Wissoker, senior executive editor, and his assistant, Kate Mullen, for their support and sincere interest in this project; and to Livia Tenzer for her exceptional work as project editor of this book. I am also genuinely indebted to the anonymous reviewers who read and commented on the first draft of this book's manuscript very generously and in great detail. Their suggestions were essential for rethinking, revising, and tightening the final version of the book.

I am very fortunate to have serendipitously found the perfect artwork for the cover of this book in the work of my dear friend, the superb Mexican artist Claudia Padilla Peña. Thanks to her for permission to use *Inspiración 3* on the cover of this book. I am deeply honored.

This book is available in an Open Access edition thanks to funding from Harvard University's Publication Grant Office, Harvard's Department of Music, and a publication subvention from the General Fund and the Iberian and Latin American Fund of the American Musicological Society, supported in part by the National Endowment for the Humanities and the Andrew W. Mellon Foundation.

Introduction: Questions About the Circulation of Knowledge at the Sonic Turn

I suppose an archive gives you a kind of valley
in which your thoughts can bounce back to you, transformed.
You whisper intuitions and thoughts into the emptiness,
hoping to hear something back.
—Valeria Luiselli, *Lost Children Archive* (2019)

As a scholar, a question I often get when presenting my research is "What is your archive?" The question seldom refers to any physical repository safeguarding documents or materials, and my response rarely mentions those kinds of archival ventures. Instead, my answer usually highlights the collections of materials, embodied practices, archival constellations, and overall performance complexes that one gathers or concocts when working on a particular project but that are very rarely confined to the boundaries of one specific repository. Terry Cook conceptualizes the difference between these two formations by using the terms *archive* versus *archive(s)*. He argues that scholars engage an archive as a "metaphorical symbol, as a representation of identity, or as the recorded memory production of some person or group or culture," while *archive(s)* refers to the "history of the archive, from [its] initial creation or inscription to its appearance in the archival reference room, [and] the internal concepts and processes that animate actual archivists working inside real archives..., or of the distinct body of professional ideas and practices those archivists follow, or of the impact all this has on shaping both the surviving record and historical knowledge."[1] Cook's explanation does not simply differentiate between the archive as a scholarly formation and the archive(s) as a physical repository that is also legible as an embodiment of the histories and epistemes that provide it with cultural capital; by addressing the history behind the physical creation of a particular archival space, he alludes to labor as one of the most significant considerations in understanding the differences between these two types of archival entities. Achille Mbembe has also written about archives in the plural to refer to "a building, symbol of a public institution [as well as] a collection of documents...kept in this building," highlighting the labor involved in "convert[ing] a certain number of documents into items judged to be worthy of preserving and keeping in a public place, where they can be consulted according to well-established procedures and regulations."[2] Both Cook's and Mbembe's descriptions of the archive and the archive(s) take place within a larger interrogation of their apparent transparency. On the one hand, this critique is aimed at finding out how scholars mediate the information kept in these archives as they use it to support and develop larger narratives. On the other, it is also about rendering visible how, by appraising, selecting, curating, encoding, and classifying records, and by developing databases and implementing retrieval strategies, the archivist determines "what the future will know about its past: who will have a continuing voice and who will be silenced."[3] Labor in both cases—in the archiving and the retrieving, in the constitution of the archive's materiality and in the construction of a narrative spell based on the documents it stores—is the powerful performative agent that transforms documents into systems of information and normativity.

The Archive and the Aural City articulates both the archive and the archive(s) and pays special attention to the kinds of labor required for their construction as material and epistemic entities. Throughout this book, I work with a general assumption that archives are collections of objects or documents that allow for the development of certain interpretations of reality or the performance of reality. Rather than ascribing any type of ontological meaning to the documents archives preserve, I take as my point of departure that such materials acquire meaning only relationally and in tandem with the agency and labor of the individuals who engage them. However, I propose that archives also have a certain agency derived from their own design and their status as disciplining and validating institutions. This archival agency often guides how individuals engage with them and understand the documents stored within them. Thus, I see archives as systems that make information legible based on particular epistemic placeholders that, as Gary Tomlinson states, "can create specific systems of information that give rise to their own internal developmental tendencies and vectors, depending on the cognitive, bodily, and environmental constraints they involve."[4] In that sense, archives are often closed or circular systems; they are meant to reproduce themselves and the epistemic placeholders that keep them together. Within this paradigm, one may understand a wide variety of cultural practices as archives.

Memes, those fair-use remixed image-icons that circulate abundantly over the internet, are very good examples of this dynamic. They work because they refer to widely shared epistemic placeholders, including ideas about teleology, essentialism, identity, complexity, and so on. Thus, one could also consider memes as archives of the values that justify those epistemic placeholders. It is precisely the circular character of the meme as a cultural practice—the way in which they are validated by but also revalidate the values that make them work—that makes it a straightforward example of one of the quintessential paradoxes of archives: the gathering and classifying of information that continues to reproduce these epistemic placeholders; in turn, these conventions prevent the production of alternative and potentially more innovative forms of knowledge. Are these placeholders the seed of the self-destruction of the archive and the archive(s) or of the dissolution of the logic that makes them useful within specific paradigms of knowledge production? Cristina Rivera Garza argues in favor of this conceptualization when she states that "with its materiality on its back, the archive frequently obstructs the linear activity of the narration, making its development problematic, raising questions that are precisely those of its own production."[5] By focusing on the way in which the archive(s)' materiality and its history tarnish the way in which the very narrative it is supposed to validate is produced, Rivera Garza seems to take aim at the archive(s)' circular logic and to infer that the key for its deconstruction lies within itself and the labor that makes it possible.

In consonance with Rivera Garza's "anarchival" dictum, the central goal of this book is to find out and explore ways to open those closed or circular systems in order to make the documents they store and the information they generate legible in productive ways that transcend the circularity of its epistemic dynamics.[6] In other words, this book focuses on how, as Kirsten

Weld proposes in the case of Guatemalan secret police archives, records produced for social control can be repurposed as "tools of social reckoning," and how "what matters most about such archives is not their supposedly depersonalized, abstract exercise of panoptical control but rather their use-value by real humans."[7] This book launches this exploration in the context of archives that store sound and the particular types of mediation that inform the production and circulation of knowledge in and about Mexico and Latin America at the sonic turn in the humanities and social sciences. I also argue that this sonic turn has motivated the development of an Aural City (*Ciudad Aural*), an urban intellectual elite that seeks to reevaluate prevalent visuo-centric and logocentric ideas about understanding and representing the world from a locus provided by sound and listening as a type of epistemic labor. Thus, *The Archive and the Aural City* sits at the intersection of archives, archival labor, and aurality to question the viability of an epistemic project that, by attending to the power of listening to perform and sound alternative types of knowledge, may be able to bypass the epistemic and political shortcomings of what Ángel Rama termed the Lettered City (*Ciudad Letrada*).[8]

The development of new sound recording technologies at the end of the nineteenth century revolutionized musicological and ethnographic research and the possibilities of documenting, storing, copying, and circulating music, speech, and sound. The establishment of the first sound archives soon after slowly allowed for an unprecedented access to musical practices from around the globe and the development of new forms of knowledge in relation to and framed by nation- and empire-building projects. A new type of relation between sonic practices and listening individuals began to develop out of the schizophonic mediations that modern sound objects—such as the wax cylinder—entailed.[9] Based on the notion that archives are never stable nor complete and that they only speak through practices of interaction, this book interrogates the work of traditional sound archives but also contributes to expanding the notion of what a sound archive could be and do.

The Archival Turn

Historians have a methodologically critical relationship with the evidence they find in the archives. As Martin Johnes explains, historians always "read between the lines, ... examine the way a source says things and consider its relationship to wider social, cultural and political contexts."[10]

Nonetheless, the interrogation of the archive initiated by Michel Foucault and Jacques Derrida in the late 1960s through the mid-1990s brought about a radical reconceptualization of the archive that had profound consequences, not only for and among historians, but also for the humanities and social sciences more generally. This move is central to the unfolding of what scholars in a wide variety of disciplines have termed the *archival turn*.[11] In his critique of structuralism, a philosophical system that neglects historical events in an effort to highlight the presumed structures, patterns, and dynamics underlying all of human activity, Foucault suggested that although "history has no 'meaning' . . . it is intelligible and should be susceptible of analysis down to the smallest detail—but this in accordance with the intelligibility of struggles, of strategies, and tactics."[12] Indeed, for him, rather than emphasizing the symbolic field or any signifying structure, critical inquiry should focus on the power relations that authenticate and render specific regimes of knowledge meaningful. As such, the archive(s) as an institution that aspires to a timeless totality is actually a reflection of the power relations that shape the regimes of knowledge that the archives aim to represent, become an icon of, and, in its attempt to render time immobile, end up reproducing.

In his continuation of this critique, Derrida also understands the archive as a place of violence. However, rather than focusing on it as an epistemic practice, he takes into account the materiality of the archives, stating that "there is no archive without a place of consignation, without a technique of repetition and without a certain exteriority," to contend that "the technical structure of the *archiving* archive also determines the structure of the archivable content even in its very coming into existence and in its relationship to the future."[13] For Derrida, the performative logic of repetition that provides the archive with its aura of memory bank and authenticity is intrinsically connected to the compulsive repetition in Sigmund Freud's death drive and thus to a sense of inevitable self-annihilation. Thus, the desire for origins that drives many into the archive(s) leads to a type of fetishization of its space, its holdings, and its sense of timeless transcendence that renders it discursively unproductive and invites oblivion rather than memory.

More than a challenge to the faith in the factual positivity of the materials held in archives, the radical character of Foucault's and Derrida's theoretical interventions lies in their questioning of the assumed stability of archives and the discursive regimes that archival labor engenders. It is in that sense that their ideas came to disrupt the work of traditional historians and their narrative assertions. This archival turn, the understanding of the archive and archive(s) as systems of rules and epistemic placeholders that inconspicuously regulate what one can and cannot say, has led historians to reflect on the discursive implications of their archival labor more consistently and systematically. If the archive and archive(s) tended to be phantasmatic presences in the work of historians, as Martin Johnes has argued, the archival turn not only rendered them visible but also placed them at the center of these scholars' intellectual conversations and made them subjects of study themselves.[14]

Through the first two decades of the twenty-first century, this reconceptualization of the archive and the archive(s) has gained traction in a wide variety of disciplines across the humanities and social sciences, making them into influential concepts and analytical tools. In this transdisciplinary reincarnation, the archive and the archive(s) have taken on broader meanings. This has made them focal points for innovative explorations of identity, belonging, memory, tradition, communication, regulation, subjectivity, borders, and so on, in many historical and cultural settings, from colonial encounters, nation-building efforts, and canonic formations to political protests, the curation of art exhibits, and the performance of expressive culture. Thus, besides their character as institutions, repositories, collections, storage spaces, information networks, and constellations, the archive and the archive(s) have also become metaphors to talk about the production, transmission, and circulation of knowledge within larger power dynamics in a wide variety of texts, practices, plots, scenarios, and objects.[15]

The Archive and the Aural City embraces both, the reconceptualization of the archive and archive(s) as well as the challenge to understand other cultural, material, and virtual formations as archives. Here, the archive works as an episteme and as a metaphor that, as Daniel Marshall and Zeb Tortorici allege, allows "material[s to get] turned into something else: evidence or loss, history or an inspiration to do history differently."[16] Highlighting this performativity of the archive and the archival labor behind its production and usage, the case studies in this book explore internet networks, musical instruments, museum exhibits, and books and the performance complexes around them, as well as institutional and alternative repositories, as material, epistemic, and metaphoric archives.

Coming back to the issue about defining one's archive with which I opened this introductory chapter may be a productive way to clarify the concerns that first got me interested in interrogating sound archives. For that matter, I reformulate the original question: What does it mean to ask someone what their archive is? What does it mean for someone to define their archive? The most normative answer to these questions would focus on outlining the borders, walls, or limits of an archive or archival project. This concern revolves around the possible identity of the archival formation we seek to study but also around the characteristics that bound together the materials and documents we seek to analyze and those we choose to exclude from that exploration. Like any other project about walls and borders, as much as this is an effort to preserve, protect, and shape certain ideas and knowledges, it is also, and foremost, a mechanism of control or, at the very least, a mechanism that creates the illusion of control. By enforcing what makes it in and what stays out of the archive, this mechanism and the gatekeepers in charge of it impose a certain sense of authority over the archives and their holdings. Once this dynamic is established, it is clear that defining an archive in such a way is all about control: control over the production of knowledge, affectivity, and order, or certain forms of order often assumed to be in the archive's nature. Nevertheless, rather than focusing on control and authority, a nonnormative answer to those questions may contemplate instead novel ways of emphasizing the agency of the documents and materials in the archive as objectifications of the power relations that make up the archive itself; such answer would also highlight how documentary ontologies are connected to these processes of objectification. That is, a nonnormative look at the archive would focus on "the capacity of documentary practices to make things come into being."[17] Here, defining the archive would be about giving up control and introducing a sense of chaos that could eventually help us deconstruct the epistemic placeholders that make the archive legible along with ideological commands that it also helps reproduce. In that sense, defining the archive in a nonnormative way would be all about trying to find ways to discover *lo inaudito* in the archive.

In Spanish, the term *inaudito* has two connotations. It may refer to something that is exceptional or unprecedented, or, more literally, it may refer to something that has not been heard. In the context of the sound archive projects this book embarks on, this double entendre is particularly fertile. It shifts the focus from finding the expected to encountering the surprising, from listening to what is there to figuring out ways to listen to the unheard, whether because it has been silenced, because it has been left out of the repository altogether, or because we have not learned how to listen to it or for it. For that reason, the central premise of this book is to find ways to access *lo inaudito*, to estrange the archive to listen to its contents from different perspectives and as part of new relational networks and constellations that make it reverberate in more productive ways, ways that transcend what its design and structure allow us to retrieve from it.

Doing archival work in Latin America is often a very different experience from doing archival work in the United States or Europe. While US and European institutions usually have the funding to preserve documents and materials and systematize access to them, the extreme financial precarity of many archival endeavors in Latin America determines a very different way of accessing documents and materials and retrieving information from these repositories. Early in my academic career I found it extremely frustrating to try to sort out my way through archives that not only could be uncataloged but also could be in such a state of chaos that sometimes finding what one was looking for could mean the discovery of holes in the archive, not only metaphorically but often quite literally: documents that were supposed to be there but were not because they may have been extracted from the archive or lost in the past, or documents whose inadequate preservation status made them impossible to handle without jeopardizing their very existence. Nevertheless, when some of these archives became institutionalized, organized, and systematically cataloged, I realized that there was something uniquely productive about figuring out how to locate and trace documents within the archive's former chaos that was lost with their institutionalization and regimentation. While the disciplining of those archives allowed for a more ordered, guided, and methodical transit through their holdings, their patrimonialization also presumed the privileging of certain documents and materials, the hiding of the archive's holes, and the advancing of certain narratives over others in the process of retrieving information. In other words, the organization and patrimonialization of these archives led to a certain restriction of their holdings' agency. The documents and materials were no longer able to point researchers toward the archive's silences and holes in the same way that figuring out paths through the chaos had allowed. In a way, the sense of exhilaration that I experienced when uncovering and hearing *lo inaudito* in those archives was lost when they became more efficient in guiding my navigation through their holdings in accordance to their specific predetermined placeholders. Thus, my exploration of strategies for the introduction of chaos, estrangement, and schizzes into archival systems in this book is informed by a desire to recapture that sense of exhilaration in discovering—even generating—and listening to *lo inaudito* that archival systematization, in its logic of preserving things, keeping them from changing, and making them immobile, often renders superfluous.

The Sonic Turn, the Lettered City, and the Aural City

Sound studies emerged as an interdisciplinary field in the late twentieth and early twenty-first centuries, advocating for a shift in scholarly focus beyond visuo-centrism that recognized the importance of sound and listening in everyday life and their potential as analytical categories. The field sought to pay serious attention to sound and aural culture to "enrich our understanding of perception and its role in situating oneself, forming beliefs, and acting upon the environment."[18] This scholarly shift in the humanities and social sciences has been termed the *sonic turn*. Tom McEnaney traces it back to the publication of Emily Thompson's *The Soundscape of Modernity* (2002) and Jonathan Sterne's *The Audible Past* (2003). He explains that the work of these scholars defined the field of sound studies by placing

> an emphasis on the detail of sound as an isolated object of study, but also sound as a more general principle of selection (rather than "music" or "speech"); a reorientation to denaturalize hearing and reconceive listening practices as historically contingent, material, and social techniques; the need for a media archaeology that links technology and technique without falling into "impact histories" or "media determinism." [As such, the study of sound is not just about] a new object of research, but a new method that acknowledges the performative character of culture without concealing the felt reality of material life.[19]

Following on this, David Novak and Matt Sakakeeny suggest that listening and hearing someone is a matter of recognizing their subjectivity and consciousness, thus proposing that the sonic turn takes sound as "a substance of the world as well as a basic part of how people frame their knowledge about the world."[20] The central place of listening in the sonic turn has led some to refer to it as a shift "where aurality becomes an epistemological issue located at the intersection between knowledge and power."[21] Indeed, sound studies is not just about sound, noises, or silences existing out there; it is about exploring how certain modes of listening help us make sense of those sounds, noises, and silences within specific cultural and historical contexts as well as how those modes of listening lead to certain understandings of the world out there.

The arrival of the sonic turn in the humanities and social sciences speaks of cultures of sound and listening that respond to a democratization of culture that, as Jean Franco puts it, tells us of "the 'invasion' of the literary text by the 'noise' from outside [that] succeeded in breaching the walls of... the 'lettered city.'"[22] Therefore the unabashed critique of logocentrism at the base of sound studies could also be read as a further sign of the decline or decentering of the Lettered City. Such a trend is also connected to larger social and cultural processes in which technology revitalizes forms of knowledge production considered premodern, especially orality, aurality, and a detailed attention to sound. In this context, I propose the concept of the Aural City not just as a contemporary sonic and aural counterpart to the Lettered City but also as a window into critically addressing the way in which several important Mexican and Latin American artistic projects have uncritically celebrated the sonic turn as the locus for a more democratic construction of and access to knowledge. Like Rama's Lettered City, the Aural City does not refer to any actual metropolis or urban center. Instead, the city stands as a metaphor for an urban intellectual elite; their relation to specific modes of producing, circulating, and accessing knowledge; and the political potential of that knowledge.

Civilizations have always been associated with specific cities. When we think about classic civilizations such as the Sumerians, the Greeks, the Toltecs, or the earliest Chinese dynasties, we immediately picture the majestic architecture of Uruk, Athens, Teotihuacán, or Chang'an Cheng. In the Western understanding of human history, urban centers and the states they have stood for have always been synonymous with civilization. Their architectural remains, which transcend the people who once inhabited and made them alive, are, in a way, archives of ancient lifestyles, belief systems, social arrangements, understandings of nature, and the like. Thus, the idea of the city has always been a crucial aspect in defining civilizing projects that take the civilization-barbarism dichotomy as their central tenet.[23] *Polis*, the Greek word for "city," was originally coined to account for the city center encompassing administrative and religious institutions. Eventually, the term was used also to refer to the city's body politics, the intellectual and political elites or groups of citizens that ruled these urban centers. Ángel Rama articulated these ideas when he coined the term *Lettered City* to describe a privileged yet amorphous, transhistorical, diverse, and multivocal Latin American urban elite that found in literary production and the control of the written word a powerful tool of representation. For Rama, the Lettered City was a metaphor for this cultural elite and their use of written culture to access cultural capital, to shape and control a number of political and cultural projects throughout the history of the region, and to assist them in their attempt to render rhetorically silent and invisible the Indigenous civilizations they encountered and the epistemic worlds these civilizations entailed. In other words, the Lettered City could be defined as an intellectual elite, a logocentric episteme and its power structure, and the strategies used by that elite to reproduce such a structure.

In *La ciudad letrada* (1984), Rama explains how the urban design of colonial cities in the Americas allowed the Spanish conquistadors a utopian assertion of rationalized order that came to symbolize their assumed Eurocentric civilizing mission. As he puts it, "Isolated amid vast, alien, and hostile spaces, the cities nevertheless undertook first to 'evangelize,' and later to 'educate' their rural hinterlands." To achieve this, the cities "had to dominate and impose certain norms on their savage surroundings. The first of these norms was an education centered on literacy."[24] This task fell on a group of *letrados* (lettered or educated individuals) who oversaw the administrative operation of the mechanisms of political power, the ideologizing of Indigenous populations, and the education of the local ruling elite. This is the specialized social group that Rama called the Lettered City: the citizens in charge of dealing in and with the written word. According to Rama, the longevity of this intellectual elite and its lasting political influence in the region, from the sixteenth century through the twentieth century, can be explained by its ability to control the production and circulation of knowledge through the written word in a largely illiterate society. During this period, the Lettered City oversaw the solidification of European colonial rule as well as its eventual collapse, the development of republican nation-states, and the unfolding of nationalistic modernizing agendas symbolically validated in the incorporation, control, disciplining, and appropriation of vernacular culture. The variety and even contradictory nature of these political and cultural projects is a witness to the fact that the Lettered City has never been a monolithic formation. Instead, there have been many Lettered Cities, often coexisting in time, and always shaped in response to specific historical contingencies and in negotiation with local circumstances and the demands of privileged as well as subaltern actors. The historical membership of the Lettered City may also be debatable and open to discussion since social mobility and class struggles within specific local arrangements of personal and political relations have always rendered its borders porous. Nevertheless, its modus operandi as a central entity mediating power dynamics and prompting the negotiation of hegemonic pacts throughout the history of Latin America is very clear.

Franco argues that the demise of the Lettered City came about because of the impact of the Cold War in Latin America and the resulting anticommunist repression and censorship of military dictatorships and civil governments alike in the 1970s and 1980s. The brutality of these regimes forced many writers and artists into exile and "ended [their] utopian dreams [as] agents of 'salvation and redemption' . . . leaving older structures, both cultural and political, in fragments. Terms such as 'identity,' 'responsibility,' 'nation,' 'the future,' 'history'—even 'Latin American'—had to be rethought."[25] For Franco, the imposition of neoliberal policies in the region after the Cold War, through the 1990s, with their pragmatist approach to economic as well as cultural production, and the weakening of the nation-state as a sovereign political formation, marked the end of the Lettered City project. The advent of the internet and other forms of media that claimed a more democratized access to information, and a shift from civic participation to the consumption of commodities as the main agent in the production of citizenship, certainly challenged the relevance of a cultural project that sought to control individuals and their agency through regulating their access to information and certain types of knowledge and cultural capital. In the ideal neoliberal city, the role of consumption is to help the economic system "reproduce the labor force and increase profit on commodities."[26] Indeed, in a social and economic arrangement that does not need civically committed citizens but instead requires avid consumers, controlling the circulation of the written word and knowledge seems secondary to generating desires and aspirations and regulating the renewal of consumption.

Just as I remain skeptical about the celebratory welcoming of sound studies as the answer to the epistemic excesses and shortcomings of logocentrism, I refuse to comply with the enthusiastic academic discourse about the demise of the Lettered City brought about by the sense of disillusionment described by Franco. Many scholars have argued against that.[27] I would also add that the civilizing mindset that transhistorically informs the Lettered City does not disappear with the avowed shift from logocentrism to aurality proposed by sound studies either. As the architectural remains and urban layout of Tenochtitlán remind us, a city does not just vanish when it is replaced by another city. Traces of Tenochtitlán lurk beneath and around the streets, corners, and buildings of Mexico City. When we learn to see and identify them, we gain a novel perspective that gives the newer city meaning well beyond the moment of its violent birth, the utopian dreams of those who founded it, and the labor of the Lettered City that has rhetorically made it part of a civilizing project for centuries. To borrow Carolyn Steedman's allegory, nothing disappears; it always leaves us its dust in the archive, which we must learn how to deal with.[28] Following on these metaphors, one of the goals of this book is to explore the epistemic connections and differences between the Lettered City and the Aural City as displayed in several sound archival projects. With that, I do not intend to claim that the Aural City has replaced the Lettered City; that would be a very reductionist interpretation. Instead, I argue that the Aural City represents a new intellectual episteme that characterizes the labor of an urban intellectual elite whose members sometimes, strategically, may travel back and forth between the Aural and the Lettered Cities. In other words, the Aural City and the Lettered City coexist, occasionally sharing spaces and intellectual concerns but often following diametrically opposed intellectual and political principles.

Rather than simply proposing the *Aural City* as a theoretical term, I intend to use it in an embodied way. I am interested in identifying and mapping out the labor and networks of specific individuals whose work with sound may qualify them as citizens of the Mexican Aural City. My way of going about it is twofold. On the one hand, I do it by looking at the archives these individuals or their networks may have crafted and left behind or by studying the ways in which they have engaged already existing archives. On the other hand, I do it by grouping together archival constellations that render visible networks that otherwise remain invisible or *inauditos* in the vastness of the infinite archive of the everyday, for which everything out there could eventually be an archival document. The possibility of collective grouping, of finding resonances and echoes in the archive, is a strategy to partially render visible portions of the Aural City and its labor in a more embodied way. The type of labor I seek to identify here could be depicted as creative, reterritorializing, institutional, or alternative; it could reproduce logocentric hegemony, challenge it, or flourish in the in-between cracks. However, rather than classifying individuals and their labor into discrete uncontaminated categories, I seek to show the many ways in which this labor escapes an essentialist characterization. It is precisely in the apparent contradictions of the projects that the citizens of the Aural City engage in that one can visualize the cultural and social complexities that the term embodies as well as its descriptive potential vis-à-vis and beyond the structures of the Lettered City.

Geoffrey Baker has challenged Rama's emphasis on the written word and the *letrados*, arguing that "music, sound, and performance ... were equally integral to [the] process of colonization and urbanization in the New World, with the ordering of the city ... conceived and enacted not only in verbal but also in sonic terms."[29] He refers to this "intersection of sound, urban form, and colonial power" as a "*ciudad sonora*, or 'sonorous city.'"[30] Baker is correct in pointing that out as a critique of Rama's Lettered City. His work, along that of scholars such as Linda Curcio-Nagy, Jesús Ramos-Kittrell, Alejandro Vera, and Leonardo Waisman, provides plenty of evidence of the ways music was used to systematize social and urban life during the colonial period.[31] Along Baker's lines, Natalia Bieletto-Bueno has proposed the notion of the *ciudad vibrante*, or Vibrating/Vibrant City, to study how "the listening modalities and strategies of the inhabitants of a particular city allow them to build specific relationships between sounds, localities, history, memory, cultural identity and senses of belonging [but also how] individuals generate opportunities to make their cities sound, how those sounds transform the space in its acoustic, social, affective, physical, and perceptual dimensions, and how this type of agency presupposes a civic and political action."[32] The practices that Baker and Bieletto-Bueno articulate largely denote the use of music and sound to do something in or develop affective relations with particular urban spaces. They refer to urban experiences rather than to the acknowledgment of an elite's listening as an epistemic practice and labor not only to understand the world around them but also to establish networks and strategies for political and intellectual influence and mobilization. Bieletto-Bueno's description insinuates the latter, but just like Baker's, her choice of the noun *city* is still meant to focus on a particular metropolitan area rather than serving as a metaphor for an urban elite and their intellectual projects. For that reason, Baker's and Bieletto-Bueno's choices of terms are suitable for their endeavors as they refer to practices that have made the city sonorous or vibrating/vibrant and deal with sound in the city as an ordering social and affective factor.

In this book, I propose the term *Aural City* instead of options like *Sonic City*, *Sonorous City*, *Vibrating/Vibrant City*, or my own previously preferred expression, *Sounded City*. Originally, I coined the notion of the Sounded City to conceptualize the urban elite and their work as described above. Adjectivized with the past participle of a verb, the idea referred to the quality of something that has been sounded out or searched for, as well as something that has been powerfully projected as the result of an action or the outcome of a certain labor.[33] Although I still consider this sounding action to be an essential feature in the relation between sound and listener that characterizes the intellectual urban elite at stake here, in the end I chose to use the term *Aural City*. The reason is that rather than a simple interest in experiencing or describing sound itself, what truly characterizes this elite is their aurality or how their listening produces a certain type of sonic and sound knowledge. In other words, the Aural City as I propose it emphasizes aurality and listening as labor. They are the type of labor needed in order to make something sound, meaningful, understandable, circulated, and reproducible as part of larger intellectual projects. Moreover, the concept of the Aural City also helps in mapping out how, regardless of the Lettered City's heralded decline, the Latin American elites that coalesced around the idea of the Lettered City continue to carry cultural and political valence at the beginning of the twenty-first century. In doing that, I am interested in taking the Aural City as a model for cultural critique and political struggle that, while articulating some of the premises of the Lettered City, transcends its transhistorical efforts and political alliances.

Ana María Ochoa Gautier has already invoked Rama's Lettered City in the context of an intensification of the aural in Latin America derived from the democratization of modes of circulation, mediation, manipulation, and reproduction of sound in the region that globalization has made possible. In *Aurality* (2014), Ochoa Gautier takes a critical stance toward the kind of unsuspecting optimism—what Jonathan Sterne has called the "audiovisual litany"—that characterizes some scholarship in sound studies by stating that "its complex relation to the political theology of orality and to alterity [continuously returns] as an obvious construction despite repeated historical deconstructions. [This is] the 'spectral politics' of modern aurality—this capacity to present itself as 'an other' when it is in fact 'the same.' . . . Without suspecting it, what we are doing is reproducing the same sensorial/expressive scheme that we are critiquing."[34] In her early work on these developments, Ochoa Gautier maintained that "under the contemporary processes of social globalization and regionalization coupled with the transformations in technologies of sound, the public sphere is increasingly mediated by the aural. These processes are, if not subverting, at least displacing the relation between the sonic and the lettered word."[35] She concluded that "this is not so much an issue of the sonic replacing the lettered, as a move from the gaze to listening as a locus of analysis and political struggle."[36] Evidently, Ochoa Gautier was aware of the negative epistemological connotations of the term *Lettered City* and thus avoided referring to this shift in terms of an Aural or Sounded City, preferring to coin the term "aural public sphere" since "the public sphere is being redefined to include forms of participation which are not channeled by the forms of debate or participation historically recognized as such by official polity."[37] Ochoa Gautier's aural public sphere emphasizes the shifts that such democratic access to sound and an inclusive construction of knowledge may make in redefining political strategies and ideologies of nation building.

There are important conceptual differences between the notion of the aural public sphere and that of the Aural City. The aural public sphere is a concept that invokes a cultural area, a situation, a condition, or a region; besides invoking an area of social and cultural life, the Aural City speaks of a heterogeneous amalgamation of individuals, clusters, ventures, and institutions; missions and commissions; strategies, policies, and, most important, a type of labor that makes these alliances and projects possible. Regardless of whether the Aural City could be associated with the type of cultural labor that characterized the Lettered City or could be understood as a reaction against it—as the case studies in this book show, depending on the particularities of the projects it is involved with, it could be described either way—it is manifested or materialized in a series of networks developed to enact the type of labor at stake. Thus, rather than just an urban space, the Aural City refers to an intellectual elite whose agency, labor, and the cultural regions they develop are established and channeled through a variety of networks. Sometimes these cultural regions may map onto specific urban geographies; however, the Aural City's labor and networking efforts often transcend these topographies. Therefore, the Aural City should be understood as a series of cultural networks developed to enact the type of labor that characterizes their strategic engagement with sound, space, and practice.

It must be clarified from the outset that individuals acting within this Aural City do not engage with sound in a uniform manner. Those involved in patrimonial projects tend to ascribe value to sound in itself, especially sounds that can be traced back to important historical figures or moments, or iconic geographic locations and cultural practices. This tendency often leads to a fetishization of the objects that store these recorded sounds as well as the archives that hold these objects. For others, sound acquires meaning only in relation to a listening subject. However, what most of these folks agree on is that listening is a knowledge-conducive practice and a way of making sense of the world. Thus, the Aural City's labor is based on the premise that listening is a "resonant power" that, as Lizette Alegre González and Jorge David García argue, invoking Veit Erlmann's work on modern aurality, resonance, and reason, "connects the different epistemic, affective, and sensorial dimensions that are part of the social framework, and the way individuals relate to it."[38] Since listening is not just an "auditory sensory phenomenon, [but] an act that recognizes the constitutive hierarchy that predates any sonic expressive genre," the labor of the Aural City revolves around the artistic and intellectual strategies that, as Ana Lidia Domínguez Ruiz puts it, "understand the individualization, socialization, culturalization, and adaptation processes that mediate our relationship with sound [and] account for the ways in which the world is configured under an aural logic."[39]

Although I understand the Aural City as a system and a type of labor rather than as a specific group of individuals or an urban space, it is still important to mention that it is individuals, their agency, their labor, and their ability to network that define this otherwise abstract entelechy. Thus, it is relevant to unveil, if not their names, at least their presence and participation in a number of cultural spaces, projects, and institutions, in order to get a better sense of how their positionality defines the character of the Aural City also as that of an intellectual elite, albeit often differently situated socially, culturally, and ideologically than the traditional elites enacting the Lettered City. Such recognition may also help in understanding the intersections of the Lettered City and Aural City projects and making sense of the mobility of individuals continuously traveling across the often-porous lines dividing them.

In the last chapter of *In Search of Julián Carrillo and Sonido 13* (2015), I refer obliquely to the Aural City when discussing the unique alternative audiences attending concerts featuring Julián Carrillo's microtonal music in the 2010s. These were audiences who "may have heard about Carrillo and *Sonido 13* [the Thirteenth Sound] through cultural channels and networks that override the elitist world of classical music, engaging instead the mystic and mysterious representations of Carrillo that circulate via Mexican popular music."[40] Although people trained within the Mexican music conservatory system were members of these audiences, most of them were folks whose intellectual interest in experimental music and sound developed outside of academic music spaces. The work of Arturo Castillo, Víctor Garay, Alfredo Martínez, and Juan Pablo Villegas with Mexican Rarities that I explore in chapter 5 clearly reflects this dimension of the Mexican

Aural City. My second encounter with individuals whose labor could place them within the Aural City was in 2012, when I was invited to teach a seminar in music and performance studies at the Escuela Nacional de Música, the music department of Mexico's Universidad Nacional Autónoma de México (UNAM). At that time, I was able to identify some of the unique features of these folks' training as well as their approach to sound and music when Carlos Prieto Acevedo, Rossana Lara Velázquez, and a few other members of an intellectual project that would lead to the establishment of the Seminario de Arte y Sonido in 2014, invited me for a collective conversation. For me, one of the most salient features shared among these individuals was that, except for Lara Velázquez, none of them were trained as musicians or musicologists. Their interest in experimental music and sound came from intellectual conversations in their own disciplinary fields: art history, sociology, philosophy, communications, anthropology, and so on, which often bypassed the ideological prejudices that had characterized music academia since its inception as a scholarly field and that still typified it at the beginning of the twenty-first century: an emphasis on the Western art music tradition, a disdain for popular music, a particular canon of "great men," and a fetishization of the musical work as a depositary of unequivocal and univocal meaning. In *Unbelonging* (2023), Iván Ramos studies a number of distinct Mexican subcultural scenes (punk, metal, sound art, computer music, electronic music, etc.) characterized by a "sonic refusal [that] fueled a generation of younger contemporary artists who were attempting to run away from historical and aesthetic legacies that tied national and ethnic identities to specific aims and aesthetic forms."[41] Although Ramos's intention is not to identify any specific artistic or intellectual elite, his study does outline the labor and mission of these subcultural and underground scenes—a rejection of consumerist and nationalist mentalities—in ways that resonate with the work of what I have called the Aural City. Ramos also identifies a number of spaces as well as formal and informal institutions that are fundamental for the existence of the subcultural projects he studies and that have also been central to the development of the Mexican Aural City's networks and the circulation of its work: El Chopo (both the museum and the *tianguis* [street market]), Museo Ex Teresa, Radio Educación, Museo de Arte Carrillo Gil, Laboratorio Arte Alameda, Centro Digital, Centro Cultural Tijuana, Museo Universitario de Arte Contemporáneo (MUAC), and so on. Although these are not the only places that the Aural City favors, they are indeed some of the institutional spaces that often welcome them and their projects. To borrow the popular

Mexican saying, *Ni son todos los que están, ni están todos los que son* (Not all of those who are there belong, and not all of those who belong are there), yet these accounts do offer glimpses into some of the names, spaces, and strategies one could associate with the Aural City. *The Archive and the Aural City* deepens the study of their shared ideals and the particular individual and collective intellectual and artistic strategies to achieve them, but rather than simply naming individuals, I provide a series of road maps to identify them and their work.

Like the Lettered City, the Aural City is undeniably a privileged artistic and intellectual elite. Although its constituency is very heterogeneous, its citizens tend to be well educated and intellectually curious, have access to resources, and enjoy a cultural capital that differentiates them from other working-class subaltern groups whose voices are largely silenced in Mexico. In that sense, the Aural City, in its inherent connection to the Lettered City, resonates with the shortcomings of what Robin James has termed the "sonic episteme." It "remakes and renaturalizes all [the political baggage inherited from Western modernity] in forms more compatible with twenty-first-century technologies and ideologies—which is exactly what the neoliberal episteme does with its calculative rationality."[42] Nevertheless, one of the main differences setting the Aural City apart from the Lettered City is precisely the former's general skepticism toward the nationalist and modernist projects whose promotion was central to the intellectual and political agendas of the Lettered City. This critical attitude also defines the Aural City's relation to national institutions. While during most of the twentieth century—certainly after the Mexican Revolution—citizens of the Lettered City were organic intellectuals who pushed for the creation of national institutions and whose political agenda was often channeled through them, the citizens of the Aural City are savvier in their articulation of these institutions. Rather than uncritically embracing the mission of these spaces, the citizens of the Aural City often engage them when their work with these bodies helps them advance their own personal and collective agendas. In other words, there are certainly moments and geographies in which the contemporary Lettered City and the Aural City intersect or overlap, especially in their understanding of listening as a point of entry into knowing the world around us. Nevertheless, as the case studies presented here show, national institutions often seek to control the way in which listening practices are conceived and enacted in order to promote top-down nationalist and modernist projects, while the Aural City is often interested in the agency behind individual listening practices as a way to empower the listener and expose the shortcomings of these nationalist and modernist agendas. So, rather than conceptualizing an "institutional Aural City" and a "noninstitutional Aural City" as a dichotomy defined in terms of particular individuals to whom one should attach those labels unequivocally, univocally, and in perpetuity, considering the porosity between them, one should think about them in relation to specific and unique types of labor that respond and eventually advance different political projects. In that sense, the Aural City's potential to disrupt central tenets of the Lettered City also resonates with what James recognizes as productive ways to "think with and through sound" in order to "avoid and/or oppose the systemic relations of domination that classical liberalism and neoliberalism create."[43]

Sound and Sound Objects as Intangible Heritage, Memory, and Patrimony

At this point in this introduction, the unique nature of the archival projects discussed in this book is evident: We are dealing with archives whose mission is to store sounds. Despite its obviousness I want to highlight this because we are dealing with sound within an epistemic shift regarding the production and circulation of knowledge. This shift impacts the authority of certain cultural elites and raises important questions about the nature and understanding of sounds as memory, heritage, and patrimony.

The fact that UNESCO has elevated some of the collections studied here to the category of Memory of the World makes this discussion particularly salient. Such recognition bestows significant importance on the institutions hosting these archives and validates them locally, which is especially significant in the precarious context that has characterized the development of cultural projects in Latin America. This precarity has become more extreme due to budget cuts imposed by neoliberal policies.

Evidently, as Brian Kane explains, sound is intangible, ephemeral, and invisible; it is a temporally contingent event, matter, or flux.[44] Recording technologies may help sound transcend its transient nature by creating a sound object out of it. However, the sound object should never be confused with the sound qua vibration. As Michel Chion states, "The sound object is something perceptual"; it is not the sound itself nor its physical source but rather how we come to perceive it in a materially mediated way.[45] Once a particular sound transcends its temporal contingency by being recorded into a sound object capable of reproducing it, our relationship to the sound qua vibration ceases. We no longer listen to the original vibration; we listen to a material imprint left in the past by its specter, frozen in time. Yet, as Jonathan Sterne argues, this "may be the moment when we credit the tool for the sound."[46] In other words, our alienation from the sound qua vibration may lead to a fetishization of the object that stores it, attributing the properties, qualities, and values of the event in flux to the technology that allows its anachronous reproduction. This dynamic is articulated in the discussion about the patrimonial rhetoric that characterizes national archives such as Mexico's Fonoteca Nacional (National Sound Archive), which is studied in chapter 2. Moreover, the political implications of this paradox are further explored in chapter 4, in relation to the type of fetishization implied by the patrimonial invocation of sound in national and imperial archival projects such as the Berliner Phonogramm-Archiv (Berlin Phonogram Archive). Differentiating the sound object and sound qua vibration and avoiding conflating them is essential in understanding the fetishization that often occurs in affective local characterizations of patrimony and how UNESCO assesses sound objects and sound qua vibration in relation to its prevailing notions of memory and heritage.

The archives discussed in this book evoke sound through sound objects. These archives do not actually store sounds; they safeguard the objects that mediate or represent sounds and often take advantage of the dynamics of fetishization that provide these materials with their aura of authenticity to further the institution's patrimonial logic. Consequently, even though these archival projects deal with intangible elements, the fact that their actual storing, cataloging, and circulating efforts focus on the material sound objects makes these collections ineligible to be classified as intangible cultural heritage according to UNESCO's definition, which describes it as the "practices, representations, expressions, knowledge, skills—as well as the instruments, objects, artefacts and cultural spaces associated therewith—that communities, groups and, in some cases, individuals recognize as part of their cultural heritage."[47] In sum, what this institution considers intangible cultural heritage are oral traditions, performing arts, ritual practices, and traditional craftmanship, not their recording in physical formats. Thus, UNESCO considers musical traditions such as Mexican mariachi, Argentinean tango, or Cuban rumba to be intangible heritage but not collections of recorded music or sound. For the most part, these collections and recordings fall within the category of documentary heritage and are thus eligible to be labeled Memory of the World. Documentary heritage is defined by UNESCO as documents comprising "analogue or digital informational content and the carrier on which [they] reside.... [They are] preservable and usually moveable. The content may comprise signs or codes (such as text), images (still or moving) and sounds, which can be copied or migrated. The carrier may have important aesthetic, cultural or technical qualities. The relationship between content and carrier may range from incidental to integral."[48] In sum, for the archives I write about, what UNESCO considers Memory of the World is actually the material objects where those sounds have been recorded, not the sounds themselves. These considerations are important for navigating the notions of memory, heritage, and patrimony that inform the different case studies in this book.

Listening to the Archive of the Aural City

Throughout this book, I repeatedly invoke science fiction and poetry as ways to jumpstart the discussion of certain matters and topics that are central to the arguments in each of the chapters. In the epigraphs that open each chapter and in the presentation of the ideas that inform my take on each of the case studies in the book, I refer to plots and imagery from literature, films, and poems. This is because good science fiction and poetry have the power to make us reflect on the human condition and offer unforeseen avenues to engage issues of memory, identity, time, and representation. These themes lie at the core of our humanity and pungently inform the discussions about how we relate to, use, and could potentially revamp archives. As José Montelongo's alter ego proposes in his novel *No soy tan zen* (2022), poetry is a way to use words to reveal the world that "hides behind the words."[49] In this case, paraphrasing Montelongo, poetry and science fiction are points of entry into revealing the world that hides behind the documents in the archive(s). For me personally, science fiction and poetry have allowed me to free myself from some of the dogmas of academic writing and thought and have afforded me an estranged perspective on archival documents and materials that I believe has been conducive to finding instances of *lo inaudito* in the archive. My articulation and understanding of these archival constellations and the questions that guide my exploration of them would not have been the same without the liberating affective and intellectual input of these artistic expressions.

Chapter 1, "Performative Listening, Writing, Reading, and the Assemblage of Archival Constellations," focuses on an archival network developed by me in order to expose the placeholders that keep together the larger archive of Mexican nationalism, one that both creates an essentialist past and argues for a teleological future. Inspired by Ana María Ochoa Gautier's admonition to conduct "acoustically tuned explorations" that help us dilucidate the ways in which certain listening practices have been entextualized into written archives, the archival constellation at stake in this chapter is the result of putting in dialogue two books and an exhibition: Daniel Castañeda and Vicente T. Mendoza's *Instrumental precortesiano* (1933), Carlos Chávez's *Hacia una nueva música/Toward a New Music* (1932–37), and the musical side of the exhibit *Twenty Centuries of Mexican Art* at New York's Museum of Modern Art (MoMA, 1940)—which I contend was partly a materialization of the ideas featured in the aforementioned books.[50] This archival constellation offers a window into recognizing the performative relation between modernity and tradition that informs the postrevolutionary Mexican nationalist narrative. Articulating this as an archival constellation offers an aural space for us to hear an *inaudito* counterpoint between the invention of the past and the imagination of the future that is key in understanding the aspirational essentialism that has informed Mexican music historiography since the 1930s. One of the goals of the chapter is to highlight how music was also an instrument used by the Lettered City to further the nationalist turn that their civilizing project took after the Mexican Revolution, which makes Chávez and the actors who took part in this project into a type of proto–Aural City.[51]

Chapter 2, "Patrimony, Objectification, and Representation at Mexico's Fonoteca Nacional," explores the foundation and mission of Mexico's national sound archive in order to understand how the nationalist nature of this project determines the type of labor informing the activities it sponsors, the sounds it deems worth preserving, and the types of listening it privileges. By analyzing the disciplining character of the soundscape and sound map projects conceived by the Fonoteca Nacional's staff, this chapter argues that regardless of the archive's rhetoric about democratization of knowledge, its activities often reproduce the civilizing project of the Lettered City as well as the aesthetic canons privileged by the proto–Aural City discussed in chapter 1. These shortcomings lead to an understanding of the circulation of information and knowledge based on a top-down model that misses the opportunity to engage the sounds that are most significant to the archive's users in their everyday lives.

Chapter 3, "*Critical Constellations of the Audio-Machine in Mexico* and the Performativity of Archiving/Archival Labor," analyzes *Critical Constellations of the Audio-Machine in Mexico*, a sound exhibit presented at Berlin's Kunstraum Kreuzberg/Bethanien as part of the 2017 CTM Festival (an annual music event in Berlin), to ponder the political implications of archival construction and deconstruction. By paying attention to how Carlos Prieto Acevedo, the exhibit's curator, introduces chaos into the archive in order to deactivate the nationalist narratives behind the Mexican music canon and to rearticulate it in novel rhizomatic ways through an active engagement with his audiences' corporeality, I explore how estrangement may open new paths for a recognition of *lo inaudito* and, in this case, a postnational reimagination of the body politic. I argue that Prieto Acevedo's kind of archiving/archival labor generates a libidinal economy that, in tune with Gilles Deleuze and Félix Guattari's theorization about schizophrenia, may provide the epistemic conditions for new socio-personal orders.[52] Invoking Deleuze and Guattari's ideas here is not gratuitous; it is a response to the critical theory background that informs much of Prieto Acevedo's curatorial work. This theoretical background is shared by many citizens of the Mexican Aural City.

The objects and documents in an archive usually tell and retell stories that performatively reproduce the larger ideological frameworks informing the dynamics between objects, documents, representations, and users. The central concern in chapter 4, "Things, Sound Objects, and Legacy at the Berliner Phonogramm-Archiv's Konrad T. Preuss Collection," is whether it is possible (and how) for archives to tell stories different from the ones they are designed to tell us. The first half of the chapter studies the collections of Náayeri and Wixárika chants recorded for the Ethnologisches Museum Berlin by Konrad T. Preuss between 1905 and 1907 (and currently housed by the Berliner Phonogramm-Archiv and the Humboldt Forum) and proposes that the way the sound objects in those collections were created responds more to Preuss's expectations regarding these Indigenous communities than to how these communities conceptualize their music and ritual practices. Based on Arjun Appadurai and Silvia Spitta's take on thing theory as well as Sandra Rozental's exploration of the uses of patrimony in Mexico, the second part of the chapter explores how Mexican anthropologist Margarita Valdovinos has engaged this archive since the 2000s.[53] It proposes that her interrogation of its constituent materials, with the end of repatriating its recordings to Náayeri and Wixárika communities in Mexico, is a model of how to ask questions of archives that force them to tell us stories different from those embedded in their design, structure, and materiality. Valdovinos's listening in detail to the archive and the transnational mobility of its sound objects exemplifies the type of performative labor that characterizes the Aural City.

The historical chronology of the archival complexes explored in chapters 1 through 4 shows the trajectory of the Mexican postrevolutionary nationalist discourse. Chapter 1 explores its development in relation to ideas about the past and the future of the Mexican state; chapter 2 investigates the logic of an archive informed by this discourse; chapter 3 studies a postnationally informed way to rearrange the objects in the archive in order to turn the nationalist discourse that gives them narrative meaning on its head; and chapter 4 analyzes a strategy that reevaluates the sound objects in an imperial archive in order to listen to the silences such an archive entails and repatriate the objects to their communities of origin, while avoiding the nationalist patrimonial efforts that often inform these types of restitution projects. Conversely, chapter 5, "Mexican Rarities, *Disco pirata*, and the Promise of a Sound Archive of Postnational Memory," explores two archival projects of postnational inspiration. Following on recent scholarship about postnational memory by Nadim Khoury and Nigel Young, this chapter studies Mexican Rarities, an archival project developed in Mexico City in 2020, and *Disco pirata*, a 2016 performance action turned sound archive, as models for a possible postnational rearrangement of the logics and dynamics informing the traditional archive(s).[54] Mexican Rarities is an analog archive focused on the identification, storage, and recirculation of Mexican alternative experimental sound and music projects that, by way of their bizarre and eccentric nature, escape the patrimonial gaze of the Mexican state's nationalist rhetoric. On the other hand, *Disco pirata* was started by French sound artist Félix Blume as a project to record everyday sounds that he identified as endemic to Mexico City. The recorded sounds were packaged and presented as a pirate CD, imitating informal-economy circulation strategies, and later featured as part of a larger sound installation at the Fonoteca Nacional that sought to encourage *chilangos* (Mexico City natives) to develop more deliberate ways of listening in detail to the sounds of their city. Eventually, Blume uploaded *Disco pirata* as an open-access internet archive available for free downloading and use. This move presented sound designers in the Mexican film industry with a significant resource when trying to re-create the sonic environments of Mexico City in film. Based on Cristina Rivera Garza's conceptualizations of *noriginales* (nonoriginals), *escrituras geológicas* (geological writings), and *escrituras colindantes* (adjacent writings), as well as her theorization about archives as the previous future of a hyperreal present, this chapter examines the potential and shortcomings of these two archives as repositories of postnational meaning in continual flux rather than as databases of fixed, static value.[55]

Mexican composer Julián Carrillo spent most of the last forty years of his life crusading for microtonal music—especially his so-called Sonido 13—as the future of the Western art music tradition. Nevertheless, Carrillo's constant invocation of nature and law in his theorization of microtonality established a universal, general prescription of Sonido 13 as a closed normative system. If the future Carrillo dreamed of never truly materialized, the presentation of his fifteen microtonal Carrillo Pianos at the Expo 58 in Brussels opened the door for his ideas to have a new life in a different future. Chapter 6, "Aurality, Materiality, and the Carrillo Pianos as Archives," takes the idea as well as the storing, retrieval, and circulation logics of an open-access archive such as Blume's *Disco pirata* and expands it to study the production, circulation, storage, and eventual reinvention of the Carrillo Pianos as metaphorical and literal open-source archive(s). As a figurative interpretative tool, this notion facilitates an exploration of these instruments as archival interfaces of futurity that, following on the work of Thor Magnusson, Carla Maier and Holger Schulze, Roger Moseley, and Alexander Rehding about instruments as archives, provides windows into how individuals can reinvent instruments according to new sonic fantasies about their own presents and futures.[56] Although these instruments were designed with specific musical goals in mind, they have the anarchist potential of becoming sources of new sounds and creative processes in line with the sonic affordances stored in the instruments' materiality. If instruments can be considered archives, one should ask what they are designed to preserve, how one can retrieve that from them, and whether one would be able to retrieve something else from them. Indeed, in the spirit of finding ways to allow archives to say something different from what they were designed to say, I explore the notion of the open-source archive in relation to how poetry, as in the work of Polina Barskova, has been described as an experience that enables the affective recovery of that which the archive has rendered invisible.[57] Thus, a poetic, aesthetic, and creative exploration of the open-source archive may provide access to *lo inaudito* in ways that resonate with Sean Williams's call for the use of poetry in ethnography "as a pathway for understanding an array of experiences in the field, raising essential issues in fieldwork for our students, or transmitting cultural knowledge through multisensory 'creative making.'"[58]

Chapter 7, "In Search of the Aural City: Collective Action and the Invisible Sound Archive," is a dual exercise. It attempts to identify a series of archival projects that reside on the internet in close to invisible fashion while providing a more tangible picture of the Aural City and its citizens. These projects live at the triple intersection of being discursively invisible, being analogically invisible, and representing an invisible nonplace. In this chapter I map out a network of these archival projects by following the actors who made them possible and by establishing connections among them. Once the voices of these individuals and their projects are identified, I let them guide the conversation about what sound archives mean for them. Archives and archival networks are not inherently visible or invisible; they may appear one way or the other depending on whether one knows how to look for them or identify their traces or not. I take this exploration of the archive as an excuse to learn how to look and listen anew. Central to this exercise is an examination of the labor of the constituencies who put together and maintain these virtual repositories as well as my own labor as the researcher identifying and articulating them serendipitously by following the actors behind them. Focusing on the routes that enable this archiving/archival labor also shows the pragmatic and savvy ways in which many citizens of the Aural City engage larger institutional spaces and projects as long as doing that furthers their individual intellectual agendas and those of their allies.

Meant as a conclusion or epilogue of sorts, "The Relevance of Archives in Times of Post-Truth: An Essay Against Nihilism in the Neoliberal Age" aims to tie together the ideas and case studies in the previous chapters by thinking about the relevance of archives at a historical moment in which their traditional value as repositories of truth is being challenged by progressive and conservative agendas alike. I argue that in times of post-truth it is imperative to understand the different ways in which these challenges to archival authority operate and what they mean in relation to notions of truth and falsehood. The epilogue closes with a call against the type of nihilism that the hopelessness of neoliberalism instills in contemporary society. If we can think of the sound archive as a nodal point that helps us make sense of the world by making sense of ourselves, then the archive can be a space for turning resignation into agency, and emotional capital into a liberating political resource. As I repeatedly argue throughout this book, to be able to encounter *lo inaudito* we just need to learn or relearn how to listen.

As a researcher, I do not intend to present myself as someone who brings an asserted objective perspective on my objects of study. Instead, I acknowledge from the outset not only that my subjectivity and positionality mediate how I approach and read my objects of study but also that my interaction with documents, stories, and practices often creates and shapes these materials into their form as objects of study. This is evident in chapters 1 and 7, where the archival constellations and archival complexes that are featured as objects of study are evidently the result of my gaze into the vastness of the archive. But it is also tangentially clear in chapter 6, where, based on an understanding of instruments as archives, I explore strategies to aesthetically estrange the retrieval of the information stored in those archives. Rather than trying to defend my position, I will argue that scholars always mediate the information and documents they work with in order to define them as their archive. Nevertheless, I have also been directly and indirectly involved in some of the projects studied here as well as some of the institutions that the citizens of the Mexican Aural City articulate in their artistic, pedagogical, and intellectual ventures. The reader could argue that such experiences, along with my own interest in listening, sound, and Mexican and Latin American artistic scenes, make me a citizen of the Aural City. That is certainly a possibility. I would not argue against that other than pointing out that participant observation, a key methodology in anthropological and ethnographic work, also seeks to integrate the researcher into the cultural milieu being researched. Some may argue that such labor makes them part of the communities they study; some may think that believing that is a rather optimistic take on the benefits of ethnographic work; while others could rightfully argue that such claims are instances of the colonial gaze that continues to inform much of Western scholarship. The particularities of my case are complicated. However, I am not especially invested in answering that question one way or the other, and thus I leave it to the reader to decide whether my intellectual labor makes me a member of the Aural City or not.

Performative Listening, Writing, Reading, and the Assemblage of Archival Constellations

. . . y porque el futuro, muchas veces dejado atrás, mantendrá siempre sus atributos.

(. . . and because the future, many times left behind, will always maintain its attributes.)

—Adolfo Bioy Casares, *La invención de Morel* (1953)

In 1941 Columbia Records released a collection of 78 rpm (revolutions per minute) recordings entitled *A Program of Mexican Music*. The repertory on those records was a selection of music that Mexican composer Carlos Chávez (1899–1978) had curated for an exhibit titled *Twenty Centuries of Mexican Art* at the Museum of Modern Art (MoMA) in New York a year earlier. Chávez's curatorial efforts had been made possible thanks to a subvention from the Rockefeller Foundation. The exhibit was the result of a joint project between the museum and Mexico's Ministry of Foreign Affairs that attempted "to spread knowledge of the rich artistic tradition of Mexico . . . bringing about a better understanding of our life, both past and present. [And so that] . . . those who are privileged to enjoy this exhibition . . . know us better, . . . [gaining] a fairer and more enlightened judgment of the cultural and artistic evolution of our country."[1] As part of this project, Chávez was tasked with organizing a series of concerts that would "present an idea, general but as detailed as possible, of the music of Mexico" and that would respond to the following question: "In what form and to what degree has a new music, characteristic of Mexico, been produced from the sum of [Indigenous, Spanish, and African] elements?"[2] Three years earlier, Chávez had been invited to conduct a series of concerts with some prominent US orchestras, and since the Columbia Broadcasting System (CBS) Symphony Orchestra premiered his *Sinfonía india* (Indian symphony) on a CBS Radio broadcast in the winter of 1936, he was already a well-known figure in the New York music scene.[3] For the 1940 exhibition, Chávez proposed a final program that included his own compositions (*Xochipili-Macuilxochitl* [1940] and two dances from his ballet *Los cuatro soles* [The four suns, 1925]) and those of José Manuel Aldana (1758–1810) (*Misa en re mayor* [Mass in D Major]), as well as arrangements of traditional Mexican music made by Blas Galindo (*Sones de mariachi* [Mariachi sones, 1940]), Vicente T. Mendoza (*Corridos mexicanos* [Mexican corridos, 1940]), Gerónimo Baqueiro Foster (*Huapangos*, 1940), and Chávez himself (*La paloma azul* [The blue dove, 1940]) as well as Luis Sandi's arrangements of music from the Indigenous Yaqui community. The program premiered on May 16 under Chávez's baton and was broadcast nationally by Blue Network, one of the radio networks of the National Broadcasting Company (NBC). A special national broadcast of the concert, courtesy of CBS, took place on May 19. In both cases, the concerts were broadcast to Latin America via shortwave radio.

The 1941 release of the record collection by Columbia Records contained four 78 rpm recordings—which allowed no more than five minutes of music on each side of the twelve-inch discs—and offered a selection of Chávez's curated program that included *Sones de mariachi, La paloma azul, Xochipili-Macuilxochitl,* "Danza a Centéotl" (Dance of Centéotl) from *Los cuatro soles,* Yaqui music, and *Huapangos.*[4] The cover of the collection featured a typical design by Alex Steinweiss, who just a couple of years earlier, in 1939, had developed the very concept of album cover art for Columbia Records (figure 1.1).[5] Ironically, Steinweiss's design problematically reproduced stereotypes about Mexico (the hats, the donkey, the poncho, the cactus, and the adobe houses) that the organizers of the exhibit sought to overcome with a more sophisticated representation of the country and its culture. However, the design was economical and efficient in its direct and easy appeal to the generalized imaginary that US audiences already had about Mexico. In that sense, the record collection, as a consumption good, articulated cutting-edge marketing ideas in the recording industry at the time.

Out of the two works by Chávez recorded in *A Program of Mexican Music,* I pay attention to *Xochipili-Macuilxochitl* due to its character as a

FIGURE 1.1. Album cover of *A Program of Mexican Music* (1941). Design by Alex Steinweiss. Columbia Records.

simulacrum of cultural authenticity that speaks volumes about the aspirations and desires that coalesced in the production of this record collection as a consumption good. Given that the goal of the concerts was to offer a historical landscape of the development of Mexican music to accompany the art exhibit featured in *Twenty Centuries of Mexican Art*, it was necessary to include music to represent the era prior to the Spanish conquest. Without a pre-Hispanic musical repertory to choose from, Chávez was forced to compose a work based on what he imagined were the constitutive elements of pre-1521 Indigenous musical practice. Chávez's response was *Xochipili-Macuilxochitl*, a work for flutes, clarinets, trombone, and copies of pre-Hispanic percussion instruments reconstructed with the financial support of the Rockefeller Foundation.[6] Chávez became familiar with these instruments thanks to Daniel Castañeda (1898–1957), who had collaborated with him on several educational and research projects since the early 1930s. Thus, in the exhibit's program, the piece was announced as "music for pre-Conquest instruments" with the added explanatory caption "XVIth Century" as a symbolic date, when the composer believed these instruments were used quotidianly. Contrary to the pentatonic simplicity with which Chávez himself described pre-Hispanic music in the program notes, in the different sections of *Xochipili-Macuilxochitl* he mixes pentatonic and diatonic collections and uses pentatonic scales in different

The Assemblage of Archival Constellations   31

transpositions simultaneously.[7] In both cases, the combination of scales generates half steps and dissonances strange to the type of pure pentatonic system that the composer describes as common pre-Hispanic Indigenous practice in the program notes.[8] Years later, in the mid-1960s, when Mills Music published the work (in 1964) and Columbia Records reissued the record collection on LPs (1965), Chávez changed the title to *Xochipilli* and went on to famously describe it as "imagined Aztec music." He even wrote a program note in which he acknowledged the fanciful character of a music he characterized as "an attempt to reconstruct—as far as it is possible—the music of the ancient Mexicans [based on the] pre-Cortesian musical instruments in the archeological museums of Mexico and of other cities, also described in codexes and reports written by the chroniclers of that time."[9]

The connections between *Xochipilli* or *Xochipili-Macuilxochitl* and the ideas about the pre-Hispanic world in Mexico during the first half of the twentieth century, especially the role of this work in the construction of an imaginary about the Indigenous in postrevolutionary Mexico, have already been studied.[10] So, rather than reviewing that, in this chapter I take the materiality of *Xochipili-Macuilxochitl* in its 1940–41 versions as specific articulations of imaginaries of past and future in dialogue that bear witness to the type of "hungry listening" that Dylan Robinson describes as settler colonial forms of perception that "bolster an intransigent system of presentation guided by an interest in—and often a fixation upon—Indigenous content, but not Indigenous structure."[11] In that sense, read relationally, the collection of 78 rpm recordings and the program of the exhibit's concert series are the materialization of ideas about the past and the future—and a type of cannibalistic construction of the Other as part of these fantasies—of a Mexican Lettered City in construction according to forms of listening whose forerunners are already evident in a variety of archival constellations immediately preceding *Twenty Centuries of Mexican Art*.[12]

Archival Constellations and the Past and Future of the Lettered City

In the prologue to his 1925 postdoctoral habilitation, *Ursprung des deutschen Trauerspiels* (published in 1928), Walter Benjamin argues that "ideas are not represented in themselves, but solely and exclusively in an arrangement of concrete elements in the concept: as the configuration of these elements," and goes on to famously postulate that "ideas are to objects as constellations are to stars."[13] By focusing on the relational aspect of grouping as the source of meaning of any given configuration of elements, Benjamin's notion of constellation places the agency of the observer, the individual who comes up with a specific arrangement, at the center of the process of signification. It is the gaze of the observer tracing a constellation that stabilizes these materials by making them into elements of specific discursive narratives. Like the unrelated stars that become fantastic mythological or astrological figures as imaginative individuals group them together, random objects do not have any social meaning until they are arranged into configurations in which they become meaningful in relation to each other. I understand an archival constellation in the Benjaminian sense, as a pseudo-arbitrarily constructed network that enables us to establish new rhizomatic relations between objects, and where new ideas can take form. I use the term *pseudo-arbitrary* because although the nodes articulated by these relational networks may seem random, they often invoke alternative logics of intentional and intuitive forms of pattern recognition as well as unconventional ways of following actors or connecting sources, events, and historical moments. Rather than arbitrary assemblages, these constellations simply respond to nonordinary grouping strategies. It is precisely this nonconventional logic that endows them with the potential to be particularly insightful interventions. As such, an archival constellation is a theory of reading that, in agreement with Andrea Krauß, "designates both the instrument and object of reading, mutually intertwined with each other in a complex interaction [that] draw[s] attention to the discursive production of objects of knowledge."[14] Thus, archival constellations are performative because they provide apparently random networks of objects with discursive meanings that make their relationship real. In that sense, archival constellations are performance complexes because they operate as archival spaces that allow for "performative processes to occur as networks of relations between experiences, events, and actors in the past, present, and future [are seen and understood anew]."[15] This creative practice could be invigorating as it estranges old objectual relations and allows us to see their constitutive objects in a different light. Nevertheless, one must be careful because with time passing and under specific hegemonic contexts, these novel constellations can also become ossified forms that may prevent us from seeing beyond the eventual normalization of those sets of relations.

In this chapter, I propose to read *Twenty Centuries of Mexican Art* and *A Program of Mexican Music* vis-à-vis two books written in the early 1930s in Mexico, Daniel Castañeda and Vicente T. Mendoza's *Instrumental precortesiano* (1933) and Carlos Chávez's *Hacia una nueva música/Toward a New Music* (1932–37), as a transhistorical archival constellation, a performance complex that reveals something *inaudito* in the sway between these two books that offers novel ways to think about the past in relation to the future in postrevolutionary Mexico.[16] The first book is an organological project that seeks to classify Indigenous musical instruments. The second book is an essay about the role of new sound technologies (electric instruments, novel sound recording and reproducing systems, radio and broadcasting methods, sound film, etc.) in the development of modern musical practices and aesthetics. The books articulate their specific subjects of study through ideological frames of reference that were in wide circulation in a variety of national and international intellectual networks in the early 1930s. Both books have been independently influential to a certain extent. For instance, *Instrumental precortesiano* was one of Robert M. Stevenson's main sources in writing *Music in Aztec and Inca Territory* (1968), a seminal study of pre-Columbian musical practices, while *Toward a New Music* remains in the imaginary of many scholars as a "subtle premonition with respect to the media and quality of the music of the future."[17] Nevertheless, the fact that both were produced under the intellectual aegis of Carlos Chávez—one written by the composer himself, the other written under the auspices of the research academies he founded during the period he was director of the National Conservatory of Music in Mexico—gives us an excuse to read them in dialogue. In the end, I argue that listening practices defined historically and contingently by different Lettered City projects articulate continuities between pre- and postrevolutionary cultural production in Mexico and that such listening practices can be used as ideological hinges for the circulation of imaginary exercises for a relational construction of the past and the future.

My intention is to focus on these two books not because they could be depositories of specific content and information but rather because, together, they constitute an archival constellation of ideas that, understood intertextually and dialogically, speaks about the interconnected processes in which inventing the past means imagining the future. Both aspects of this dialogic creative process were essential for the modernist and modernizing agenda that was urgent to implement in postrevolutionary Mexico, an agenda that articulated the cosmopolitan through the local. In this particular case, reading *Twenty Centuries of Mexican Art, A Program of Mexican Music, Instrumental precortesiano*, and *Toward a New Music* as a transhistorical archival constellation allows us not only to establish new, unforeseen rhizomatic relations and networks among material objects, practices, and individuals across time but also to understand how these individuals perform themselves as subjects through the discursive exchanges that give historical, ideological, temporal, and spatial meaning to their intellectual, political, and artistic networks of socialization and identification. Thus, an archival constellation is a grouping of materials and practices that not only refer to a larger archive but could also, by establishing new relations, be constituted into micro-archives themselves. Therefore, the point of entry into exploring this larger transhistorical archival constellation is an exploration of the historically situated micro-archival constellation made of *Instrumental precortesiano* and *Toward a New Music*. In the case of these books, I suggest that by approaching the ideas and listening practices that a transtextual reading of these texts puts in evidence, rather than the facts contained in the texts, we can generate a new epistemic archive. This new archive is a constellation that articulates and gives meaning to a new way of understanding the world, and at the same time, in its transhistorical unfolding, it is a constellation with the potential of becoming a simulacrum that could end up being perceived as the reality in which we live.

Practices of Entextualization and Transduction: Writing Is Archiving

Listening practices are always informed by cultural and ideological networks. Nevertheless, they also reproduce the values that inform those particular networks. In this epistemological loop, the continuous process of signification between individual engagements of sonic reality and collective understandings of that sonic reality, listening practices have the potential of becoming performative as they naturalize specific ways of making sense of the world. In other words, the particularities of an apparently subjective listening experience could be generalized as an effectively affective way to apprehend reality if those experiences are understood to be unmediated acts of accessing the materiality of the world. When individuals are convinced that their listening provides access to the only sonic reality of an object, practice, or moment, it is because the impression that listening may be an unmediated way to engage reality creates a powerful sense of authenticity in relation to that listening experience. Nevertheless, as a culturally and socially informed practice, listening is always informed by the positionality of the listening subject. As such, the perceived authenticity of the listening experience results more from the cultural and social network that informs the listening subject's ways of navigating the world than from an access to any type of sonic real. As Sarah Weiss suggests in the case of the musical listening to Otherness, "authenticity is a malleable concept, within and between individuals as well as over time.... Listeners imagine they are listening to another musical world, but they are hearing themselves, as their own musical world is used as a filter."[18]

Here, I focus on the ways in which writing about musical instruments or musical technologies is an inscription of the authors' listening practices and experiences, and the ways in which two particular entextualizations (*Instrumental precortesiano* and *Toward a New Music*) reveal ideas about Indigenous Otherness and projects of modernity in 1930s–1940s Mexico. I am interested in seeing the act of writing as a way of documenting certain experiences and epistemologies and, as such, as the creation of archives. In this, I follow Nuraini Juliastuti's reflections on the linguistic homology between "archive" and "document" in the Indonesian language, to argue that the documents produced in the act of writing are themselves archives of the very experience of writing them.[19] This "know-show" function of documents, the knowledge that is produced in the practices of showing, copying, circulating, and archiving them, provides them with their epistemic power, as Cait McKinney argues.[20] Thus, as documents and archives, these literary texts not only report on specific moments or practices but also store techniques of entextualization ("the process by which an object... becomes a specific type of text") that speak about the epistemes that made sense of the materials reported in the historical document, imply certain storing and transmission technologies, and lend themselves to the exploration of possible retrieval strategies.[21]

Focusing on *Instrumental precortesiano* and *Toward a New Music* as archives leads us to seriously think about them in terms of information activism. As such, it is important to keep in mind that the processes of encoding and decoding are fundamental actions in the relational production of knowledge that makes archives important. In the aural context that made these books possible, entextualization, as the process that transduced these listening experiences into written texts, is the encoding action that creates the document, and an acoustically tuned reading of the text, as proposed by Ana María Ochoa Gautier, that works as the necessary decoding strategy. In *Aurality* (2014), Ochoa Gautier describes what an "acoustically tuned exploration of the written archive" may be when conceiving of historical texts as collections that speak about "how the idea of a valid aural expressive genre was constituted depending on the listening practices or 'audile techniques' through which it was constituted. What is revealed by such a disjuncture is that many of the acoustic dimensions of the colonial and early postcolonial archive are not presented to us as discrete, transcribed works or as forms neatly packaged into identifiable genres. They are instead dispersed into different types of written inscriptions that transduce different audile techniques into specific legible sound objects of expressive culture."[22] Fernando de Sousa Rocha explains that both the written word and recorded sound are inscriptions that produce "a reusable object [and that] issues related to the mastery and control over sound recording and transmission through writing are transferred to modern sound recording and transmitting technologies."[23] Thus, my contrapuntal reading of *Instrumental precortesiano* and *Toward a New Music* articulates the intersection of transduction, acoustically tuned reading, and technologies of entextualization that produce reusable sonic inscriptions. In this case, I take these two books as indexes of transduction and sonic inscription that enable us to recover and reassess voices as well as performative listening practices. Here, the performative is understood as the production of a sense of reality through the enunciation or execution of a very specific practice, in this case, listening. Taking the Spanish definition of *escuchar* (to listen) as paying attention to what one hears, in this reading I pay attention to what the authors of these texts listened for and how that listening had the possibility to create new worlds of aesthetic and political meaning.[24] My interest in these texts lies in reading between the lines to reveal how a proto–Aural City hides behind the strategies of the Lettered City. My goal is to discern how these texts articulate, build, and imagine a proto–Aural City in the making, its motivations, and its aspirations. In sum, the intention of this study is to understand the raison d'être of an eventual Aural City that, as an intellectual endeavor, transduces the project of the Lettered City into an epistemology that pays attention to what is being listened to and to how that act of listening does something at the discursive and representational level. In other words, the purpose of this chapter is to map out and analyze the traces of performative listening in the archival constellation that is produced in a dialogic reading of *Instrumental precortesiano* and *Toward a New Music*, and in the larger transhistorical constellation produced when reading the books vis-à-vis the MoMA exhibit *Twenty Centuries of Mexican Art* and the Columbia Records collection *A Program of Mexican Music*.

From the Lettered City to the Proto–Aural City

The notion of the Lettered City was coined by literary critic Ángel Rama to describe the privileged Latin American elites that found in literary production and the written word a source of cultural capital and access to political power. Throughout Latin America, these elites were fundamental in the construction of national projects. The work of these types of intellectual elites is evident in the development of cultural and political projects in the newly independent Mexico of the nineteenth century as well as in the postrevolutionary Mexico of the early twentieth century: from the liberal crusade that led to the 1857 Constitution to the educational campaign of José Vasconcelos and the irruption of artistic groups and movements like Los Contemporáneos and Estridentismo in the first half of the 1920s, and from the nativist *indianismo* (Indianism) of the end of the nineteenth century to the nationalist *indigenismo* (Indigenism) of the 1930s state policies. Since the mid-nineteenth century it is possible to clearly identify several Mexican writers who, to a certain extent, took listening as a serious tool for the articulation of larger cultural and intellectual projects. One can think of the *costumbrista* musical chronicles or the Orientalist texts written by Rubén M. Campos (1876–1945) as well as the occasional descriptions of dance and music culture in the works of Guillermo Prieto (1818–97), Ignacio Manuel Altamirano (1834–93), Marcos Arróniz (fl. 1850s), Federico Gamboa (1864–1939), or Amado Nervo (1870–1919) as examples of this move.[25] Nevertheless, it is with the arrival of a new generation of young postrevolutionary intellectuals, especially those who coalesced around the projects of Carlos Chávez as a cultural broker, that we can speak of a more systematic interest in listening for the nation-state; in other words, an interest in the production of knowledge and its political and ideological articulation in a nation-state project based on music and sound.

The serious and engrossed listening to not only musical but also broader sonic expressions that supported political projects of national scope by a growing group of artists and researchers, including those in Chávez's inner circle, gives us the opportunity to think about the development of a Mexican intelligentsia that could be described as a proto–Aural City. It is evident in the artistic and educational projects put in place with the public intervention of thinkers, politicians, writers, and cultural brokers like Gerónimo Baqueiro Foster (1898–1967), Rubén M. Campos, Julián Carrillo (1875–1965), Daniel Castañeda, Francisco Domínguez (1897–1975), Alba Herrera y Ogazón (1885–1931), Vicente T. Mendoza, Pedro Michaca

(1897–1976), Augusto Novaro (1891–1960), Antonieta Rivas Mercado (1900–1931), Jesús C. Romero (1893–1958), Gabriel Saldívar (1909–80), and José Vasconcelos (1882–1959), or the members of the Grupo de los Nueve and later the Ateneo Musical Mexicano, among others, that the aural as an intellectual and political exercise acquired an unprecedented importance precisely in the 1920s and 1930s.

Setting Up the Stage: Carlos Chávez as an Organic Intellectual

In 1928, after almost two years living in New York City, Carlos Chávez returned to Mexico City with the fervent desire to shake up and modernize the city's music scene. Soon after his arrival, the Sindicato de Filarmónicos del Distrito Federal (Federal District Musicians' Union) and a patronage committee organized by Antonieta Rivas Mercado invited him to lead the restructuring of the Orquesta Sinfónica Mexicana, which would later be called the Orquesta Sinfónica de México and would eventually become the Orquesta Sinfónica Nacional.[26] Also that year, Antonio Castro Leal, director of the Universidad Nacional de México (National University of Mexico), invited Chávez to direct the Conservatorio Nacional de Música (CNM), which at the time was an academic unit affiliated with the university. It was through his transformative pedagogical, creative, and promotional activities as leader of these two institutions that Chávez was able to push for a comprehensive program intended not only as a curricular reform but also as a platform for the education of new audiences, echoing the ideas that were beginning to shape the modernizing and nationalist agenda of the postrevolutionary regime. It was also his work at these two institutions that propelled his career as an organic intellectual, meaning, as a public servant and artist who sought to play a role in the construction of a new hegemonic pact and to have a positive political impact on his country's everyday life.[27]

One of the most salient examples of Chávez's transformational work as an organic intellectual is the organizational and curricular mark of his administration as director of the CNM between 1928 and 1933. Chávez arrived in this position at a critical moment in the history of the institution, just as the National University was struggling to obtain its autonomy from the Ministry of Public Education. In this context, with the excuse that "Mexico did not need doctors nor graduates in music but rather good band, orchestra, opera, and ballet musicians, as well as good middle school music teachers," Chávez opted for the conservatory (which he renamed the Escuela de Música, Teatro y Danza in accordance with the type of technical-school profile he argued for) to stay connected to governmental agencies that could help with coordinating the work among the different art schools under a common mission, different from the goals of the university's science and humanities departments.[28] "Letting the Conservatory stay [as a unit of] the autonomous university does not mean it will fail. But it does mean that its development will be slower and more difficult for those who come to direct it," Chávez wrote in one of his many apologies in favor of splitting from the university, arguing that only "its organization within similar institutions and under state sponsorship, as it has always been, will allow for its development to accelerate."[29] Evidently, Chávez was interested in guaranteeing the Escuela de Música, Teatro y Danza's financial health as well as its curricular cohesion and independence, which he believed the university's autonomous status would endanger. For him, the ideal institutional niche for the school was the Department of Fine Arts, an agency of the Ministry of Public Education. Finally, Chávez's activism was successful when the new National Autonomous University's Ley Orgánica (Organic Law) detached the Escuela de Música, Teatro y Danza from the new university organization and assigned it to the Department of Fine Arts.[30] This development generated a schism by a group of professors who had remained very critical of Chávez's project and who decided to create a music department within the new National Autonomous University.[31]

Although this split was a hard institutional blow for the conservatory, it also paved the way for Chávez to put in place the program of curricular and organizational reforms he had been fighting for since his arrival at the school. The situation became especially favorable since the split meant that a group of professors and students who did not agree with his ideas and had tried to remove him from office were no longer an obstacle to his agenda.[32] Chávez's reforms sought to modernize a curriculum that he believed was "antiquated and dogmatic, that privileged theory over practice, that was copied from systems used by old European academies, and that did not respond to the authentic needs of the place where they were implemented."[33] The goal of these curricular changes was not only to make the instruction of musicians a more practical, efficient, and professional task but also to create a number of instrumental ensembles—of Western as well as non-Western music—and establish a free concert series. These changes also attempted to provide students with the possibility of receiving stipends for their performances and of having access to a musical repertory that transcended the Eurocentrism that characterized the institution.[34] Chávez presented the problem of the institution's exacerbated Eurocentrism in the following terms: "If it has been said that European music was known, what was known was exclusively that continent's music during the eighteenth and nineteenth centuries. Twentieth century [music] was unknown and [music] before the seventeenth century was completely unknown.... Studying the major and minor diatonic scales and the science of harmony and counterpoint that can be deduced from this system is not enough; it is indispensable to know all the so-called exotic and primitive scales, the Greek and liturgical modes, and all of what those systems generate."[35]

It is precisely as a response to that problem that Chávez founded the research academies at the conservatory. Organized after the labor division proposed by Guido Adler's *Musikwissenschaft* (musicology), the working groups were divided into the Academy for Historical and Bibliographic Research, the Academy for Research into New Musical Possibilities, and the Academy for Popular Music Research (or Academy for Mexican Music Research), which respectively encompassed Adler's historical musicology, systematic musicology, and comparative musicology (what would later be known as ethnomusicology).[36] These working groups were populated with conservatory professors as well as students. Being part of these groups allowed professors to fully devote themselves to research without having to worry about teaching, which was an unprecedented luxury in the Mexican music scene. Curiously, the creation of these academies shows that, contrary to what one may think after reading Chávez's arguments for detaching the Escuela de Música, Teatro y Danza from the National University, the composer was truly interested in the intellectual aspect of music studies, its impact on the everyday practice of performers and composers, and the ways these concerns could help audiences develop a new understanding of the notion of national culture.

Instrumental precortesiano and *Toward a New Music*: An Archival Constellation in Contrapuntal Sway

One of the projects of the CNM research academies was collecting traditional Mexican instruments and music. These were made available to interested composition and performance students with the goal of putting together a "Mexican orchestra" that combined Indigenous and modern instruments.[37] This was the academic context in which Daniel Castañeda and Vicente T. Mendoza's organological research project, which would lead to the publication of *Instrumental precortesiano*, unfolded. Castañeda had started his musical training as a child; nevertheless, he enrolled at the CNM only in the 1920s, after he had finished his studies in civil engineering. Always informed by his background in the hard sciences, he became a frequent presence in the musical scene of Mexico City on two fronts from the mid-1920s: first, as a member of the Grupo Nosotros (a group of artists and intellectuals interested in public debate about the fine arts with whom he participated in the organization of two national congresses of music in 1926 and 1928) and, second, as a founding member of the musicology section of the Ateneo Musical Mexicano (a cluster that included the members of Grupo de los Nueve), a group that had questioned the feasibility of theorist and composer Julián Carrillo's microtonal ideas in a very public debate in 1924.[38] Castañeda's interest in Indigenous music was clear in the papers he presented at the aforementioned music congresses. In 1926 his presentation dealt with the use of microtonal scales for credible transcriptions of Indigenous music; in 1928 he presented a paper about methodologies for the better study of musical folklore in Mexico.[39] Both lectures were written in collaboration with musicologist Gerónimo Baqueiro Foster, one of the first and most loyal supporters of Carrillo's microtonal theories in the mid-1920s.[40] In 1930 Castañeda was named professor at the CNM as well as director of the Academy for Mexican Music Research.

Vicente T. Mendoza (1894–1964) grew up in a family of church musicians in the state of Puebla but moved to Mexico City at age twelve, continuing his musical training at the Academia de Bellas Artes and later at the CNM.[41] Along with Baqueiro Foster, Mendoza was one of the most vocal supporters of Carrillo's microtonal crusade, although their relationship became strained after Mendoza's participation in the 1926 National Congress of Music, at which he presented a paper about the use of the overtone series as a source for a microtonal system whose aims and goals differed from those of his mentor.[42] In 1929 Mendoza was named professor of music theory at the CNM, thus becoming a close associate of Chávez in his educational reform at the school. In 1930, when the research academies were created, Mendoza was commissioned to the Academy for Mexican Music Research. Years later, both Mendoza and Castañeda would collaborate with Chávez in the process of curating *Twenty Centuries of Mexican Music* at the MoMA.

As part of their job at the Academy for Mexican Music Research, Castañeda and Mendoza developed an organological research project focused on documenting pre-Hispanic percussion instruments. They called it a project of "musical archaeology."[43] Although the project came about within the Academy for Mexican Music Research, it received the sponsorship of the Museo Nacional de Arqueología, Historia y Etnografía. Therefore, Castañeda and Mendoza's final product was first published in five parts in the museum's journal, *Anales del Museo de Arqueología, Historia y Etnografía*, in 1933. It was not until 1934 that, as an extension to that edition, the CNM published it as a single volume through the Ministry of Public Education. Originally, they planned the project to lead to a multivolume publication. Nevertheless, the project was truncated, and only the first volume, devoted to percussion instruments, was published.[44] The text, as it appeared in *Anales* as well as in *Instrumental precortesiano*, is divided into five parts: "Teponaztlis in Pre-Cortesian Civilizations," "Pre-Cortesian Percussive Instruments," "Huehuetls in Pre-Cortesian Civilization," "Small Percussive Instruments in Pre-Cortesian Civilizations," and "Pre-Cortesian Percussive Instruments (Appendix)."[45] Each of these parts is divided into several chapters according to the relation of the materials with the main theme of the section. Thus, the materiality of the instruments—their material characteristics, physical dimensions, and acoustic properties—is described in detail, providing extensive information about the places the instruments come from and where they are kept. They also provide information—based on how they are described in historical chronicles as well as traditional poetry and oral traditions—that helps the reader understand and place the instruments in their proper historical, cultural, and mythological contexts.

The editions are practically identical, except for minor typographic and presentation differences. The first part of the book, "Pre-Cortesian Percussive Instruments," which gives the project its title, not only describes and contextualizes very well the conditions in which the research project developed but also delineates very clearly the philosophical and ideological background that informs Castañeda and Mendoza's endeavor, both the purely taxonomic aspect and the interpretative framework behind their classificatory exercise. Curiously, "Pre-Cortesian Percussive Instruments" was not the first but rather the second article in the series published in *Anales*. The text has an explanatory character that one would expect to find in the first article of the series. This apparently innocuous fact could indicate that the project may have originally started as a study of teponaztlis (slit drums), since the article about these instruments is the one that opens the series in *Anales*, and that only later was it thought of as a more comprehensive organological study.

Although it is difficult to affirm with any degree of certainty, since the only published volume of the project is entirely dedicated to percussion instruments, it appears that the taxonomic principle of the project may have been influenced by Victor-Charles Mahillon's system of instrumental classification—based on four large groups: strings, winds, drums, and other percussion instruments—rather than by the most recent and meticulous system by Erich von Hornbostel and Curt Sachs. The fact that Castañeda and Mendoza do not mention Hornbostel and Sachs's terms, such as "idiophones" and "membranophones," as they describe their instruments, preferring to use the term *percutores* (percussive instruments), is evidence that they were thinking about more general taxonomies than the type of thorough discrimination based on the place in the instrument where sound vibrations originate. Nevertheless, this does not mean that Castañeda and Mendoza were not interested in matters of the origin of sound. To the contrary, their discussion about sound as "air in movement" and the need of "lo hueco" (a hollow, an emptiness, e.g., a soundbox) to make it perceptible is evidence that they were interested, even if more in empirical terms than in purely acoustic ones, in the identity of sound.[46] In principle, Castañeda and Mendoza's organological project seems to stand in between the system used by Mahillon to curate the Musée Instrumental du Conservatoire Royal de Musique (Museum of Instruments of the Royal Conservatory of Music) of Brussels in 1879 and the one Erich von Hornbostel and Curt Sachs published in *Zeitschrift für Ethnologie* in 1914.[47] The latter would eventually become the standard system of organological classification in the world but had not been translated into Spanish or English when Castañeda and Mendoza published their findings.

The central body of the collection of instruments studied in *Instrumental precortesiano* comes from museums in Mexico City, Puebla, Morelia, and Toluca. Direct contact with these materials allowed the researchers to conduct the exhaustive measuring job that characterizes the organological descriptions throughout the book (figure 1.2). In the book these types of illustrations are accompanied by very extensive descriptions of and theorizations about the physical and acoustic characteristics of each instrument that account for the sounds, intervals, and scales they produce, and the relation of these features to the material dimensions of the instrument. Castañeda's training in engineering is on full display in the descriptive tables and their complementary transcriptions as well as the mathematical formulas and equations that explain the acoustic relation between the pitches available in each instrument, the overtone series they belong to, the partials they emphasize, and the intervals and scales one can play on them. Figure 1.3 illustrates Castañeda and Mendoza's acoustic theory of the teponaztli. It includes the sounds available for each of the fourteen teponaztlis they studied: the fundamental pitch of the soundboard (in whole notes), the first partial it generates (in eighth notes), and the actual sounds of the drum tongues (in thirty-second notes). This detailed acoustic, almost functionalist description of the instruments was possible thanks to Castañeda and Mendoza's comparative study of the ancient instruments and their modern copies, which allowed them to reconstruct many of them. In fact, the publication of the measurement tables and the design prints was meant as an archive to guide the future reconstruction of these instruments. Thus, Castañeda and Mendoza's entextualization indexes how their listening transduced the pre-Hispanic instruments into quantitative data on which a variety of narratives could be imposed. As such, theirs is essentially a type of hungry listening that bypasses Indigenous meaning and takes Indigenous content as the basis of a modernist fantasy in which only science and technology can make the past live again in the present and the future. It also opens multiple avenues to describe the Other in relation to social evolutionist narratives supportive of novel forms of domination.

Based on the intervals that the teponaztlis were able to produce (major second, minor third, major third, perfect fourth, and perfect fifth), Castañeda and Mendoza concluded that pre-Hispanic scale systems must have been pentatonic. Their statement was in line with the early scholarship about pre-Hispanic music in Peru by Leandro Alviña and Daniel Alomía Robles, as well as the subsequent book by Raoul and Marguerite d'Harcourt, *La musique des Incas et ses survivances* (1925), which argued that the structure of this music was fundamentally pentatonic.[48] Although slightly problematized in the work of Andrés Sas, Theodoro Valcárcel, and Carlos Vega, these evolutionist ideas were gaining traction among Latin American scholars interested in South American pre-Hispanic music.[49] Although this type of social evolutionism is evident in *Instrumental precortesiano*—in the continuous references to "primitive" and "civilized" musics, the bibliography quoted (Frederick Starr, Hans Brüning, and Stephen-Charles Chauvet, besides the d'Harcourts), and the mentions of Indigenous pentatonicism throughout the book—one can identify how Castañeda and Mendoza tried to reconcile the contradictions between this ideological frame of reference and the larger political agenda that informed their work and sought to present pre-Hispanic Indigenous cultures as "civilized peoples." This is in evidence when the authors argue that

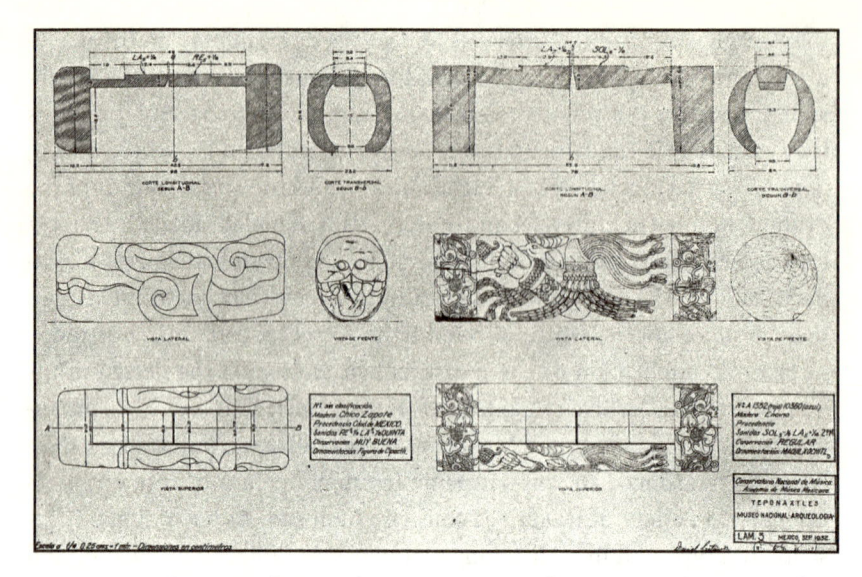

FIGURE 1.2. Drawings showing the measurements of two teponaztlis, in Daniel Castañeda and Vicente T. Mendoza, *Instrumental precortesiano*, vol. 1 (1933).

FIGURE 1.3. Castañeda and Mendoza's acoustic theory of the teponaztli, in *Instrumental precortesiano*, vol. 1 (1933).

based on the use of pedals on intervals other than the perfect fifth in teponaztlis, one could affirm that "our primal music *was not primitive*, but one could rather esteem that it was rigorously civilized and parallel, in terms of musical possibilities, to the music of great cultures from China, Egypt, and Mesopotamia."[50] Although Castañeda and Mendoza's statement was typical of social evolutionism, it tweaks the essentialist argument to reposition its object of study at a superior level that, nevertheless, prevents it from escaping the primitivist logic their scholarship reproduces. At the same time, the use of geometric and mathematical arguments in their analyses, and the description of the acoustic materiality and character of the objects under study, is an effort to validate positively and scientifically the culture that produced these instruments.

The systematic, comparative, and social evolutionist character of Castañeda and Mendoza's project has a marked influence from European folklore studies of the time, especially its emphasis on the study of local social groups. In the case of *Instrumental precortesiano*, the fact that they focused on immemorial local musical cultures gave the project the appearance of a rescue endeavor rather than a simple report that, as Arturo Chamorro would put it, "was about the search for a prestigious past" on which to build the foundation of the modern nation.[51] Nevertheless, it is evident that the project reproduces some of the internal colonialist overtones that that European folklore studies of the time were criticized for. Although Castañeda and Mendoza were very diligent in explaining the geographic provenance of their instruments, they define the materials and practices studied in their book in a general way as simply "pre-Cortesian" or "Indigenous," which fetishizes and homogenizes a large and diverse variety of Indigenous musical cultures that exist and have existed in Mexico in the present as well as in the past. In that sense, through a triangulation between their scientific method, the essentialist construction of the Other, and their performative listening experience, the sounds of those instruments are transduced into a monolithic imagination of the national past that tacitly disciplines them into a discourse of cultural and historical homogeneity. This type of hungry listening is not surprising if we consider the intellectual and political context that informed this research project. Marina Alonso Bolaños has documented how Mexican popular music patrimony had been invented since the early 1920s as a way to develop a unified identity for the politically new nation-state.[52] It is in the construction of that discourse and its eventual naturalization that a dialogue between the primitive and the civilized, the new and the old, modernity and tradition, takes place. It is a dialogue forged as a process of mutual construction in which one does not exist without the other. For that reason, it is relevant to read Castañeda and Mendoza's work in a polyphonic counterpoint with the ideas that inform Carlos Chávez's *Toward a New Music*. The intellectual give-and-take between the two books provides a unique space to recognize the configuration of an incipient Mexican Aural City.

The first three articles of what would become *Hacia una nueva música* were published in the newspaper *El Universal* in the summer of 1932. Nevertheless, the book (with five extra chapters) was published in English first with the title *Toward a New Music: Music and Electricity* (1937). In fact, the book would be reprinted in English in 1975 before El Colegio Nacional published it in Spanish for the first time in 1992. In that sense, the influence the book may have had among the Mexican intellectual elite during the twentieth century seems to be minor. Nevertheless, it is important to study the ideas in the book because, although they may not have circulated in printed form in Spanish, they are fundamental in the world of senses that informs the development of a Lettered City on the articulation of sound and in relation to the modern nation-building project that such an intellectual elite imagined and pushed for in the 1930s.

Chávez spent the first quarter of 1932 in the United States, where he traveled to oversee the rehearsals and premiere of HP (1926–32) under the baton of Leopold Stokowski. Chávez had started collaborating with painters Diego Rivera and Agustín Lazo on this ballet since 1926, although the music was not finished until 1931. During this stay, Stokowski took Chávez to visit the Bell Laboratories in New York and the RCA Recording Studios in Camden, New Jersey, with whom the US conductor had very close relations. The Bell Laboratories were the place where Joseph Maxfield and Henry Harrison developed the Westrex electric recording system in 1924, a system that had revolutionized the recording industry by replacing the acoustic recording system in vogue since the early twentieth century. With the adoption of this system by the most important record labels, including Columbia and the Victor Talking Machine Company, which had licensed the technology since 1925, the acoustic era came to an end.[53] The first commercial electric recording of a symphonic orchestra was done by the Philadelphia Orchestra, performing Camille Saint-Saëns's *Danse macabre*, op. 40, under the direction of Stokowski and recorded at the Victor studios in Camden on April 29, 1925. This was only four years before the company was sold to the Radio Corporation of America and its name changed to RCA Victor. Stokowski's relation with RCA Victor was very important for both the prestige and commercial success of the Philadelphia Orchestra and the label's Red Seal collection (the company's classical music series) as well as the Bell Laboratories' development and tryout of sophisticated cutting-edge recording technology. During the early years of the electric era, Stokowski and his orchestra were always willing to collaborate and experiment with new formats and techniques, including the 33 1/3 LP and the high-fidelity (or hi-fi) recordings that expanded the audio spectrum up to ten thousand hertz, allowing for the better capture of higher frequencies.[54] Since Chávez had been fascinated with the reproduction and communication possibilities that the radio and other modern electric technologies enabled, it is natural that Stokowski would invite him to visit these locations.[55]

Touring the Bell Laboratories and speaking with the company's engineers and sound technicians left a powerful impression on Chávez. On his return to Mexico, he arranged for the Ministry of Public Education to commission a report on what he had seen and learned there.[56] Chávez presented the ministry with a study consisting of three chapters: "Música y física" (Music and physics), "Producción y reproducción musical" (Musical production and reproduction), and "Los instrumentos eléctricos de reproducción musical" (Electric instruments of music reproduction). The study was published as three independent articles in *El Universal* on July 22, August 4, and August 16, 1932. Chávez's descriptions of acoustic phenomena in "Música y física" are reminiscent of Castañeda and Mendoza's discussion about sound that would be published in the introduction of *Instrumental precortesiano* one year later. In both cases, the texts refer to an empirical and almost philosophical search for the identity of sound. In Chávez's case, the description of the materiality of sound highlights the composer's purely artistic concerns when he affirms that "vibrations or physical matters, when grouped together, acquire aesthetic properties. Sounds grouped together in a set according to the specific sensibility of a man become music, and acquire an artistic meaning and expression."[57] In the rest of the article, Chávez features a deterministic narrative in which he argues that the progressive complexity of musical scales throughout history is directly connected to the physical possibilities of the musical instruments available at a given time. This teleological model—which moves from the primitive to the civilized—would also inform Castañeda and Mendoza's organological work. However, in Chávez's case, instead of being used to retrospectively reconstruct the past, it is used to speculate about the future and about musical modernity based on the ability to control and manipulate sound allowed by electric technology.

The same type of deterministic, teleological argument permeates Chávez's discussion about music production and reproduction in the second article of the series. There, Chávez establishes a nexus between reproduction technologies and the complexity and difficulty of musical works. Thus, he argues that it was the invention of musical notation that allowed for the development of "complex" musical styles—which he exemplifies with the music of Richard Wagner and Igor Stravinsky—that would not have been possible without a tool that guaranteed their efficient reproduction.[58] Although Chávez tacitly acknowledges musical notation as a tool of entextualization, as a means to transduce an aural/sonic experience into a visual/written one, his argument clearly shows a frame of mind that privileges the mapped visuality of the literary over the seemingly elusive character of the purely aural. In the version of this article that appeared in *Toward a New Music*, Chávez adds a long introduction that expands the discussion about the parallel development of these "complex" musics and the musical instruments capable of acoustically materializing them. This argument concludes in the third article, "Los instrumentos eléctricos de reproducción musical," which expands the discussion to include the development of mechanical and electric musical instruments as the medium that materializes the ideas of the composer. The article ends with a celebration of the recording technologies he witnessed at the Bell Laboratories as the ideal media for the faithful reproduction of music and its circulation into the future.[59] The version of this article published in *Toward a New Music* is much longer and details the development and technical characteristics of the electric phonograph as well as the possibilities it enables to internationally circulate musics from the remotest corners of the planet. In that sense, Chávez celebrates the anthropological use of technology that, in his opinion, allowed the opportunity "to keep an exact image of the successive stages of traditional music" and in a way anticipates and welcomes the process of musical massification that Michael Denning describes in *Noise Uprising* (2015) when the Mexican composer states that "this great musical wealth ought to be spread from one country to another. It ought to bind humanity together. A universal diffusion of music, realized in the double sense of geography and history, is a thing only the phonograph could achieve."[60] Chávez's celebratory tone articulates both the archival and the museographic labor of academic institutions as well as the commercial and market-oriented work of music labels. However, if we lend a critical ear to these passages, we can also clearly identify the Eurocentric neocolonialist overtones informing the composer's ideas. The teleological and rescueist rhetoric as well as the reifying, alienating, and disempowering character of the project argues for the recording, circulation, and global consumption of local musics that in the process are uprooted and decontextualized in an effort to make them into consumption goods of arbitrary purchase-sale value. Such rhetorical strategies reflect a neocolonial, almost extractivist ideology at the core of Chávez's listening of the imagined future (as promised in cutting-edge modern technology) as well as the imagined past (as retroactively fantasized in the sounds of traditional local musics).

The remaining chapters of *Toward a New Music* center on sound film, radio, and electric musical appliances. The chapter about sound film focuses on the technical aspects of sound recording and postproduction. Here, Chávez is especially interested in the transduction of sound into light as part of the optical sound technology that made the sound-on-film format possible—exemplified here with the Western Electric (vitaphone) and the RCA (photophone) systems (figure 1.4). Besides that, Chávez also takes time to describe cinema "as an organization for spreading ideas and feelings [whose] power is unequaled. In this sense, the movies are an institution comparable only to the Church in its best times.... Their social value will constantly be taken advantage of more directly in favor of the particular educational or political tendencies of governments."[61] It is no surprise that Chávez's interest in cinema was stronger than ever in those years, especially in terms of ideological propaganda—as expressed in his book and as evidenced in the quinquennial cinematography plan he proposed to the Department of Fine Arts at the time; this interest was in line with the modernizing ideological project of the Mexican proto–Aural City that coalesced around his figure as cultural broker.[62]

If the interest in technology as the factor guiding and shaping a future is clear in these chapters, the proleptic character of the modernizing project is very evident in the chapter about electric instruments. In this text Chávez focuses on the potential of these instruments to control physical and acoustic variables (pitch, intensity, duration, and timbre) that he considered to be the very essence of sound.[63] If these ideas remind us of the empirical discussion about sound in Castañeda and Mendoza's *Instrumental precortesiano*, the ideological connection between the authors and their books is even more palpable in Chávez's teleological discussion about scales at

FIGURE 1.4. Transduction: inscription of sound through light with the vitaphone. Drawing by Antonio Ruiz, in Carlos Chávez, *Toward a New Music* (1937).

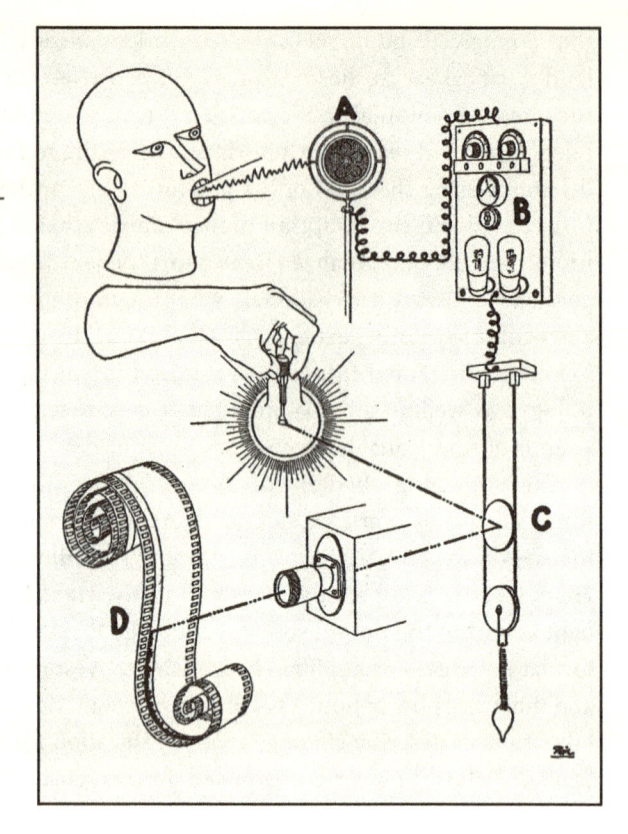

the end of this chapter. Here, Chávez not only reproduces Castañeda and Mendoza's social evolutionist argument about pentatonicism and "primitive" music but expands it to include a discussion of chords that feature the highest partials in an overtone series to argue for a teleological interpretation of complex harmonies as in "a constant movement toward intervals smaller than the whole tone."[64] Furthermore, in a move reminiscent of Castañeda and Mendoza's apology for the "rigorously civilized" character of pre-Hispanic music, Chávez concludes his argument with a daring paternalistic statement that "the so-called exotic musics of the Arabs, the Hindus, and many American Indigenous are a manifestation of their intuitive search for the musical expressions of the intervals corresponding to the high harmonies of the scale of concomitant sounds."[65] For Chávez, the only way to faithfully reproduce the nuances in these tuning systems is through electric instruments capable of meticulously controlling basic sound parameters of pitch and intonation that acoustic instruments cannot completely control.[66]

While Castañeda and Mendoza invoked the scientific method as a path to retrospectively construct the Other, Chávez's argument revindicates the performativity of Otherness of his colleagues in order to imagine a future in which science offers the only possible way to understand an epistemological enigma conceived and imposed over the past also in a retroactive way. A contrapuntal reading of these two books shows that for the postrevolutionary Mexican Lettered City—as well as the proto–Aural City of the 1930s—science appears as a tool for understanding the social as an engine of technological progress. In this sense, science is seen as an instrument of control of both past and future. In an attempt to read *Toward a New Music* in relation to Chávez's flirtation with futuristic ideas in the 1920s and 1930s, Gloria Carmona's introductory study in the book's first Spanish edition focuses on the creative possibilities of new sound production and reproduction technologies in the 1920s and 1930s.[67] Nevertheless, Carmona's aestheticist take misses the ideological-political framework in which Chávez's ideas acquired wider cultural and historical meaning as well as the very specific language in his book that reveals it. Contrary to Carmona, I consider that an acoustically tuned reading of Chávez's book, one that listens in detail for those resonances, provides a point of entry into understanding not only Chávez but also the world of senses and structures of feeling that he inhabited. By establishing a connection between his and Castañeda and Mendoza's ideas, one creates an archival constellation that reflects and reveals the intersection between intellectual and political labor in Mexico in the early 1930s.

Back to the 1940s: The Future and Its Everlasting Attributes

Listening is a process in which certain habits, objects, and physical and epistemic repositories conform a data storage that, as Ann Cvetkovich describes an archive of feelings, "[is] encoded not only in the content of the texts themselves but in the practices that surround their production and reception."[68] It is with that conceptualization about storage and retrieval in mind that I return to the exhibit *Twenty Centuries of Mexican Art* at the MoMA, to *Xochipili-Macuilxochitl*, and to the recording collection *A Program of Mexican Music*. Listened to through the archival constellation generated in the contrapuntal sway between *Instrumental precortesiano* and *Toward a New Music*, the exhibit, the concert, and the LP can be understood as the condensation and materialization, eight years later, of the ideas that the books phantasmagorically shelter.[69] In other words, *Twenty Centuries of Mexican Art* and *A Program of Mexican Music* would also be a core part of this archival constellation because they can be heard in the sense of Cvetkovich's description of an archive of feelings. Chávez's and Castañeda and Mendoza's very act of writing is a technology of entextualization and storage that generates an archive since both the empirical descriptions in *Instrumental precortesiano* and the teleological argumentation in *Toward a New Music* speak of an attempt to capture an aural authenticity through inscriptions of future reproduction. Taken as part of that constellation, the exhibit is a space for the invention of a sonic past based on Castañeda and Mendoza's instruments, their reconstruction thanks to the financial sponsorship of the Rockefeller Foundation, and Chávez's use of them to compose his "imagined Aztec music." That space, the materials that conform it, and their dissemination—through radio broadcasts as well as up-to-date marketing strategies—made a present reality into the futurity one hears in the polyphonic reading of *Instrumental precortesiano* and *Toward a New Music*. Listening in detail to it also puts in evidence the type of hungry listening that informs the whole endeavor.

Although it is true that the epistemic frame of reference and the ideas in *Instrumental precortesiano* and *Toward a New Music* can be found in a more dispersed way in a variety of writings by several national and international artists and intellectuals of the time, what makes this contrapuntal listening relevant is that it provides us with tools for thinking the space that Chávez generated out of his institutional positionality as a project in which artistic, intellectual, and political labor worked hand in hand. Thus, the "acoustically tuned" exploration of the written archive leads us to listen to how the project of the Lettered City is transduced into a proto–Aural City that generates the canonic indigenist/modernist/nationalist narrative that fueled the imagination of the past and the future of the nation throughout the second half of the twentieth century. Thinking about this particular concatenation of narratives in terms of hunger/devouring as metaphors for the colonial cultural encounter is telling because it forces us to understand the colonialist and modernist projects and the surpluses they generate as a continuum. If the colonial overtones are clearer in *Instrumental precortesiano* than in *Toward a New Music*, that is because, while Chávez's project seems to simply and innocuously point toward the future, Castañeda and Mendoza's devouring of the Indigenous Other in order to produce a past that would work as the foundation for a possible future is clearly an extractivist endeavor. However, when we establish a link between the books and the MoMA exhibit as a transhistorical archival constellation, the type of hungry listening informing Chávez's project also becomes evident. If, as Oswald de Andrade argues in his "Manifesto antropófago" (1928), the devouring of European colonial culture is fundamental in the production of local cosmopolitan modernist projects, Chávez's articulation of the Other in the context of the MoMA exhibit provides an unexpected twist to an avant-gardist call often read as anticolonial.[70] This archival constellation puts in evidence the cannibalistic character of modernism as a project that does not discriminate when it comes to selecting those to be sacrificed at the altar of its futurist utopia.

This chapter opens with an epigraph from Adolfo Bioy Casares's classic science fiction novel *La invención de Morel* that is relevant to further understanding the performative work of the archival constellation I have proposed here. Reflecting on the technology that creates the hyperreality he experiences on the desert island he is hiding on, the Fugitive, the unnamed main character in Bioy Casares's book, states, "That would allow us to feel we are always in a new life, because there will be no other memories of the projection than those that take place in the moment of recording, and because the future, many times left behind, will always maintain its attributes."[71] Bioy Casares's story presents us with an archive of memories that, by evading any real past, have currency only as triggers of futurity. It is only in the future that the Fugitive imagines and continuously reconfigures on the basis of the projections he has access to that what he experiences as reality makes sense. The future maintains its attributes because of the gaze of the observer and the way they conceptualize a narrative of reality according to their own desires. The archival constellation I examined in this chapter presents us with a similar situation. In the impossibility of having real access to the past, the objects in the archive and the modernist labor and practices that make them into archival materials provide an entry into imagining the past in relation to the aspirations and desires for the future. And as Bioy Casares suggests, exercises that imagine the future always imply a libidinal postponement that makes that future's imaginary features into everlasting attributes. As such, visits to the archive that provide us with tools to dissect how that future was imagined, independently of whether it became a reality or not, will always be the delightful source of a peculiar and uncanny intellectual concupiscence.

The articulation of the books *Instrumental precortesiano* and *Hacia una nueva música/Toward a New Music*, the MoMA exhibit *Twenty Centuries of Mexican Art*, and the Columbia Records LP *A Program of Mexican Music* as an archival constellation or performance complex in this chapter is threefold. First, it introduces the reader to a Mexican intellectual community, a proto–Aural City of sorts for whom aurality and sound became points of entry into articulating a teleological narrative behind the revolutionary regime's nation-building project, thus foreshadowing the episteme of what we could more formally label a Mexican Aural City at the turn of the twenty-first century. Second, it enables the reader to become familiar with a critical take on the canonic nationalist narrative that the Aural City projects studied in chapters 2 and 3 of this book engage with and react against in productive and reproductive ways. Third, it shows the labor of researchers as central in the assemblage of their own working archives and illustrates the ways in which these can be articulated, rendered visible, and productively activated through the rhizomatic constitution of relational performance complexes. This strategy and operation are focal in the identification of the type of invisible archives discussed in chapter 7.

Patrimony, Objectification, and Representation at Mexico's Fonoteca Nacional

El universo estaba justificado,
el universo bruscamente usurpó
las dimensiones ilimitadas de la esperanza.

(The universe was justified, the universe suddenly usurped the unlimited dimensions of hope.)

—Jorge Luis Borges, "La biblioteca de Babel" (1941)

On March 21, 2013, I took a group of students from the Universidad Nacional Autónoma de México (UNAM) for a guided visit of the Fonoteca Nacional (National Sound Archive) in Mexico City. I had heard a lot about this relatively new governmental project and wanted to take advantage of being in Mexico City teaching a sound studies seminar at UNAM's Escuela Nacional de Música to visit the Fonoteca and learn more about their goals and the kinds of projects they sponsor. The tour guide took us first to the Murray Schafer Auditorium, where she proceeded to inform us about their weekly lecture and concert series as well as their temporary sound art exhibits, the ideas that led to the creation of their weekend programs Caminatas Sonoras (Sound walks) and Rodadas Sonoras (Sound bike rides), and explained the trajectory of their soundscape series and the free distribution of the resulting CDs among Fonoteca visitors. After a brief question-and-answer session, she took us to the Fonoteca installations and explained the projects they are used for. At the Octavio Paz Audiotheque, she demonstrated how to search, locate, access, and listen to the sound files and recordings in their remarkable collection.[1] At the Thomas Stanford and Henrietta Yurchenco Halls, we learned the details of the many workshops and academic activities they plan and program.[2] And at the Salvador Novo Library, we were able to go through some of the books in their book collection.[3] She also explained how the archive came to be, due to numerous acquisitions as well as unsolicited donations, and how it has grown exponentially since the founding of the Fonoteca in 2008, making it difficult for the archive workers not only to keep up with the classification and cataloging of materials but also to simply store them and make them available to researchers. She mentioned that many of these unsolicited materials are personal collections of homemade audio and video recordings and memorabilia documenting domestic or semidomestic everyday activities that may hold dear sentimental value for their owners but for which the Fonoteca has no use.

Amid the scratchy sounds of the oldest recording kept at the Fonoteca archive—a message from Mexican President Porfirio Díaz to Thomas Alva Edison recorded in 1909 on an Edison wax cylinder, which our guide proudly retrieved digitally for us to hear at the audiotheque—and the uncanny experience of listening to György Ligeti's *Lux Aeterna* (Eternal light; 1966) through the multichannel equipment at the Jardín Sonoro (Sonic Garden), I could not help but notice that the congratulatory tone of our guide somehow collided with my students' skeptical attitude (most of them had previous experiences visiting the Fonoteca).[4] While the guide celebrated the seeming ontological transparency of a project that seeks to "safeguard the country's sound patrimony [and] to give researchers, teachers, students, and the general public access to the sound heritage of Mexico," most of my students adopted a very defensive attitude toward this rhetoric.[5] Later, in the classroom, they acknowledged their lack of enthusiasm, and I figured out it was the result of a healthy skepticism when it comes to government-sponsored institutional projects that may seem too good to be true as well as their own previous experiences at the archive, when they often ran into numerous bureaucratic snags while trying to retrieve sound files and use them as part of their academic projects. Their reserved mindset resonated with my own concerns regarding some of the political and philosophical implications of an otherwise seemingly noble project. My main interest in this chapter is to explore the space at the intersection of the democratic access the Fonoteca project celebrates and the civilizing project it stems from within the specific context of the rise and consolidation of a Mexican Aural City at the start of the twenty-first century. In doing this, I discuss some of the ontological shortcomings of this sound archive project and offer possible ways to respond to them.

One could consider the Fonoteca as both a cultural and a political project. As a cultural project, the Fonoteca seems to be a response to a phenomenon Ana María Ochoa Gautier describes as a public sphere "increasingly mediated by the aural" and an "intensification of the aural that has come about due to the technological, economic and structural transformation of the modes of circulation of sound in the past two decades."[6] In Mexico the increasing relevance of sound culture within a sector of the educated middle and upper classes—an interest that tends to bypass national institutions traditionally thought of as the gatekeepers of sound culture in Mexican society, such as music or folklore academic programs and centers—seems to point toward a new epistemological model, one in which sound and aurality become as important as the written word in trying to make sense of the natural, social, and cultural worlds around us. Access to technology and information has allowed the transmutation of professionals like graphic designers, communication majors, philosophers, and art curators from passive consumers into active producers and participants in a number of alternative sound and music initiatives and scenes—from contemporary art music, noise, electronic dance music, or Sonido 13 scenes to acoustemology, acoustic ecology, urban sound design, and sound art and installation projects. It has also allowed the creation of sound-based collectives and activist initiatives that encourage new ways of engaging everyday social, economic, cultural, and ecological relations in the urban spaces the individuals who constitute these collectives live in order to propose novel ways of understanding and dealing with the complexities and contradictions of neoliberalism and late-capitalist urban life.[7] Although many of these individuals were trained in traditional music, musicology, and composition programs, a large majority actually come from other disciplines, and their contact with and understanding of sound was not mediated by the orthodoxy of music academia. Moreover, their important noncanonic sonic cultural capital (from their extensive knowledge of cosmopolitan alternative music practices and histories of experimental art to their participation in transnational sound ecology networks interested in addressing pressing global problems), creativity, and technological savviness do not merely make up for their lack of formal training in music or sound. In fact, their information and skills give them access to sound archives, auralities, and listening strategies often neglected and marginalized in music circles closely linked to the Lettered City project. These folks and their aural practices and labor are central to the configuration and shaping of the Aural City. For them, knowledge about alternative sound practices and participation in alternative sound scenes have become markers of cosmopolitan intellectual distinction.

As a political project, the Fonoteca is a response to the Mexican federal government recognizing the importance of sound heritage and the need to create an institution to safeguard such a legacy in the cultural program of its 2001–6 Plan Nacional de Desarrollo (National Development Plan).[8] Nevertheless, these noble patrimonial reasons also intersected with more worldly economic concerns, especially the government's growing awareness of the potential of culture in diversifying and galvanizing Mexico's tourism industry, which was also recognized as a key factor in the country's economic development in the 2001–6 Programa Nacional de Turismo (National Tourism Program).[9] As becomes clear in the following sections, several of the Fonoteca's institutional programs and projects were born out of this concatenation of patrimony and tourism.[10]

Contrasting the larger cultural and institutional projects behind the inception and subsequent endeavors of the Fonoteca Nacional offers a window into exploring what happens to the aesthetic and political principles of the Aural City as they intersect with official national projects. One could use the term *institutional Aural City* to refer to the labor of individuals in the Aural City who may choose to channel their interest in sound and aurality through projects sponsored by national institutions like the Fonoteca Nacional. However, the borders between this institutional Aural City and other, noninstitutional subsets of the Aural City are in fact much more porous than a dichotomy like this could clarify. While national projects provide a clear identity to the work of state institutions, the individuals who make these institutions often savvily move in and out of them to articulate projects that may be at odds with the mission of those institutions. This becomes evident in chapter 7 when following the noninstitutional work of several individuals who have been involved in the Fonoteca project. Thus, the notion of the institutional Aural City, rather than denoting specific individuals, refers to their labor under the specific conditions of larger institutional undertakings. It is precisely the seeming contradiction between the local labor of this institutional Aural City and the contestatory project of the larger Aural City, with its origin in alternative aural epistemes circulated transnationally, that this chapter addresses and assesses critically.

A Sound Archive for the Nation

At the end of the nineteenth century, the invention of the daguerreotype and the phonograph as well as the development of subsequent forms of recording technologies that allowed for the preservation of image and sound generated new types of delicate historical documents in need of protection, conservation, and archiving. Austria's Phonogrammarchiv, founded in 1899 by members of the Kaiserliche Akademie der Wissenschaften (Imperial Academy of Science), was the first sound archive in the world. The early large-scale sound recording collections gathered under the Phonogrammarchiv's aegis included field recordings from Greece, Brazil, Croatia, Papua New Guinea, Palestine (Jerusalem), and South Africa, retrieving music, religious rituals, and languages among other sonic materials, as well as recordings of local historical figures. These activities, with their emphasis on "capturing," studying (through a variety of transcribing methods that often imply acts of cultural translation and transduction), and documenting sounds and practices from "faraway and exotic" places as well as "great local individuals" (mostly men) for preservation, reveal the encyclopedic, civilizing, and largely imperialistic/nationalistic character of the project. In 1999, after being nominated by the Austrian government, the Phonogrammarchiv's historical collections (recordings from 1899 to 1950) were included as part of UNESCO's Memory of the World Register.[11] Regardless of the various goals of particular archival projects, it is precisely the civilizing patrimonial attitude informing these two moments—the archive's foundation and its recognition as world heritage—separated in time by almost a century, that has characterized the work of sound archives in national institutions throughout the world through the beginning of the twenty-first century.[12]

In Latin America, sound and audiovisual archives developed in the twentieth century as branches of larger national or regional archives and as leftovers of various kinds of larger private cultural enterprises—recording companies, radio broadcasting stations, television networks, and so on. However, centralized efforts to develop archives specifically devoted to sound are rather recent. Two of the most important sound collections in Brazil's Arquivo Nacional—the country's national archive, not a center focused on sound or audiovisual archives—are donations from private entities, the Casa Edison (one of the earlier recording companies, active in the country from 1900 to 1932) and Rádio Mayrink Veiga (a radio station founded in Rio de Janeiro in 1926).[13] The Argentinean audiovisual patrimony is scattered across a variety of nonspecialized archives, including the Archivo General de la Nación (which houses newsreels from private companies, TV collections, and government documents), the Biblioteca Nacional (newsreels), and other TV and film archives.[14] In Bolivia it was not until the late 1990s that the first audiovisual archives, intended to preserve the country's linguistic diversity, were opened after a governmental initiative.[15] In most cases, the idea of preserving a national patrimony has been the main motivation for the creation of larger state-sponsored initiatives to share in or take over the archiving of those materials from private institutions. In other words, the impetus for the creation of these archival centers has almost always been a patrimonial attitude within a larger nationalistic vision.

The case of Mexican sound archives is no different. The Fonoteca del Instituto Nacional de Antropología e Historia (Fonoteca del INAH; Sound Archive of the National Institute for Anthropology and History) and the Filmoteca de la UNAM (Film Archive of the UNAM), both created in 1964, were the first audiovisual and sound archives in the country. They were both founded under the spell of nationalism and patrimony as ideological frameworks. The Fonoteca del INAH was started in response to the formation of the field of ethnomusicology in Mexico, under the influence of folk studies, and thus the project came about largely as an archive of traditional musics, ethnographic interviews with musicians and practitioners, and oral histories. Its avowed goal was to gather the "sonic memory of a people."[16] The Filmoteca de la UNAM was created as an institution in charge of "rescuing, restoring, preserving, and disseminating moving images as well as their sound elements, written and iconographic documents, and cinematographic devices. This is the national and international heritage preserved by the University."[17] The creation of both institutions evokes the rhetoric about memory, patrimony, and nation that gave birth to the Austrian Phonogrammarchiv. With these projects, both the INAH and UNAM reproduce the enlightened ideas behind the Austrian Phonogrammarchiv project while validating the same type of political institution, the nation-state—albeit at a more local but nonetheless equally patronizing, colonizing, and exoticizing level.

Plans for the establishment of a Fonoteca Nacional in Mexico began in 2001, when Lidia Camacho, at the time director of Radio Educación, the radio station of the Instituto Politécnico Nacional, presented Consejo Nacional para la Cultura y las Artes (CONACULTA), the country's national culture and arts council, with a proposal for the creation of a Fonoteca Nacional.[18] One of the most salient arguments in favor of founding the

Fonoteca was that the country's sound memory was in real danger of disappearing due to the fragility of the formats in which these historical recordings were made. Thus, as Perla Olivia Rodríguez Reséndiz argues, the Fonoteca was conceptualized from the beginning as "the repository of the sounds of the nation."[19] It took seven years for the project to come to fruition as it was necessary to find the right building, prepare it for storing the fragile materials, gather and digitize the collections, develop a system for public use, and put together a team of specialists trained to work with sound archives. Under Camacho's directorship, the Fonoteca Nacional finally opened in Casa Alvarado, the colonial building that hosts it, in Mexico City's Coyoacán neighborhood on December 10, 2008.

Sandra Rozental argues that "the Spanish term patrimonio condenses both its English equivalents 'heritage' and 'inheritance' while also indexing the patriarchal power relations that have historically ensured the existence of the Mexican state. [The 'pater' of 'patria'] implies the Mexican state's political sovereignty over certain lands and their assets, as well as over its citizens."[20] Thus, Rodríguez Reséndiz's conceptualization of the Fonoteca as "the repository of the sounds of the nation" is a clear pronouncement of the patrimonial character of the Fonoteca project. From the outset, its raison d'être was not only to safeguard but also to define and shape the sonic property of the nation. Thus, gathering sound objects for the Fonoteca archive was one of the early challenges of the institution. A curatorial committee formed of musicians, researchers, specialists in the preservation of sound archives, and other intellectuals evaluated a series of collections and recommended acquiring them: purchasing them, borrowing them temporarily, offering space to keep them safe, establishing exchange programs with national and international institutions, or accepting donations from private or public entities and individuals.[21] Central to the initial conformation of the archive were the Thomas Stanford Collection (consisting of over five thousand ethnographic recordings of a wide variety of ethnic groups throughout the country), the Radiopolis Collection (which includes over twelve thousand recordings of radio shows produced by XEW, one of the most influential Mexican radio stations of the twentieth century), and the Rebeca Rangel and the Armando Pous collections (made of historical 78 rpm records), as well as smaller collections of ethnographic field recordings by neurologist and ethno-cinematographer John C. Lilly Jr. and ethnologist Alfonso Muñoz Güemes.[22] In 2020, twelve years after its formal opening, the Fonoteca housed more than half a million sound documents organized in over two hundred collections and compilations. Eight of the

Fonoteca's core collections (the Thomas Stanford Collection; the Raúl Hellmer Collection; the Henrietta Yurchenco Collection; the Baruj "Beno" Lieberman, Enrique Ramírez de Arellano, and Eduardo Llerenas Collection; the *De Puntitas* Collection; the *El Foro de la Mujer* Collection; the Estudios Churubusco Collection; and the Encuentros de Música y Danza Indígena Collection) were included in UNESCO's Memory of the World Register between 2010 and 2019.[23] Nevertheless, more than two-thirds of the Fonoteca's total holdings remained uncataloged at the time, which speaks to the institution's limitations when trying to handle the sheer amount of materials it was able to amass in such a short period of time.[24]

Recycling the Sound Archive: Sound Map and Soundscape Projects

One of the main differences between the Fonoteca Nacional and other similar institutions in Mexico and Latin America lies in the fact that central to its task is what Perla Olivia Rodríguez Reséndiz calls "making the sound archive alive," by generating a culture of sonic awareness and artistic sound recycling.[25] Rodríguez Reséndiz explains that the objective of the Fonoteca "is not only preserving part of our identity but also reutilizing [the sounds] in educational, cultural, and scientific realms."[26] Thus, the institution was born out of the idea of not only providing a space for the safeguarding of historical sound files but also transcending the borders of the archive and offering a number of educational and creative programs that allow for the development of a "listening culture" among general audiences as well as for artists to use these files to create novel sound art pieces, installations, and exhibits.

Rodríguez Reséndiz argues that these programs have succeeded because "sound culture is deeply embedded in [Mexican] society. I would say that although it does not have the preeminence of visual culture, [sound] catches people's attention; it is very present [in their lives]. You just need to provoke them a little bit, and they get really interested."[27] Central to the Fonoteca Nacional project is the issue of democratization of sound that Ochoa Gautier has identified as the "intensification of the aural" in Latin America, as discussed earlier. This is evident in the design of the institution's goal of fostering participation by audiences and artists, an apparent response to this intensification. Thus, one of the keys to the success of the Fonoteca Nacional has been its managerial staff's ability to identify their audiences' desires and aspirations and respond to them accordingly. Besides the edu-

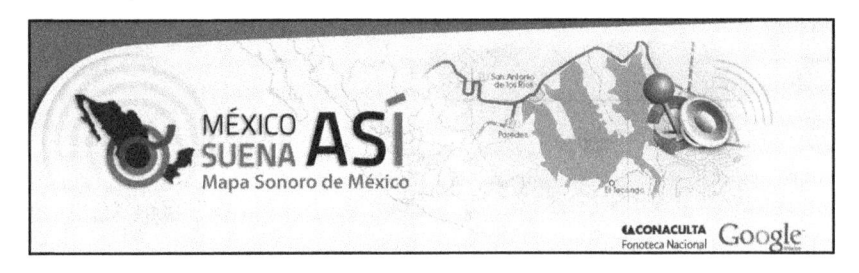

FIGURE 2.1. Official iconography of the project México Suena Así: Mapa Sonoro de México (Mexico sounds like this: Sound map of Mexico), a collaboration of the Fonoteca Nacional with Google. Screenshot, October 14, 2014.

cational workshops, guided sound walks, sound bike rides, and sound art offerings, the most salient Fonoteca projects are the sound map of Mexico and the series of soundscapes from different Mexican states. México Suena Así (Mexico sounds like this), their sound map project, is a participatory internet initiative developed in conjunction with Google (see figure 2.1).

Started in 2010, the project seeks to encourage people from all over the country to record and upload the sounds they identify as part of their close environment in an attempt to "map out the sonic geography of Mexico.... [I]t is an invitation to listen to what Mexico sounds like and develop a community of listeners interested in capturing [the country's] sonorities for the appreciation and delight of all of its visitors."[28] The website instructs users to upload up-to-ten-minute-long files in specific formats; the website also warns users that their sound files should qualify as "field recordings, meaning they should have a testimonial character that accounts for 'a here and now'[;] unlike studio recordings they are done on-site and recover natural or urban everyday sounds that show specific circumstances, such as natural environments, traditional village festivities, people's everyday speech, trades, and a variety of musics, such as street music or that which accompanies a ritual."[29] By the end of 2014, 380 sound files had been uploaded into the map (figure 2.2).

A large majority of those files represent sounds from urban areas, with more than half (224 samples) corresponding to sounds from Mexico City. Most of these urban files were uploaded by private users, while a majority of the audio samples from rural areas or snippets from Indigenous communities (both music and speech) tend to have been uploaded by Fonoteca staff. It would be a mistake to dismiss the Mexican countryside as technologically unsophisticated and assume that a perceived absence of internet access or internet habits among people from these areas may

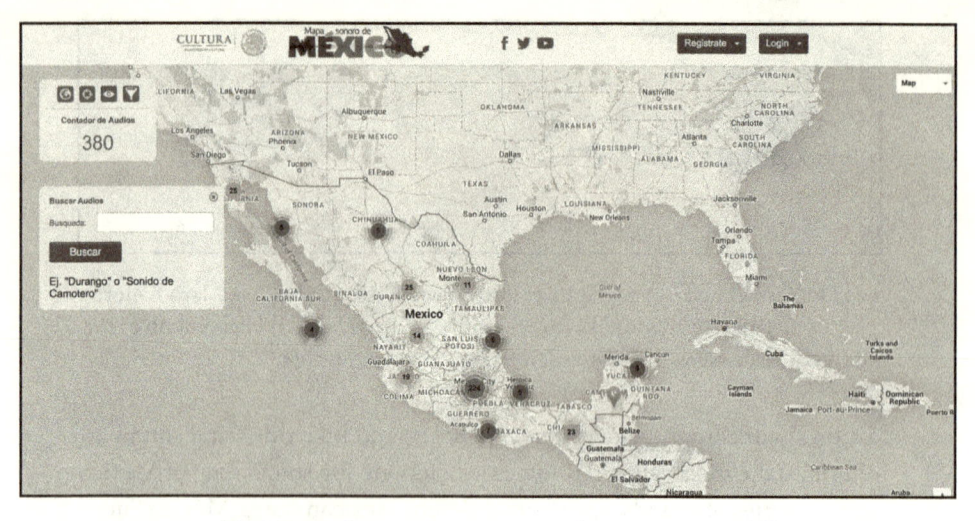

FIGURE 2.2. The México Suena Así project map, showing 380 sound files uploaded. Screenshot, October 14, 2014.

have prevented them from participating in the Fonoteca project. Increasing access to technology and modern forms of communication is fundamental in the lives of people in the Mexican countryside in the context of emigration to the United States, which has characterized these areas for decades. The use of these technologies allows for an emotional and inexpensive connectivity between those who stay and those who leave.[30] As such, the abundance of urban sounds is revealing of the elite that controls México Suena Así as a representation project. Furthermore, as Natalia Bieletto-Bueno suggests, the dominant presence of sounds from upper-middle-class Mexico City neighborhoods (such as Coyoacán, where the Fonoteca is located), to the neglect of lower-class districts, signals a "deficient acknowledgment of the existence of these sonic environments and thus of the people who live and make them possible."[31] This preeminence of the urban (especially affluent areas of the urban) over the countryside is a curious trend that acquires more relevance when read vis-à-vis the trajectory of the Fonoteca's soundscape project.

In November 2010 the Fonoteca Nacional hosted the IV Encuentro Iberoamericano de Paisaje Sonoro (Fourth Ibero-American Soundscape Conference). The occasion presented the Fonoteca with an opportunity to launch its series of soundscapes with the presentation of two CDs, *Paisaje sonoro de San Luis Potosí* (Soundscape of the state of San Luis Potosí; 2009) and *Paisaje sonoro de Veracruz* (Soundscape of the state of Vera-

FIGURE 2.3. CD covers for *Paisaje sonoro de Veracruz* (2009) and *Paisaje sonoro de San Luis Potosí* (2009). Paisaje Sonoro de México Collection. Fonoteca Nacional.

cruz; 2009) (figure 2.3). The soundscape project was a continuation of an older endeavor started by Lidia Camacho, director of the Fonoteca, at her previous post at Radio Educación. There, she had entrusted the legendary Mexican electronic and ethnofusion musician and sound artist Jorge Reyes (1952–2009) and German sound engineer Peter Avar (b. 1962) with developing a series of sound pieces based on the recording of emblematic sounds from the states of Michoacán and Chiapas, as well as Mexico City. On Camacho's move to the Fonoteca Nacional, the project also migrated. At her new home, Camacho incorporated Francisco "Tito" Rivas (b. 1977), a sound artist formally trained in communications and philosophy, first as an assistant and later, after Reyes's death, as the person in charge of the project. Except for the *Paisaje sonoro de Veracruz*, which was recorded and produced by Avar, Rivas, and Arturo Jiménez, the rest of the soundscapes presented in 2010 (of San Luis Potosí, and the ones produced earlier by Radio Educación: of Michoacán, Chiapas, and Mexico City) were all recorded, edited, and assembled under the artistic direction of Reyes. Later, Rivas continued collaborating with Avar, and together they produced *Paisaje sonoro de Oaxaca* (Soundscape of the state of Oaxaca; 2012), *Paisaje sonoro de Guerrero* (Soundscape of the state of Guerrero; 2013), and *Paisaje sonoro de Chihuahua* (Soundscape of the state of Chihuahua; 2015).[32] According to Rivas, the first five soundscapes in the series were made with an artistic perspective rather than an archival one; he states that "originally there was no attempt at developing a soundscapes project; it was all about making artistic pieces based on field recordings."[33] Once the project was taken over by the Fonoteca Nacional and Rivas became part of the producing team, recording sounds for the conformation of the Fonoteca's own sound archive also became a priority. Thus, all the sound files recorded for each of these soundscapes (about sixty hours per project) were cataloged, archived, and made available free for listening to Fonoteca users.

Recording in the Field, Performing the Field

The personalities of Reyes and Rivas, neither of whom was trained in ethnomusicology or anthropology, defined how the different soundscapes of the series were approached and recorded. Being a more intuitive person, Reyes approached field recording by means of continuous improvisation. According to Rivas, "The way it was done with Jorge [Reyes] was freer. It was like 'Let's see what we find. When we get there, we start asking questions [of local people], and let's see where that leads us.' It was a little bit chancier and more circumstantial.... There was no predetermined plan; it was more like an adventure."[34] When Rivas took over the coordination of the project, he started devoting more time to preproduction in order to take better advantage of the little time that, due to budget constraints, the recording team was able to spend in the field. For Rivas, one of the goals of his field trips was to gather sounds for the Fonoteca archive, not just to find sounds for the creation of an artistic soundscape. Thus, it was necessary to know exactly what types of sounds they wanted to record in order to ensure they were in the right place at the right time. Preproduction made the project more efficient but also created unforeseen problems. According to Rivas, researching what to record is difficult because when looking at travel guides for each of the places visited, "you have to imagine what is left untold on the basis of what [the travel guides] say, and more or less think whether [those places] may be useful from a sonic point of view."[35] To better imagine the sounds in relation to situated spaces they had not experienced, the production team had to rely on the knowledge of local folks. However, the epistemological gap between researcher and locals also proved problematic because, as Rivas states, "When you ask someone in a town, they answer: 'What do you mean, what do we hear here? What do you mean, where is something good to listen to?' ... In the end it is an issue of making them aware because it is not in their discourse.... [I]t implies explaining to them what a soundscape is, what the value of those sounds is, and giving them examples."[36] It should be noted that both Reyes and Rivas relied on the experience of locals to find sources for the sounds they wanted to find; as Rodríguez Reséndiz states, "The local artist would help us make a selection based on important historical, social, and cultural criteria."[37] In all cases, Reyes and Rivas sought the collaboration of local artists and regular folks in order to identify the places and moments that offered the possibility of sonic markers of local identity. Nevertheless, the type of epistemological gap described by Rivas discloses the character of an endeavor that strives to present itself as a democratic, bottom-up project but is rather informed by complex top-down power dynamics.

A central difference between Reyes's creative and Rivas's systematic approaches to field recording was their interaction with the environment. As Rodríguez Reséndiz informs, "Reyes was an active participant within the soundscape not only because he took part in the selection of locations ... but also because he interacted with the landscape. [He actively] generated sounds from the environment, and thus the sounds provoked by the artist are an addition to the soundscape."[38] In a nod to the illusory idea of low-impact recording, Rivas attempts to erase himself from the field by simply recording the sounds he encounters during his field trips. Although this approach is practical since the systematization of his preproduction phase allows him to have a clearer idea of the types of sounds he will find once he is in the field, it also predetermines the field regardless of Rivas's attempt to negotiate his absence/presence. Therefore, while Reyes was ready to create the sounds he needed and those he imagined while in the field, Rivas preimagines the sounds he wants and enters the field with the goal of finding them. These two approaches speak of two very different understandings of how the field could be performed: as a process open to creativity, invention, and imagination in which the producer unabashedly admits to being part of the field or as a process of identifying fragmented pieces in a larger preimagined sound puzzle. Regardless of the patently different approaches to the field, both experiences refer to idealized soundscapes whose existence in the imagination, either as a predetermined sonic blueprint or as an improvised sonic process, guide both the act of recording and the montage of the final sonic product. This idealization of the field as a space that one must listen to in either a creative or an arguably self-erased way presupposes that the field exists before the arrival of the recorders. However, as Mark Peter Wright argues, "as it turns out, [the field] never existed in the first place. Instead, the indexical nature of geology and organic life stratifies recordists into its polyvocal stories across time and space. In other words, recordists are always part of the landscape's voice, whether audible or not," and thus perform the field with their actions.[39] Furthermore, one could argue that the imaginary character of the resulting soundscapes works as a type of simulacrum in Jean Baudrillard's sense, an idealized representation made of fragmented "memory banks" that takes over reality.[40] A good example of this dynamic is the reaction of Rivas and his team to the setbacks experienced while doing fieldwork for *Paisaje sonoro de Guerrero*. When the production team arrived in the field, they realized that it was impossible to follow their preproduction plan due to the extreme narco- and military violence that has affected the state of Guerrero since the early 2000s. Rivas explains that when they tried to access some of the places they had planned to visit, the military discouraged them, arguing that they would not be able to guarantee the safety of the researchers.[41] The reaction of Rivas and the team, who avoided going to these areas, may have been a practical decision based on understandable safety concerns. Nevertheless, it also reveals a fixed ontology regarding the sonic world they expected to hear and a refusal to listen to the sounds of violence that folks in Guerrero experienced as the real soundscape of their everyday lives—or at least a refusal to include them as part of the soundscape they sought to "preserve." It may be understandable that a state-sponsored project would intend to suppress the sounds that reveal the crisis of that very state, choosing instead to celebrate a romantic and even touristic notion of what a particular place should sound like. It could not be otherwise within the patrimonial logic at the core of the Fonoteca project. However, this is not a minor shortcoming given that the bursts of Kalashnikovs or AR-15 assault rifles not only allow people living in these areas to assess their surroundings but also dictate their everyday behavior. Removing the sounds of violence from the more touristic and sanitized soundscapes celebrated by the Fonoteca project clearly leaves out a central element in how people from Guerrero understand their world and their sense of survival. Although it would be an irresponsible act of cynicism for a state-sponsored institution to include these sounds of violence in the highly contested current political and social moment of the country, the decision to leave them out also contradicts the very raison d'être of these soundscapes. Under these conditions, the soundscape cannot be the sonic window into particular subjectivities and epistemologies it is meant to be. The basic conundrum that a soundscape project like this one faces is that the social and political conditions in which it unfolds make it an oxymoron; they make it an impossibility and a contradiction in and of itself. A closer look at the very conceptualization of soundscapes as cultural artifacts and practices further accentuates their shortcomings in relation to their implementation at the core of the Fonoteca's program of activities.

Soundscapes, Listening, and the Body

The term *soundscape* was coined by American urban planner Michael Southworth in an article published in 1969. There, he used the concept to refer to aspects of a sonic environment that mark and identify a city. Southworth affirms that these types of sonic environments are defined on the basis of "the uniqueness or singularity of local sounds in relation to those of other city settings and... their informativeness or the extent to which a place's activity and spatial form [can be] communicated by sound."[42] In other words, Southworth's interest in soundscapes lies in the exploration of how sonic experiences are essential to the way in which individuals recognize and identify with the urban spaces they move through in their everyday lives, and how they find in and derive from them a type of aural enjoyment.[43] As such, the way in which Southworth understands soundscapes is intimately related to the bodies of those who listen, their pleasure, and their urban routes—or, to be more precise, the way in which the sonic experience informs the way in which those bodies develop, use, and enjoy those urban routes. For Southworth, the soundscape is not just about the sounds per se; it is mostly about the bodies that listen and their aural affectivity.

Nevertheless, the concept of the soundscape comes to us today largely through a different genealogical route, that of sound studies, acoustic ecology, and the so-called ecomusicology, which recovers the concept out of its reinvention by Canadian composer and environmentalist R. Murray Schafer (1933–2021) in the context of the early 1970s World Soundscape Project. Schafer's aesthetic and political posture is clear in his book *The Tuning of the World* (1977), which summarizes and interprets the results of that project. There, Schafer identifies noise pollution as a major world problem due to the indiscriminate proliferation of sounds he deemed vulgar and undesirable, which, rather than being diminished, should be systematically resisted. To accomplish that, he proposes a program that leads us to ask ourselves, "Which sounds do we want to preserve, encourage, multiply? When we know this, the boring or destructive sounds will be conspicuous enough and we will know why we must eliminate them. Only a total appreciation of the acoustic environment can give us the resources for improving the orchestration of the world soundscape."[44] Schafer points out that a soundscape can be "any acoustic field of study. We may speak of a musical composition as a soundscape, or a radio program as a soundscape or an acoustic environment as a soundscape."[45] But it is evident that his agenda privileges acoustic environments as objects of study. It is precisely in that sense that Schafer offers a program to uncover the most meaningful features in any given soundscape based on the identification of key sounds, signals, sound marks, and archetypal sounds. He argues that to be able to discern those features, one must pay attention to the geography and climate of the landscape, to the socially codified "acoustic warning devices," to the sounds that local communities consider special and thus deserving of being "protected," and to the "mysterious ancient sounds" that possess a specific type of cultural symbolism.[46]

Regardless of an avowed but timid nod to the tolerance of aesthetic and multicultural diversity, Schafer defines soundscapes in direct relation to a strictly ecological and environmentalist frame of reference. In this context, sound acquires a patrimonial character that makes it worthy of being preserved. In contrast to the body's central role in Southworth's earlier conceptualization of soundscapes, in the type of sound economy detailed by Schafer, listening and the listener's body occupy a secondary role since what is important for his agenda is protecting those sounds that have been ascribed value a priori rather than understanding the performative potential of the aural experience. These types of exercises not only conceptualize listeners as abstractions and neglect their bodies but also render discursively invisible the mediation of the bodies of the individuals in charge of recording the sounds. Thus, the listening of the sound recorders, which in fact mediates a natural sonic environment and the possible aesthetic soundscapes such recordings may eventually generate, is naturalized, and the ideological framework that informs their actions is rendered discursively transparent. In this context, the notion of soundscape seems to privilege a disciplined type of listening whose process of instruction responds to aesthetic, political, and cultural values that are made into units of universal assessment criteria by rendering invisible the body of its sonic mediator. This dual dynamic of negation and invisibility presupposes a type of listening that, beyond the recognition and identification of the sonic features that make a soundscape meaningful, as described by Schafer and as naturalized in the specific recording projects, unfolds as a rather contemplative experience. Tim Ingold claims that such an approach to soundscapes objectifies sound and obscures the experiential character of listening.[47] Furthermore, given that this definition implies a type of listening tasked with recognizing specific sound features, it creates the need to train listeners in what they should listen for and how. This situation makes apparent a contradiction between the patrimonial character of the soundscape project and the liberalizing and democratizing mission of the sonic turn that sound studies invokes: The type of disciplined listening that the former demands contrasts with the type of performative listening—the phenomenological experience that performs the listener's social and cultural environment—that the latter promotes. As Stefan Helmreich states, although Schafer's "pastoral conception ... no longer characterizes most mobilizations of soundscape—comfortable with urban worlds and broadcast space—contemporary treatments continue to approach soundscapes as things in the world, waiting to be tuned into."[48]

It is no coincidence that the Fonoteca's main auditorium is named after R. Murray Schafer. The Canadian's work has been an important influence in generating the objectives and developing the programs and activities of this institution; his environmentalist and patrimonial mission reverberates with the Fonoteca Nacional's soundscape project—in the theoretical framework it presumes, the types of sounds it values, and the type of listening it privileges. For that reason, dissecting Schafer's conceptualization of the soundscape is important, as such an exercise recognizes the notion's historical and epistemological limitations, which may help us problematize the apparent utopian transparency of the Fonoteca Nacional as an archival project.

Listening to the Fonoteca Nacional's Soundscape and Sound Map Projects

As mentioned, there seems to be a certain transparency in the rhetoric about the Fonoteca Nacional's soundscape and sound map projects. They invoke the democratic character of access to sound and the aural's challenge to the lettered model of modernity that early theorizations about a more inclusive aural public sphere celebrate. However, a closer look at the dynamics informing these projects may challenge their apparent transparency and show instead a number of institutional mediations that idealize the production, naturalize the representation, control the circulation, and shape the consumption of these sonic products.

México Suena Así, the country's sound map, was a project based on the participation of people throughout the country. At the outset, the very existence of the resulting sound map relied on the active participation of an audience willing to determine what their most significant places sounded like and able to record them. Google went so far as to give away digital sound recording devices at the launch of the project to ensure the ability of an interested audience to capture their most beloved sounds. However, by the end of 2014, sounds from urban areas heavily dominated the map, showing that for one reason or another the representation of Mexico's sonic space was de facto controlled by urban folks. Furthermore, the fact that the rural and Indigenous sounds present in the map were mostly uploaded by the Fonoteca team also suggests the likely possibility that the rural representation of Mexico featured in the map may in fact be the idealized fantasy that urban folks have of the countryside—the abundance of Indigenous speech and ritual dances among these sounds makes one believe that such urban idealization also reifies the Mexican countryside in an essentialist way as premodern and "authentic." The problematic arrangement of voices in this map may lead us to also ask about the voices that the soundscape project allows to speak and those it unintentionally silences.

The rhetoric about democratization of culture informs the soundscapes project as much as it does the sound map project. The production team is always quick to point out that each soundscape was a collaborative endeavor that involved the working together of the Fonoteca team and a local crew. Both Jorge Reyes and Tito Rivas relied on local folks to direct them to areas that were significant for them as sources of sonic affect. One could argue that the resulting soundscapes are oral palimpsests that reveal a multiplicity of voices (in preproduction, planning, selection, recording, etc.) behind the montage of the final product. Nevertheless, Reyes's and Rivas's authorial voices as well as their imagination dominate not only the final product but also the process leading to that final product, as exemplified in each of the Fonoteca soundscape CDs. From Reyes's creatively performing the field to Rivas's imagining sounds out of travel guides, the leading principle behind the conception and realization of the soundscapes is an idealization of sound that in both cases operates like a personal simulacrum that overtakes reality. This is more evident in the case of the Guerrero soundscape, when problems in accessing the field could have led the production team to question the validity and authenticity of their preimagined utopian soundscape. However, that did not occur, and not just because of the danger in attempting to sample the everyday soundscape that Guerrero citizens listen to but also because even considering the possibility of such a soundscape is out of the question; it does not fit the imaginary desires and fantasies of the individuals who make the Fonoteca or its local institutional partners. In fact, the type of idealized imaginaries that control the production of these soundscapes vividly resonates with the rhetoric behind Pueblos Mágicos (Magical Towns), as seen in figures 2.4 and 2.5. This was a federal government program launched in 2001 through the Secretaría de Turismo (Ministry of Tourism) to encourage an influx of visitors by advertising Mexico's towns as "sites with symbols and legends, villages with history that in many cases have been the stage of transcendental historical moments for our country, they are places that show the national identity of each of its corners, with a certain magic that emanates from its attractions; visiting them is an opportunity to discover the enchantment of Mexico."[49] At the intersection of magical realism, strategic essentialism, and the preservation of patrimony, both the Pueblos Mágicos program and the soundscape project of the Fonoteca Nacional exploit simulacra in which Otherness rearticulates local pride and cosmopolitan aspirations within the dynamics of neoliberal globalization. The process of sound reification encompasses the entire soundscapes project but is particularly evident in the recording and use of Indigenous speech and language. Here, the sounds of language are completely deprived of their semantic value and are instead cherished as endangered sounds. As such, these sounds enter the simulacrum machine of unlimited semiosis to become something only in relation to the desires of the listening Other.

Another problematic feature of the soundscapes project is evident in the epistemological tension between researchers and local folks. Rivas acknowledges that usually there is a knowledge gap between him and the people he seeks advice from to facilitate his entry into the field. I do not wish to focus on the researcher's notorious paternalistic tone as Rivas explained to local folks "what a soundscape is and what the value of those sounds is" (which illustrates the power struggles at play in the field); instead, I want to highlight the epistemological implications of such a statement. This incident not only shows the top-down, civilizing character of the soundscapes project and the tacit imposition of the researcher's perspective but also speaks of how this perspective may unintentionally pursue a fragmentation and compartmentalization of the sensorial experience that local folks may not understand or undergo in such a way. Rivas's interest in sound leads him to ask local folks if there is anything "good to listen to" in their communities, forcing them to focus on just the listening aspect of an

FIGURE 2.4. Official iconography advertising the Pueblos Mágicos (Magical Towns) program of Mexico's Secretaría de Turismo. Screenshot, October 14, 2014. Source: Operadora Turística Corporativa.

FIGURE 2.5. Official Pueblos Mágicos map as of 2014, showing twenty-eight new towns added to the original eighty-three. Screenshot, October 14, 2014. Source: Secretaría de Turismo.

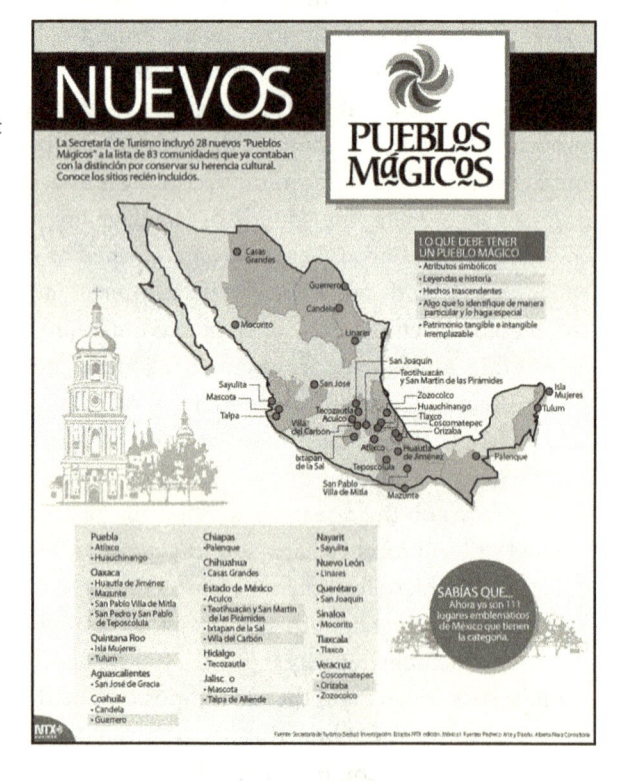

experience that, for all that matters, may be more multisensorial for them. In such cases, the soundscapes project may seem to aid a problematic epistemological transformation of the sensorial experience, a transformation that reverberates with well-known modernist Western projects of rationalization and colonization. Furthermore, the Fonoteca's patrimonial rhetoric along with its existential and essentialist search for sounds of local identification paves the way toward a naturalization of identity. This may work on the strategic level to attract the gaze of the outsider (as in the Pueblos Mágicos program), but it also inevitably ties the individuals referred to and their aspirations to specific spatial, geographic, and ecological realities. In a way, these types of dynamics essentialize Indigenous and rural folks, making them the de facto Other, in a move resembling what Susan Campos Fonseca has labeled *interior microcolonialism*. Campos Fonseca uses this term to describe how, in the case of the Costa Rican Noise scene, members of an experimental music elite "make sense of themselves through a possible internal Other; understanding it ultimately as a source of aural Otherness" established on the basis of "ethnographic materials from Indigenous communities so as to generate a local color understood as Costa Rican."[50] In the Fonoteca case, this interior microcolonial attitude translates into a series of strategies and activities that aurally reify Indigenous and rural folks and their cultural practices, making them into sources of authenticity and objects of desire to be possessed in an attempt to generate a rhetoric of national patrimony. This objectivizing dynamic is particularly problematic when read vis-à-vis the fact that efforts to repatriate many of the sounds in the Fonoteca collections would be futile given the nature of these collections as copies of earlier archives in which sounds were severed from their cultural context. Furthermore, in many cases, the details regarding the origin of these sounds and the particular dynamics informing how these archives were generated in the first place have been lost in the process of them changing hands over time.[51] This situation points toward a process of patrimonialization in which sounds are extracted from their cultural context, made into essentialized objects of cultural consumption, and ultimately repossessed by the nation-state through the validating power of its political institutions. The pairing of patrimonialism and identity in the logic of institutions such as Mexico's Fonoteca Nacional exposes the devious dynamic of dispossession and reification that often informs nationalist projects.

    The operating principle behind the Fonoteca's articulation of the notion of patrimony as a "repository of the sounds of the nation" follows on Sandra Rozental's definition of *patrimonio* in Mexico as the state's co-option and ownership of the Mexican people's heritage and its commodification into national cultural property.[52] Although *patrimonio* legislation is careful to avoid giving this cultural property any exchange value, it does endow it with the affordance to become a good for cultural consumption. As Rozental suggests, "Patrimonio works as a kind of public property that belongs to all Mexicans alike, even though, by law, the sole body responsible for administrating, caring for, and enforcing property rights over patrimonial things is the Mexican state."[53] The Fonoteca case shows that when it comes to patrimony, in principle it is also up to the state to control its visual and sonic representation.

Beyond the Walls of the Institutional Aural City: Surplus, Absence, and the Epistemic Possibilities of the Everyday

Regardless of the democratic intention and celebratory tone of institutional projects like the Fonoteca Nacional, some of their features and ideological underpinnings exacerbate the type of power relations that characterize the Lettered City precisely because the activities they promote stem out of a patrimonial attitude. This stance and its objectifying logic permeate the Fonoteca's archival mission as well as its avowed goal of "bringing to life the sound archive."[54] As examined here, the soundscape and sound map projects privilege sounds as objects of aesthetic contemplation as well as objects that are ascribed with and embody certain values. It could not be otherwise given that an understanding of the archive as a depository of worthwhile objects is what characterizes any patrimonial project. This episteme marks an ontological contradiction within the project itself, a paradox that opposes a notion of listening that implies a disciplinary strategy against listening as the intimate sonic delight that bodies experience through their interaction with and recognition of environments that shape their understanding of the world. In other words, while the rhetoric of the Aural City stresses the importance of listening as an experience that creates knowledge and shapes understandings of reality, the institutional Aural City highlights an education of the ear that would allow for the recognition of values ascribed to sounds on the basis of predetermined aesthetic and political criteria. Most of the activities sponsored by the Fonoteca Nacional respond to the institution's need to discipline the ear of its audiences so they can place value on sound objects that represent aural experiences in the process of extinction or sound objects that represent markers of local identity that can be incorporated into the larger nationalist patrimonial rhetoric. In this process of fetishization of the sound object and in the patrimonial logic of the institutional archive, sound recordings are not taken simply as indexes of an unattainable reality in the past but somehow, disregarding the processes of transduction that mediate the sound object, they come to be considered as almost mystical gates into the real past. An example of this is the moment when our tour guide proudly played for us the Fonoteca's oldest recording, Porfirio Díaz's message to Thomas Alva Edison. Although it would seem that she did it due to its significance as a historical document, what was really at stake was the recording's allure as a medium able to bring back the specter of Díaz. The uncanny fascinating character of this recording comes out of its capacity to deceive and mystify us into believing that we have actually heard Díaz himself. Following on that, one could read the fetishization of the Fonoteca recordings as a consequence of a self-delusional desire for the simulacrum to take over reality. However, in all the activities programmed by the Fonoteca, there is always a surplus that transcends the institution's prescribed listening mode; there is always something that escapes its disciplining. Often, that surplus is to be found in the many gaps and silences that listening in detail helps us recognize and locate in the archive—such as the absence of the sounds of violence in the Guerrero soundscape or the minimal participation of countryside folks in shaping the representation of their sonic surroundings in the sound map project.

One of the activities regularly programmed by the Fonoteca that has greater potential for the identification of these surpluses is the Caminatas Sonoras. Sound ecologist Hildegard Westerkamp, working within Schafer's World Soundscape Project, defined *soundwalking* as "any excursion whose main purpose is listening to the environment. It is exposing our ears to every sound around us no matter where we are."[55] For Westerkamp, soundwalks were a way to establish a dialogue between people and their everyday sonic environment that, as Andra McCartney suggests, "shift[s] power relationships between artists and audiences, acknowledging the varied listening experiences and knowledge of audience members."[56] In 2009 the Fonoteca Nacional hosted a meeting of the World Forum of Acoustic Ecology in Mexico City. Westerkamp, one of the founders of the forum and its president at the time, visited the city for the occasion and organized what many consider to be the first soundwalk in Mexico.[57] After the forum, the Fonoteca adopted soundwalks, under the name Caminatas Sonoras, as one of its regular offerings. Needless to say, the concept of the soundwalk had to be slightly adapted to the larger mission of a national institution like the Fonoteca. Thus, the Fonoteca website describes its Caminatas Sonoras as "silent sound excursions to listen to the sounds of a place and appreciate them as if we had never heard them before. [This activity] takes place under the guidance of a specialist who teaches the audience how to recognize, appreciate, and value the surrounding sonorities."[58] In agreement with its pedagogical mission, the Fonoteca makes the soundwalk project into a "program of guided sensibilization" that values the disciplining of its audience members' listening more than their "varied listening experiences and knowledge."[59] Thus, the patrimonial character of the Fonoteca's mission trumps the raison d'être of soundwalks, preventing its audiences from experiencing them fully and freely. What could these audiences miss by having a guide tell them what and how to listen for instead of simply immersing themselves in the experience?

The last activity of the sound studies seminar that brought me to the Fonoteca in 2013 in the first place was precisely a soundwalk. To prevent my biases from influencing my students' experience, my only instructions before the soundwalk were for them to walk in complete silence, refrain from pointing out to their peers any of the sounds that might grab their attention during the activity, and try to pay attention to the sources of those sounds. The exercise lasted a little over two hours. It took us from the classrooms and halls of UNAM's Escuela Nacional de Música to the streets of Coyoacán, the neighborhood where the school is located, and back. We walked through commercial areas with lots of vehicular traffic as well as less busy residential zones; we went from hidden bucolic squares and churches to popular touristic areas and crossed crowded markets and plazas before heading back to the classrooms and halls of the School of Music. A deep silence filled the space for several minutes on our return to the lecture hall. Finally, one by one, we started sharing our impressions, the things that surprised us and those that met our expectations, the physical sensation of the activity as well as the affective and emotional reactions we experienced. As the conversation developed, it became evident that rather than the sonic identity markers of any given place, what impressed us the most were the moments in which, when walking by the neighborhood houses' open windows or doors, sounds from domestic life leaked into the public sphere. In a way, we found it uncanny how the everyday intimacy of those sounds disrupted the tacit and conventional understanding of what a public space should be and what types of behaviors and experiences should give it meaning. These sounds of domesticity and everyday life are the types of surpluses hidden in the Fonoteca activities that have the potential of transcending and challenging the institution's prescribed listening mode and the episteme on which such an aural regime is based.

One of the early goals of the Fonoteca Nacional was to encourage private folks to contribute to the archive by donating personal materials. According to the tour guide who showed us the Fonoteca installations and explained the projects' offerings to us, this created a problem because many folks began showing up with materials that were of no interest to the institution, including mixtapes, sound and video recordings of everyday life events, and so on; they were mostly materials belonging to or meant to be used in the private sphere. The tour guide stated that at some point the Fonoteca team had to start rejecting these donations. This anecdote shows that there is an evident tension between what the institution considers worthy archival materials and what regular folks may consider valuable and significant. The absences in the Fonoteca archive and the untapped surpluses in its sponsored activities both point toward the potential epistemic possibilities of the everyday. These types of everyday absences and surpluses put in evidence how institutionalized archives (which may lack the infrastructure to classify and store these kinds of sound materials in systematic and meaningful ways) and their archival logic (the fact that these archives were created under epistemological-bureaucratic models that neglect the kind of knowledge such materials purport) conflict with the archival needs of regular folks that imply a circumvention of the "symbolic annihilation" at the core of an institutional archival logic like the Fonoteca's.[60] Paying serious attention to these absences and surpluses may also help break the logic of sound objects in the archive as delusional fetishes of the past, instead understanding them as powerful records that give us access to an affective understanding of the world around us.

Recognizing the importance of domestic everyday life, Julio Arce suggests that in the context of the virtual digital pervasiveness we currently live in, it is very important to pay attention to the digital technologies that mediate our interaction with the world; register our everyday practices of consumption; record our wishes, desires, and fears; and regulate the development of our social and professional networks. For him, the big data we generate for multinational technology platforms like Google, Amazon, Facebook, Instagram, YouTube, or Spotify in our daily internet interactions are the largest archive of everyday life in human history.[61] By focusing on the actions that generate the internet data stored in these archives, the types of curatorial practices that allow for that storage, and the active or passive ways in which technology allows their retrieval, Arce proposes a way to analyze the relevance of this documentation of everyday life. Another way of doing that is by listening in detail to the work of artists who bypass the ideological restrictions of archives like the Fonoteca by appealing directly to the virtual archives Arce focuses on. A salient and powerful example is *Vis.[un]necessary force_1* (2014), a sound installation by Mexican transdisciplinary artist Luz María Sánchez Cardona (b. 1971) presented at the exhibition/symposium *Sound Art and the City: Mapping Sound and Urban Space in the Americas* at Cornell University on October 23–25, 2014. This work is part of *Vis.Fuerza[in]necesaria. El sonido del México post-nacional* (Vis.[un]necessary force. The sound of postnational Mexico), a larger research-creation series in which Sánchez Cardona explores "the violence provoked by the war against the narco in contemporary Mexico and its consequences for the everyday life of the civil population."[62] To address the type of violence that Mexicans experience in their everyday lives, *Vis.[un]necessary force_1* uses seventy four Caracal-pistol-shaped portable speakers that, on activation by the audience, play independent soundtracks made with sounds of shootings from the war against drugs. These sounds were recorded with cell phones by civilians caught in the crossfire during confrontations between narco gangs and law enforcement. These people later uploaded them to YouTube "to communicate their particular experience within this context of explicit violence."[63] This is where Sánchez Cardona accessed them. It is telling that it was precisely in one of the virtual archives of everyday life that Arce discusses that Sánchez Cardona was able to find and retrieve the very sounds that the Fonoteca expunges from its sanitized soundscapes of Mexico. Other forms of creative storage, retrieval, and activation of these sounds of everyday violence in Mexico can be found in the work of performance artists and playwrights Enrique Ježik, Hugo Salcedo, and Felipe Osornio (aka Lechedevirgen), as well as Teatro Línea de Sombra.[64]

The massive proliferation and uninterrupted expansion of the types of materials and practices that Arce discusses can only be understood in terms of big data. This leads us to a critical conundrum: how to approach and deal with the potential infinity of this archive. The situation is further complicated when we think about how the types of analog technologies and predigital everyday materials that the Fonoteca has rejected in the past would substantially expand the vastness of an archive that, for practical purposes, is already infinite. In this particular context, the archive is no longer the simulacrum that threatens to take over reality; reality itself, its totality, is the archive. Everything in the world—every action, every place, every object, every word, every sound, everything there is—is part of the archive. There is nothing beyond the archive. Jorge Luis Borges prefigured a similar notional eventuality and its affective repercussions for human beings when, in his short story "La biblioteca de Babel," he imagined an "unlimited and periodic library" made of "an indefinite number, maybe even infinite, of hexagonal galleries" that "is so enormous that any reduction to human origins is infinitesimal."[65] In his fiction Borges describes people's reaction when it was proclaimed that the library contained all possible books as one of extreme optimism because "there was no personal or world problem to which an eloquent solution did not exist; . . . the universe suddenly usurped the unlimited dimensions of hope."[66] However, this was quickly followed by a turn to an extreme sadness on the intolerable realization that "some bookshelf in [one of the library's] hexagons stored precious books but that those precious books were inaccessible."[67] Indeed, the only possible response to the exhilaration of the archivist who is given the impossible opportunity of archiving everything is the intense depression of understanding that in the incommensurable chaos of an archive that includes everything, nobody should expect to find anything.

By definition, an archive that includes everything would defy the very goal of any such repository. Rather than granting access to information, the vastness of such an archive would make it unmanageable and thus prevent any systematic and regulated retrieval of its contents. It would seem that an archive that includes everything is a useless archive. However, the chaos of such an archival universe offers the industrious opportunity of approaching its contents in novel rhizomatic ways and the possibility of finding *lo inaudito* within it. Rather than aspiring to an absolute archive, one must try to make sense of this totality through the specific ways in which the objects and materials it contains could be connected according to unpredicted rhizomatic networks that allow us to see and hear different archival constellations. The performance complex studied in chapter 1 and the virtual archive articulated in chapter 7 are clear examples of how archival constellations can be built based on rhizomatic relations prompted by specific intellectual questions or rhetorical enunciations. Mexico's Fonoteca Nacional is also one such constellation. Regardless of the rhetorical turns that present the Fonoteca as an archive defined by the national specificity of its contents (e.g., sounds of the nation), the relationship between contents and archive actually works the other way around; it is the institution's mission that prescriptively leads to the rhizomatic identification and grouping of these sonic materials as sounds of the nation. The Fonoteca is the result of a rhizomatic network of sound objects whose grouping is developed out of the specific nationalist gaze and rhetoric that informs the archive's patrimonial mission. And as in any rhizomatic constellation, what it contains is as important for its identity as that which is left out. The mixtapes, the homemade audio and video recordings, and the domestic or semidomestic everyday memorabilia that are not there—that cannot be there—make up the pungent surplus of the archive. Furthermore, it is a surplus that also speaks about materials in the archive that are there but that we do not know how to listen for or recognize.

These absences, invisibilities, and surpluses raise questions as to the very nature of the archive, its social structuration as such, and its ontological dependence on continuously changing listening practices and histories of the ear. At the very least, they reinforce the idea that the sound archive is not a series of sonic documents but rather a type of fragmentary knowledge in process, as Miguel García suggests.[68] In responding to these critical issues it is important to consider seriously the potential of the chaos implied in the total archive as a foundation on which the rhizomatic unfolding of a wide variety of deterritorialized archival constellations is not only plausible but in fact highly desirable. In that context, it would not be terribly gloomy that the Fonoteca project shows how the Aural City may continue to reproduce the top-bottom dynamics that have historically characterized the logics of knowledge circulation throughout the Lettered City. It would not be such a pessimistic realization because in that universe there would always be the potential to counter this institutional Aural City with alternative archival constellations developed out of an understanding that the patrimonial sounds of the Fonoteca can also be significant and meaningful beyond the institution's reductionist and essentialized nationalist rhetoric. For those alternative constellations to be viable, the development of a postnational ear is certainly indispensable. As we see in the following chapters, that postnational ear may be on full display in some of the noninstitutional projects developed by individuals for whom occasionally collaborating with the Fonoteca may be a savvy way of furthering larger, more progressive intellectual and political agendas.

Critical Constellations of the Audio-Machine in Mexico and the Performativity of Archiving/Archival Labor

But your advantage in forgetting is that
you'll forget to write yourself off
as a lost cause.
—Jonathan Nolan, "Memento Mori" (2001)

In his 2020 science fiction film *Archive*, British filmmaker Gavin Rothery presents us with a story in the not-very-distant future, the year 2038. There, a roboticist is working on three androids. The three reflect different stages of development of an AI cyborg, with every updated version being a step closer to the roboticist's final goal, a true human equivalent to which he can completely upload his deceased wife's consciousness, which has been kept alive digitally in a cyber archive. Because the archive has an expiration date, the roboticist is running out of time to design the ideal prototype to keep his wife's essence alive. The third model seems to be the perfect candidate since its personality software, designed through samples of the digital archive that keeps his wife's memory alive, has been able to develop a type of consciousness capable of evolving and learning independently. However, when the android finds out the roboticist's true intentions and realizes its reason d'être is simply to be a vessel for the consciousness of his creator's wife to remain alive, it threatens to kill him. This takes place precisely at the split second the archive is reaching its expiration time. It is at this climatic moment that the cinematic narrative moves us quickly back and forth between a series of short and fragmented dreamlike images: nostalgic memories of the roboticist's past life with his wife, conversations between the roboticist and his wife's digital consciousness as kept alive in the archive, and a scene in which the technician in charge of the archive delivers the news of the technology's expiration to a woman. This visually dizzying maelstrom reveals the film's final twist: The roboticist and his wife did have a fatal car accident, but it was not the woman who was killed; it was the roboticist. Thus, it was his consciousness that had to be kept alive in a cyber archive, and it was this archive that finally reached its expiration date. Thus, it is the wife who mourns the loss of her husband's digitized essence, which is now gone forever, and not the other way around.

I read Rothery's film as an allegory of the ways in which archives can become discursively and intellectually unfertile accumulations of data. Here, the roboticist is literally an archive that tries to performatively reproduce itself by keeping alive the narratives through which his life made sense in the past. He is so obsessively consumed by this project that he is unable to realize that, in fact, he is already dead, unproductive, and sterile. His is a futile, neurotic struggle ultimately destined to fail.

Archives are collections of materials and data that have meaning only when they are organized, consciously or unconsciously, following specific ideological criteria, thus allowing for the interpretative production of narratives. Under the appropriate historical and political circumstances, these interpretations of the archive can take it over and, through their repetition, become performative iterations that shape our understanding of reality. When hegemonies are negotiated in specific historical contingencies, these archival narratives have the potential of becoming naturalized. It is at that moment that we start seeing them everywhere; they become our guiding principle to engage and make sense of a wide variety of experiences, practices, behaviors, and other, more informal and open archives we encounter in our everyday lives. It is as if the archive could only make sense when read in that specific way. In a move reminiscent of Jean Baudrillard's notion of the hyperreal, the narrative about the archive, its ideological representation, and its symbolic reordering become the archive itself for most people who engage it.[1] We accept it, and it works because it makes our chartings of reality simpler and our quotidian lives easier.

In the introduction to the English edition of Gilles Deleuze and Félix Guattari's *Anti-Oedipus* (1972), Mark Seem explains this stage of interpretative complacency by invoking a furtive longing in all of us to be led, to have someone else provide rules, to be fed an interpretation, and to disengage from processes of active knowledge production.[2] For Deleuze and Guattari, this is a fascist attitude that reveals the processes of repression that people go through in their everyday lives via modern territorialities, "the forces of social production, reproduction, and repression" that naturalize the dynamics of individual exploitation involved in capitalist labor relations and the extraction of surplus value.[3] As they argue, in this complex chain of representation and signification, "the real is not impossible; it is simply more and more artificial."[4] Deleuze and Guattari propose a way out of this conundrum through the production of schizzes, detachments that cut into the flow of these chains and introduce "signs from different alphabets" that have the potential to deterritorialize the spaces these chains tend to normalize.[5] The liberatory character of these schizzes lies in the fact that they cannot be enacted by one person alone; instead, they are always dialogic in nature; they can only be the result of processes of collective conversations where feedback is generated and shared understandings are negotiated. In that sense, schizzes can lead to a redistribution of the sensible that, following Jacques Rancière, constructs "material rearrangements of signs and images, relationships between what is seen and what is said, between what is done and what can be done."[6] In other words, these schizzes are all about possibilities, about providing new ways of understanding old arrangements of objects and the discourses that give them aesthetic and political meaning.

I argue that archives, as institutions that often validate digested narratives and the artificiality of the hyperreal as natural, operate as the types of modern territorialities described by Deleuze and Guattari. As such, to overcome their eventual status as discursively and intellectually unfertile accumulations of data, they need to go through processes of estrangement that could detach them from the ideologies they come to represent and stand for in the simulacrum chain. As Polina Barskova poetically puts it, it is the labor of the productive archivist to "transport the souls from one folder to another, from a folder where no one will ever hear, to one where someone will—at least for a short time."[7] One such opportunity is indexed by a dramatic but apparently innocuous scene in Rothery's *Archive*. The fact that the android destined to host the consciousness and memory of the roboticist's wife threatens to kill him when it uncovers its master's plan reminds us that one of the features of every system is that the codes built into them to guide and allow their reproduction may also provide tacit instructions for their self-destruction. In his poignant analysis of the apologetic rhetoric about the fire that destroyed Mexico's Cineteca Nacional (National Film Archive) in 1982, Javier Villa-Flores concludes that "the possibility of loss is never cancelled with the creation of an archive, as the risk of erasure and obliteration is always attached to the exteriority of the place."[8] Villa-Flores tacitly refers to the fact that at the time of the fire, the film archive temporarily housed a number of nitrate film collections waiting to be restored and copied into "safety films." It is believed that these highly flammable materials were the origin of the fire that destroyed the archive. I take the idea of the self-destructing archive a step further and argue for the potential for the archive to also house its ideological self-decimation within it. This conceptualization is ripe with possibilities when read vis-à-vis the production of the schizzes needed to counter the intellectually sterile performativity of archival ventures that continually reproduce hegemonic narratives—such as the nationalist rhetoric that informs the Fonoteca Nacional archival project explored in chapter 2.

This chapter analyzes *Critical Constellations of the Audio-Machine in Mexico* (CCAMM), a sound exhibit presented at Berlin's Kunstraum Kreuzberg/Bethanien as part of the 2017 CTM Festival (an annual music event in Berlin), to ponder the political implications of archival construction and deconstruction within the logic of schizzes proposed by Deleuze and Guattari. I pay attention to how Carlos Prieto Acevedo (b. 1973), the exhibit's curator, introduces chaos into the archive in order to deactivate fundamental aspects of the nationalist narratives behind the Mexican music canonic fantasy and to rearticulate them in novel rhizomatic ways that afford *inaudito* hearings of its constituent elements. In doing that, I show how estrangement may open new paths for a reimagination of the body politic based on a new postnational performativity. In other words, central to this chapter is an examination of the logics and strategies of a postnational ear.

An Exhibit in Berlin and a Canon That Is Everywhere and Nowhere

On January 27, 2017, CCAMM opened at Berlin's Kunstraum Kreuzberg/Bethanien, an art space organized by the Friedrichshain-Kreuzberg District Council (figure 3.1). The event, organized and presented by Mexican curator Carlos Prieto Acevedo as the exhibition of the annual CTM Festival, was described in the event's program as highlighting the "history and current state of electronic music and sound art in Mexico, guiding visitors through various musical styles and sound experiments that have emerged

FIGURE 3.1. Program of the exhibit *Critical Constellations of the Audio-Machine in Mexico* (CCAMM) at Kunstraum Kreuzberg/Bethanien in Berlin, January 28–March 19, 2017.

in the country since the beginning of the 20th-century."[9] The festival and its exhibition are organized every year around a specific theme. The 2017 edition took the notions of fear, anger, and love as points of entry into exploring "radical forms of musical expression and dissonant emotions found in and through music [in order to examine] the diverse strategies that are applied to unleash or harness them."[10] Prieto Acevedo was trained as a philosopher at the Universidad Nacional Autónoma de México, and his career as a curator and cultural broker has tackled an examination of sound from a perspective different from that of his colleagues trained as musicologists or ethnomusicologists. His world is not that of music scholars or sound anthropologists. He belongs to the art world, and in a way, his career as a prominent sound expert represents the positive side of the democratization of the listening and sonic experience brought about by the advent of sound studies. As such, "by staging a variety of objects and documents in relation to music and sound in a series of self-contained yet transversal constellations, [Prieto Acevedo] opens up a multiplicity of readings on Mexico's contemporary transformation, constantly constructing and destroying its own utopias."[11] In doing that, the curator's practice is able to bypass the nationalist discursive conventions of the traditional gatekeepers of the institutional Aural City. In fact, CCAMM offered more than just the "history and current state of electronic music and sound art in Mexico," as the promotional leaflet of the CTM festival advertised. Read against the canon of Mexican music developed in a twentieth century permeated by nationalist ideologies, Prieto Acevedo's conceptualization and execution of the exhibit actually works as a strategy that turns that canon on its head.

As in most canon formations, the nationalist rhetoric and canonic fantasy of twentieth-century Mexican music operates as an epistemology beyond the list of composers and works it often refers to. This canon is everywhere and at the same time nowhere to be found. There is not a single institutional place or policy one could refer to when trying to retrieve it or trace it back as a single entity. Instead, one finds it informally throughout Mexican mainstream culture and more specifically by conscientiously looking in the cracks between and across a series of historical texts that have shaped and reproduced our understanding of the trajectory of Mexican music in the twentieth century. More than the composers and works that reappear once and again in these texts, what truly characterizes this canon as an epistemology is the ways in which it patrimonially reproduces narratives about the historical trajectory of the nation-state, the symbols it has embraced musically in order to perpetuate these narratives, and the excuses behind the presences and absences that characterize it—the reasons why something or someone belongs or does not belong in the canon.

Otto Mayer-Serra's *Panorama de la música mexicana* (1941) could probably be described as the first comprehensive modern history of Mexican music. Written in support of the hegemonic discourse of the postrevolutionary intelligentsia, Mayer-Serra's book follows a teleological understanding of history that explains nationalism as the result of the maturity reached by Mexican music and musicians through their affiliation with the nationalist ideology of the revolution. As such, the book became an apology for the hegemonic nationalist discourse being developed in Mexico at the time, allowing for the presence of composers who engaged that discourse while excluding or marginalizing those who, for different reasons, distanced themselves from it. That Yolanda Moreno Rivas's *Rostros del nacionalismo en la música mexicana* (1989), written more than fifty years later, reproduces Mayer-Serra's interpretative framework within the specific coordinates of Mexican nationalism is not surprising given its subject of study. What is more telling is that Rivas's *La composición en México en el siglo XX* (1994) puts in evidence another teleology, one about modernism and the avant-garde as reactions to (and consequences of) the postcard jingoist excesses of the second wave of Mexican nationalist composers in the 1950s. This type of teleological interpretative framework is evident in one way or another in a wide variety of writings about twentieth-century Mexican music, from the foundational texts studied or referred to in chapter 1—Daniel Castañeda and Vicente T. Mendoza's *Instrumental precortesiano* (1933), Miguel Galindo's *Historia de la música mejicana* (1933), and Gabriel Saldívar's *Historia de la música en México* (1934)—to more comprehensive historical surveys like Guillermo Orta Velázquez's *Breve historia de la música en México* (1970), and even Dan Malmström's *Introducción a la música mexicana del siglo XX* (1974).[12] The seeds of this ideology are already present in the earlier work of scholars such as Alba Herrera y Ogazón, whose critical discussion of the origins of Mexican music in the violent encounter of Indigenous and European civilizations in *El arte musical en México* (1917) follows on her desire to establish a foundation for a national and cosmopolitan Mexican musical art. Furthermore, the legacy of this epistemology continues to inform and shape the organization of content and the narrative of more recent editorial projects, including Aurelio Tello's *La música en México* (2010) as well as Ricardo Miranda and Tello's *La música en los siglos XIX y XX* (2013).

The leitmotifs behind this epistemology are recurrent but often contradictory: (1) an attempt at recovering Indigenous pre-Hispanic musical cultures and translating them into Western epistemologies; (2) the articulation of these Indigenous traditions along with other popular and folk musics as sources for the development of an "authentic" national music; (3) a faith in progress and modernization as the engines of a teleological understanding of history; (4) postrevolutionary life and society as sources of a cohesive modern national identity based on the notion of *mestizaje*; (5) cosmopolitanism as a way for individuals to transcend the cultural boundaries and shortcomings of the nationalist rhetoric; and (6) a historical narrative arc divided into clearly separate periods: a pre-Hispanic world, a colonial period, the postindependence nineteenth century, the nationalist phase after the revolution, the avant-garde break with nationalism in the 1960s to 1980s, and the "postmodern" scene of contemporary Mexico. If anything, these narratives could be understood as a canonic formation unified by a tension between tradition and modernization that tacitly tries to balance fantasies about the past with aspirations for the future.[13] Prieto Acevedo's CCAMM identifies the contradictions behind this libidinal economy and uses them as the constitutive building blocks that guide the assemblage of this deconstructive exhibit.

Anarchy, Chaos, and Productivity in the Archive

The CCAMM exhibition is an extension of a larger, unfinished multivolume editorial project that Prieto Acevedo began working on in the 2010s. The project, called *Variación de voltaje* (Voltage variations), takes as its point of departure a series of interviews with Mexican electronic music practitioners to develop an encyclopedia with a fanzine-like format that seeks to demonstrate that "there has been a type of contamination between the world of nonacademic cultural production and the work of composers from the Conservatory and National School of Music. The thesis being that modernity is improvised out of nonacademic and academic knowledges about music following a preconception of what Mexican identity ought to be."[14] At first glance, this premise may seem self-evident. However, in the highly restrictive hierarchical order that has characterized the Mexican Lettered City as well as the institutional Aural City, which has often led to the fragmentation of knowledge along unequivocally patrolled and controlled coordinates of class and privilege, Prieto Acevedo's project stands as a distinct culturally and socially disruptive intervention.

Not only is the fanzine format central to the formal outlook of *Variación de voltaje*, but it also regulates the project's production and regulation of knowledge as well as, to a certain extent, its circulation. Fanzines are low-budget, DIY underground editorial projects written and produced by and for fans of particular cultural scenes. They tend to move back and forth across a wide variety of issues and ideas while staying away from the type of conventional writing found in mainstream journalistic and academic works (often mocking ironically these fields' attempts at "objective" reporting). This writing and editorial attitude translates into a chaotic character that juxtaposes topics and knowledges in a polystylistic fashion. As such, fanzines de-emphasize authorial voices and privilege a collective or collaborative production of knowledge that takes place as writer and readership develop strategies to navigate a chaotic stylistic landscape. This is the spirit that characterizes *Variación de voltaje* as an editorial project, with its transduction of the oral and aural experiences of his interviewees into the over-codified semantics of written language. According to Prieto Acevedo, the book seeks to "produce noise" and "feedback" among voices that are impossible to "equalize."[15] It is in this feedback, in the screeching that results from the looping of the output signal returning to the input of the same device, that Prieto Acevedo locates a site for the introduction of chaos into the larger discourses of modernity and national identity that have shaped a constellation of fragmented experiences into seemingly coherent narratives about music and sound in Mexico.

The fanzine-like spirit that informs *Variación de voltaje* also found its way into the conceptualization and realization of CCAMM as an archive to be put on display and aurally interpreted as a type of epistemological feedback that defies equalization. Prieto Acevedo states that "given the audiovisual and narrative richness of the materials [in *Variación de voltaje*] I was forced to think about more creative ways to give life to the ideas [and connections behind them].... I realized that the exhibition format was very appropriate to pose problems that would be very difficult to present and bring to a large audience in a written essay."[16] In 2015 Prieto Acevedo and his partner, artist and filmmaker Bani Khoshnoudi, were awarded a residence at Centro Cultural Border in Mexico City; there, he began working on the exhibit's curatorial structure and on mapping its soundscape (figure 3.2). The residence allowed Prieto Acevedo to foster a workshop and think tank along two central axes: (1) how to conceptualize sound art within the logic of museums and (2) how to use documents and sound objects to develop a museological soundscape. An impressive list of Mexico-based sound artists, composers, experimental musicians, and curators collaborated in a residence that culminated in a month-and-a-half-long creative workshop whose goal was to come up with drafts of the exhibit's design.[17] As Prieto Acevedo explains, "The result was a sort of visual screenplay, a sort of score. In a hall, with just clippings, with phrases and ideas, I built this large fresco [laying out] the narrative and episodes of the exhibit."[18] The curator states that the development of this curatorial scaffolding came about only after carefully considering the works, sounds, music, and materials he wanted to feature in the exhibit.

Soon after the end of the residence, Prieto Acevedo met Carsten Seiffarth, a German curator and the founder of Berlin's Singuhr Hoergalerie, who was in Mexico working on *Entre Límites/Zwischen Grenzen* (In-between limits), a binational project cosponsored by the Goethe-Institut Mexico and his gallery that sought to establish an exchange program of residencies, commissions, and exhibits for Mexican and German sound artists.[19]

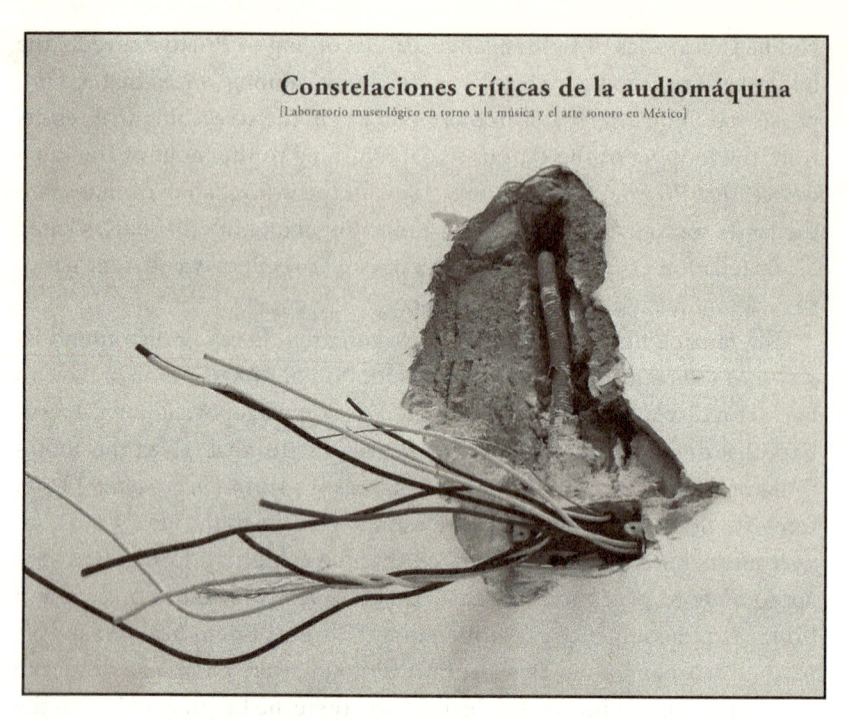

FIGURE 3.2. Flyer for *Constelaciones críticas de la audiomáquina*, a residence and think tank hosted by Centro Cultural Border in Mexico City, 2015.

The project was developed as part of the Germany/Mexico Cultural Year 2016–2017, and with that as an excuse, Seiffarth invited Prieto Acevedo to present *Variación de voltaje* in Berlin. The book presentation took place in November 2016 as part of a series of conferences about Mexican sound art at the Singuhr Hoergalerie. As a side project, Prieto Acevedo proposed to Seiffarth the possibility of CCAMM as a larger exhibition of Mexican sound art, following on the ideas developed at the Centro Cultural Border residency. The project was already structured in detail, and Seiffarth, who was working with Kunstraum Kreuzberg/Bethanien to present one of the final iterations of *Entre Límites/Zwischen Grenzen*, pitched the idea to the leadership of the CTM Festival. They were particularly excited about CCAMM since the festival had been featuring projects about experimental and electronic music in peripheral countries for a few years.[20] Furthermore, the Germany-Mexico Cultural Year proved to be the perfect contingency for the idea to flourish since José Wolffer, music adviser for Mexico's Ministry of Foreign Affairs, was also looking for projects to sponsor in the context of the dual Cultural Year.

To physically put together CCAMM, Prieto Acevedo resorted to "staging a variety of objects and documents in relation to music and sound in a series of transversal constellations" that allow visitors to be immersed in a world of sound and its visual representations.[21] Here, the curator's understanding of constellations is Benjaminian as it designates a "theory of reading[,] an interpretative procedure that draws specific attention to the instable conditions of this interpretation."[22] Instead of providing a chronology of sound art or history of music, the exhibit was an attempt to survey and question the trajectories and representations of sound culture in Mexico in relation to important questions of identity. As such, Prieto Acevedo's reading is a reordering of the sounds that make up this narrative at a moment in which the project of nation that permeated the country's history during the past ninety years found itself in a serious political, social, economic, and even moral crisis. In order for these feedbacks to happen, Prieto Acevedo consciously attributes "new meanings" to the musical works exhibited and establishes "new kinds of relations and hierarchies among them."[23] In doing that, this exhibit not only engages questions about what the Mexican nation may have been but, most important, operates as a dialectical sounding that creates a utopian map of possibilities to reread the past and nostalgically reimagine the future.[24] In order to curate such a project, Prieto Acevedo draws on an archive that addresses both official and semiofficial canonic formations, as well as alternative and marginalized sources. By placing them alongside each other, the exhibit allows the production of feedback loops and noise that may help us conceptualize the crisis of symbols of the nation-state as a potentially prolific state of anarchy or chaos. In stating this, I do not intend to create a gratuitous polemic about the symbols of the nation-state, but rather I wish to emphasize that as the national project that gave meaning to the Mexican constellation of twentieth-century sounds collapses, the very sonic symbols developed to support it—as well as those made into icons of resistance—are also decentered. Thus, as sounds and sonic cultures are emptied of naturalized meaning, the narrative of struggle that gave birth to the sonic fantasy of the Mexican nation-state also loses its significance for both Mexican individuals socialized to the nation-state narrative and foreign audiences for whom that narrative informs their own imaginaries of the exotic Other. In that sense, Prieto Acevedo's labor as a curator works as a *point de capiton*; it creates an archive by putting in evidence its anarchic state, identifying and reevaluating the conventional fragmented leitmotifs that glued together traditional fantasies about the modern Mexican nation-building project, and performatically and performatively reorganizing this field of sonic signifiers into new dramatic arrangements that prompt the audience to develop new narratives.[25] In doing this, the curator offers new ways to transhistorically put in dialogue pasts, presents, and imagined futures. In sum, Prieto Acevedo's curatorial labor is not about sound objects but rather about aurality. Because schizzes are dialogic and not the labor of one person alone, Prieto Acevedo requires from his audience not simply a disposition to listen actively and conscientiously; his labor demands new ears in order to hear old sounds anew and to marvel at *lo inaudito*, the strangeness of new noises and marginally conceived sonorities. The feedback this exhibit seeks to generate operates not only in the newly arranged constellation of archival items but, most important, in our bodies as we move through the spatial disposition of the archive and listen to and reimagine its sounds in new relational arrangements. Figure 3.3 shows images of the exhibit's preparation.

The most productive aspect of CCAMM is not the revamped archive it offers to our ears but rather the way in which Prieto Acevedo's curatorial efforts transcend their own material contingency by providing an opportunity to optimistically explore the feasibility and productive potential of anarchy in the archive. His chosen paths show us that a more fruitful and creative relation between user and material is possible precisely because the archive is in a state of disorder, and its lack of disciplining allows for novel ways to relate moments, characters, places, and their sounds. Prieto Acevedo's approach to the constellation of twentieth-century Mexican sound did not start with a blind belief in the narratives that the Mexican nation-state created for its own nationalist propaganda, but he does delve into that archive in order to see and hear the repetitive leitmotifs of its decadent discursive narrative. He is aware that although outdated and self-indulgent, these leitmotifs are easily recognizable, thus providing his audience a sense of familiarity necessary for the process of estrangement of that familiarity to take place. The reinvention of these leitmotifs into new narrative bricks provides the perfect opportunity for the production of the necessary schizzes to deterritorialize apparently natural epistemologies. Thus, the four topical axes of the exhibit, indio-futurism, the cosmopolis, the monstrous, and the posthuman body, are reinventions and rearrangements of those familiar leitmotifs. However, rather than being thematically "sound" as topics in normative histories and narratives usually are, they are instead shaped and reshaped by continual interpenetrations. Exploring and contextualizing the archives and constellations in each of these leitmotifs

FIGURE 3.3. Preparing the CCAMM exhibit at Kunstraum Kreuzberg/Bethanien. Photos courtesy of Carlos Prieto Acevedo.

and the way they leak into each other ideologically (in their conceptualization) and performatically (in their museographic staging) provides an entry into understanding the logic of Prieto Acevedo's postnational ear and how the aforementioned schizzes work within and in relation to specific narratives and the materials, practices, and sounds that support these discourses.

Leitmotif I: Indio-Futurism

The idea of the Indigenous played a fundamental role in postrevolutionary imaginaries of Mexican nationality. This was not new to the revolutionary regime; *indianismo* had already played an important part in representations of the national during the last part of the nineteenth century. Nevertheless, it was after the Mexican Revolution that the notion of the Indigenous was co-opted to validate the new regime racially and culturally. From José Vasconcelos's *raza cósmica* (cosmic race) to *indigenismo* as a state policy permeating social, political, and cultural life during President Lázaro Cárdenas's administration in the 1930s, a fantasy of the Mexican Indigenous world took over the representation of the nation.[26] This idea was so powerfully instilled that it filtered through decades of discourses about Mexican modernization, surviving in one way or another through the end of the twentieth century. Nevertheless, this fantasy of Indigenous culture was not concerned with the actual Indigenous communities that continue to precariously inhabit the national territory; instead, it celebrated an idealized and romanticized past of pre-Columbian splendor and used it to validate an identity toward the future of the nation. Music played a very important role in developing these fantasies about indigeneity. From *El fuego nuevo* (The new fire, 1921), Carlos Chávez's impressionistic ballet, which was never premiered as a ballet but still made its way into the canon of nationalist Mexican music, to Candelario Huízar's Symphony No. 4 "Cora" (1942), a type of invention of the pre-Hispanic Indigenous sonic world came to dominate the discourse of Mexican national identity.[27] As explored in chapter 1, this was a move that looked into the past in order to invent the present and imagine the future. Nevertheless, such representations left out many other ways to imagine a relation between indigeneity and the aspirations of cosmopolitan modernity that dominate Mexico's twentieth century.

In CCAMM, Prieto Acevedo looks back at that archive, recognizes its valence as a space for an imagination of the future to be negotiated—thus labeling it indio-futurism—and puts it in dialogue with more recent and forgotten moments that equally invented indigeneity in an attempt to fulfill other aspirations to cosmopolitan belonging. In this exhibit, the nationalist indigenist episteme meets the poetic evocation of the words and sounds captured in the pages of *Pedro Páramo* (1955), Juan Rulfo's classic harbinger of magical realism, and enters into dialogue with Roberto Morales Manzanares's exploration of alterity within traditional Mexican and Latin American Indigenous musical instruments via AI in *Nahual II* (1990–2017); the historical ethnomusicological field recordings of Raúl Hellmer as reinvented in the avant-garde sonic imaginary of Carlos Jiménez Mabarak's musique concrete ballet *El paraíso de los ahogados* (The paradise of the drowned; 1960); the New Age sonic re-creations of Antonio Zepeda's "Ondulantes serpientes de agua" (Undulating water snakes) from his 1986 cult cassette *La región del misterio* (The region of mystery); the mathematical transduction of Mexica mythology into orchestral music in Guillermo Galindo's *Ome Acatl* (Two arrows; 1994); and the deep future electronic dance music (EDM) loops of Javier Estrada's *3Ball prehispánico* (Pre-Hispanic tribal), an imagined rave to mingle, drink, and dance with Mexica and Maya gods. In the space of the exhibit, the Indigenous becomes a cyber-shamanic ambience evocation that opens the archival record to a large variety of discursively marginalized retro-futurist inventions of indigeneity.

This leitmotif allows Prieto Acevedo to "look back at things that are a bit stigmatized or stereotyped regarding that period of Mexico—indigenism—[and realize that the function of] the primitive in combination with the avant-garde is fascinating. [Here] indio-futurism is a laboratory of Mexican forms."[28] This recognition allows for the identification of a diverse array of sonic manifestations that dwell on the Indigenous to imagine the future beyond the specific nationalist-indigenist contingency of the 1930s—which is important in a postnational moment in which all-inclusive, homogeneous discourses of nationality fail to engage local and regional experiences of, and individual ways to affectively relate to, an ever-changing idea of the motherland. However, the indio-futurist refraction of the Indigenous as pure ambience or sonic objectification as featured in the exhibit's archive also puts in evidence that novel articulations of this ideology continue to operate in a fetishistic and co-optive way. As such, indio-futurism could be read along the lines of Afro-Pessimist scholarship as Indo-Pessimism, that is, a hopeless structural racial violence that denies Indigenous subjects their presence as human beings as a necessary state of affairs for current forms of fundamentally repressive social, economic, and political organization to continue existing.[29] This interpretation is not explicit in Prieto Acevedo's production and design of the exhibit; however, its spatial arrangement allows for this feedback loop to happen in the ear of the visitors. This potential to trigger noise in the bodies of the exhibit's guests may be the most productive aspect of Prieto Acevedo's indio-futurist rearticulation.

Leitmotif II: The Mexican Cosmopolis

An obsession with modernization has been a feature of the Mexican political, intellectual, and artistic elites since the nineteenth century. This fixation has generated a wide variety of personal, collective, and national projects that sought to imagine Mexican individuals—as well as the body politic—as equal actors in the international concert of nations. This cosmopolitan attitude is present throughout the country's convoluted history, cutting across political ideologies and historical contingencies to animate the progressive liberal ideas of politicians like Benito Juárez (1806–72) and intellectuals like José Vasconcelos (1882–1959) as well as the actions of conservative militants like Miguel Miramón (1832–67) or ideologues like Manuel Gómez Morín (1897–1972). As such, this attitude also informs an ideologically diverse variety of aesthetic projects by a cosmopolis of artists, musicians, and cultural brokers—a Lettered City of cosmopolitan intellectuals—through the twenty-first century. If many Mexican artists retroactively invented the Indigenous as a way to forge a rhetorical path toward modernity, many more decidedly embraced the call of the avant-garde to imagine their place in a world beyond the boundaries of the nation-state. This futurist constellation is full of utopias that never happened. It could not be otherwise; art is not science, and in its liberating creativity, it has provided us with a large archive of objects of desire for futures that never came to be. The avant-garde presupposes a firm and somehow delusional belief in the idea that the most radical artistic experiments of the present will only be properly appreciated in the future, when they will be quotidian practices.

The end of the armed phase of the Mexican Revolution in 1920 brought with it the collapse of older institutions; the development of new cultural, social, and political networks; and a promise to deliver the modern nation in the not-so-distant beautiful future. Cosmopolitan artists found inspiration in the revolutionary rhetoric of the avant-garde and developed a number of futurist and modernist-oriented projects. Although this rhetorical move about the potential of a perennial postponed future allows nationalism and the avant-garde to productively bounce off each other, it also has the potential to generate a type of epistemic isolationism, a rapturous evasion of the present in order to dwell on the promises of that future. One of the consequences of this cosmopolitan spirit is the production of "unique and lonely creations that are very isolated. They are like small worlds; solitudes that are [constantly] struggling against the mainstream."[30] Prieto Acevedo focuses on these manifestations and rearranges their archive to explore the delirious yet utopian aesthetic and political potential that these postponements entail.

In the 1920s, Manuel Maples Arce (1900–81), Germán List Arzubide (1898–1998), Fermín Revueltas (1901–35), and Arqueles Vela (1899–1977) became defiant *estridentistas* (the strident ones) and sang their praises, via words and colors, to a world of machines, robots, electric power, and technological progress in poetry that Luis Mario Schneider has described as "created in the imagination and not copied from reality."[31] Tina Modotti's (1896–1942) photos fragmented and froze bodies and objects into geometric yet antiformalist images that inhabit "the dim region between the optical and the haptical, form and touch, [as well as] the iconic and the indexical."[32] Julián Carrillo (1875–1965) explored the cracks between the sounds of the Western art music tradition to come up with the idea of Sonido 13, his microtonal music for a "future that never was."[33] In the 1940s, Conlon Nancarrow (1912–97) invoked the perfection of machines (player pianos) in order to realize the complex metric and temporal structures of his own musical utopia. These are some of the voices featured in this episode of the exhibit. They speak with the fervent cosmopolitan desire that shaped the Mexican experiences of modernization. As such, they all sang of a beautiful future in which technology would make us one with our cosmopolitan brothers and sisters of the world; in their worldly anthems, they sang the beauty of a future of mechanistic equality that never was. They are objects and practices that, by breaking traditional semiotic expectations and the lines of communication they entail, become "tradition[s] against themselves," as Octavio Paz would put it, the fortresses of solitude constantly rubbing against the mainstream that Prieto Acevedo refers to.[34] Their inability to become present speaks volumes about the trajectory of science and technology and their failure to deliver the future that the nation-state promised. However, by placing these artistic manifestations alongside more recent incarnations of the Aural City, Prieto Acevedo provides a space for new metaphoric rubbings to happen and for a new archive to arise that tacitly but loudly problematizes the fetishistic gaze that makes technology and its cyborgs into the "sounding historical monument[s] that attempted to put [Mexicans] in tune with the arrival of progress."[35]

This episode of CCAMM, showcased in three contiguous rooms, listens to the past through the technological cyborgs that make it viable in the present. Every instantiation of a sonic object from the past is mediated by a technological process that in a way makes it into an aesthetic automaton. Thus, Conlon Nancarrow's player piano music becomes a twice removed techno-phantom that refers not only to the ghost in the machine that Sergio Ospina-Romero invokes in his reading of the uncanny behind the "invisible player" of the pianola in Gabriel García Márquez's *Cien años de soledad* but also to the spatial-temporal reinvention of those already phantasmatic sounds in Hugo Esquinca's *Nancarrow Music Glossary* (2017), a quadrophonic generative composition made with excerpts from Nancarrow's Studies for Player Piano Nos. 3 to 36.[36] Manuel Maples Arce makes an appearance also as a cyber-ghost in Miguel Molina Alarcón's (b. 1960) 2004 sonic realization of his "T.S.H. El poema de la radiofonía" (T.S.H. The radio broadcasting poem; 1923), which transforms the poet's ode to the radio, destined to live as sound only ephemerally, over the waves of a live radio broadcast from the early 1920s, into a more permanent early twenty-first-century radio/sound installation.[37] Finally, Julián Carrillo's Sonido 13 is invoked in the exhibit through its mediation by *Improvisations 1*, a work for microtonal harp in harmonic tuning by Pilar Nava Calderón (b. 1988) and her father, Armando Nava Loya (b. 1957). Here, Carrillo's equal temperament (ET) microtonality is transfigured into just intonation in the name of the composer's own unfulfilled enigmatic aspirations to the "purification" of the Western art music system.[38] The Navas' just intonation improvisation operates like a proleptic simulacrum, a representation that takes over Carrillo and his Sonido 13 to make it real in the present even if slightly modified proleptically.

If the archive of this episode largely features reinventions of the artists' voices in technologically mediated simulacra that estrange the past to make it audible in the present, Prieto Acevedo's inclusion of the work of Rodolfo Sánchez Alvarado (b. 1937) presents an opportunity for those metaphoric vocalities to rub against the phonographic archive of Voz Viva de México (Living voice of Mexico), a collection of hundreds of sound recordings of prominent Mexican artists and intellectuals reading their own texts for which Sánchez Alvarado acted as sound engineer. In the confined space of the rooms housing this episode of the exhibit, the recordings of these voices metaphorically and literally press against these futurist fantasies and provide a sharp bodily contrast that, to the sensitive and informed listener, puts in evidence the isolated character of these fantasies and their need to be technologically mediated in order to become socially and culturally legible cyber-bodies in the present. Prieto Acevedo's transhistorical double estrangement provides a space for the recognition of the inhabitants of the cosmopolis and their projects, beyond their own discourses of identitarian authenticity, as cyber-bodies, machines without organs that regardless of their nature as isolated subjects, objects, or practices continue to reproduce mainstream ideologies. It could not be otherwise since their fetishized status as anomalies outside discourse works as the lubricant needed by the status quo to continue working properly. The exceptionalism of the isolated cosmopolitan machine provides the placebo of difference as an escape valve for the uniformity of the national fantasy to continue working as the only viable option toward the future.

Leitmotif III: Becoming Monstrous

Becoming monstrous is an episode that explores the sonic remnants of the nation in the postnational; the estranged sounds and rhythms that reverberate beyond the discursive border of the nation-state once their sonic symbols collapse. If the most successful representations of modern Mexico throughout the first half of the twentieth century were born out of the unexpected marriage of the Indigenous and the machine, one could interpret that representation as a type of discursive cyborg, a monster that, in its abject hybridity and unnatural beauty, dominates the Mexican imagination. Thus, the monstrous in this episode refers to the crisis of the symbols that this nationalist cyborg engendered for a new generation of artists and the way these creators reacted to them. Prieto Acevedo puts together an archive that borrows from very dissimilar aesthetic and artistic projects. At first sight—or hearing—it would be difficult to find similarities between Mario Lavista's (1943–2021) electronic collage *Contrapunto* (1972); the experimental pop of Interface's "Arrullos de la barraca" (Lullabies from the barrack; 1989); the electronic noise of Julio Estrada's (b. 1943) *Eua'on* (Flying away; 1980); Álvaro Ruiz's (b. ca. 1970) "El futuro más acá" (The future over here; 2003), a bricolage that combines lounge, soul, and tropical music; and the improvisatory character of Manuel Enríquez's (1926–94) visually stunning music notation. Nevertheless, these images, sounds, and objects "carry a secret inside that at the same time produces [them], turning [them] into monsters."[39] All of these projects share a common attitude; they all respond in one way or another to an essentialist discourse of national identity that no longer represented the desires and aspirations of Mexicans at the end of the twentieth century. From the direct rejection of such discourse—epitomized in José Luis Cuevas's infamous notion of the *Cortina del Nopal* (Cactus Curtain)—by avant-gardists like Enríquez, Estrada, and Lavista in the late 1950s, 1960s, and 1970s to the antinationalist and antimilitaristic commentaries in Interface's pastiche music thirty years later, at the beginning of the twenty-first century, the sonic mosaic prepared by Prieto Acevedo under the rubric of Becoming Monstrous creatively responds to an idea of national identity that has slowly become more and more irrelevant to bring a sense of unity and cosmopolitanism to the people living in the Mexican territory.[40]

Furthermore, the idea of the monstrous articulates "the delusions of the state about using science to improve and push the country forward, [a state that ends up] measuring poverty, hunger ... measuring failure instead."[41]

Thus, the monstrous refers to the great state utopia turned into dystopia. Here, the body politic, previously symbolized by the abstract notion of the people as represented and disciplined into larger collective modernization projects—in the rhetoric of José Vasconcelos; the public murals of Diego Rivera (1886–1957), José Clemente Orozco (1883–1949), and David Alfaro Siqueiros (1896–1974); the ballets of Carlos Chávez (1899–1978) and Carlos Jiménez Mabarak (1916–94); and the music of Silvestre Revueltas (1899–1940), José Pablo Moncayo (1912–58), and Blas Galindo (1910–93)—becomes an amorphous mass symbolized by the "undisciplined" multitude at the 1971 Festival Rock y Ruedas de Avándaro (Rock and Wheels Avándaro Festival), the Mexican Woodstock. This multitude, in agreement with Michael Hardt and Antonio Negri's theorization, "is an internally different, multiple social subject whose constitution and action is based not on identity or unity (or, much less, indifference) but on what it has in common."[42] Thus, the multitude rejects its postponement as body politic to a utopian future and asserts itself as a productive actor in the present, with "the potential . . . to sabotage and destroy with its own productive force," in this case, the reductionist hyperreal simulacrum of Mexican nationalist discourse.[43]

To articulate the sense of failure of the nation-state, Prieto Acevedo places Lavista's *Contrapunto*—a chaotic pastiche juxtaposing brief quotations from the national anthems of Mexico, the United States, and Japan; music from the Japanese Noh theater and Gagaku traditions; Gustav Mahler's Symphony No. 9; and fragments of rock music by The Beatles, Janis Joplin, and the Rolling Stones—alongside the sensationalist tabloid photographs by Enrique Metinides (1934–2022) of an electrocuted man, a woman hit and killed by a car, and a derailed train, and images of the Avándaro rock festival taken by Graciela Iturbide (b. 1942). Ana Alonso-Minutti argues that *Contrapunto*'s juxtaposition of these materials, especially elements from non-Western music traditions and popular music, is reminiscent of Karlheinz Stockhausen's *Telemusik* (Telemusic, 1966) and *Hymnen* (Hymns, 1966–67) and thus partakes in both the homogenizing neocolonial longings of this aesthetic and its antinationalist potential.[44] Metinides's crude images, which were originally taken to accompany *nota roja* (yellow journalism) pieces, became cult objects appreciated for their own artistic and aesthetic value as records of the absurdity of everyday life.[45] The result of this juxtaposition of sounds and images is a chaotic pastiche that generates another pastiche, an antinationalist sonic collage as the basis of a new curatorial audiovisual puzzle. This discursive maze expands the music's abstract horizon of interpretation through a gore montage that exacerbates its dystopian character. It works through an abjection that, as defined by Julia Kristeva, prepares the audience to go through a "great demystification of Power."[46] This new chaotic discourse also merges Lavista's rock references with Iturbide's rock images to metaphorically invoke the retreat of the people and the advent of the multitude as the new Mexican body politic, with its potential for a new "earthly city" beyond the Lettered City and the institutional Aural City.[47]

Leitmotif IV: Emanations of the Body

The CCAMM program closes with a sonic panel in which sound artists like Mario de Vega (b. 1979) and Angélica Castelló (b. 1972) rub shoulders with the visuo-writing interventions of Verónica Gerber Bicecci (b. 1981), natural engineering projects by Ariel Guzik (b. 1960), Arturo Márquez's (b. 1950) pre–Danzón No. 2 artistic experiments, and the progressive electronic music of Vía Láctea (active ca. 1970s–1980s) and Murcof (Fernando Corona, b. 1970). Again, as one looks at the type of interventions and the musical aesthetics favored by these artists and composers, the elements forming this particular archival constellation seem rather arbitrary. Nevertheless, the curator puts together this archive by focusing on how these sound and music practices avoid the stereotypical representation of the Mexican, engaging instead discourses and aesthetic visions that attempt to make sense of chaos. The putting together of the exhibit's last constellation speaks not only of crucial artistic strategies that help the artists avoid being labeled as Mexican and thus having their work's significance reduced to a series of geographic coordinates but also of the ways in which anarchy and chaos could be invoked and engaged in order to create a variety of archives that transcend the teleology of nationalist discourse and affect.

Emanations of the body deal with the contemporary arrival of ecopolitics in the Mexican music scene as well as a series of issues that move away from the obsessive search for identity that characterized the country's twentieth-century art and intellectual scene. Prieto Acevedo explains that the artists featured in this episode of the exhibit are no longer concerned with "very situated and contextualized aspects [but rather] a concern with issues that are part of the agenda of today's contemporary art: the invisible, fear, life, nature, and eco-philosophical speculations."[48] Thus, Murcof's intelligent dance music (IDM) loops, inspired by the music of Arvo Pärt, Sofia Gubaidulina, and Henryk Górecki—composers who have been identified with a certain minimalist aesthetic that dwells on slow gestures and an overall sense of spirituality—and based on orchestral samples taken from some of their music, "seek to transcend the [politics of identity] and reconcile with the sacred space of music.... [I]f [his attitude] is not postpolitical, it is certainly posthistorical."[49] Here, *posthistorical* does not refer to its existence out of history but rather to an existence beyond the hyperrealism of the historical narrative that attempted to give meaning to the Mexican body politic as collectivity in relation to the nation-building project.

The visuo-writing interventions of Verónica Gerber Bicecci use negative space to destabilize the meaning of preexisting texts or images in an attempt to appropriate the materiality of the artifacts and read them anew, in dialogue with her readers or audiences. For the CCAMM exhibit, Prieto Acevedo commissioned from Gerber Bicecci a graphic intervention inspired on Luis Villoro's philosophical essay *La significación del silencio* (1996). Here, Gerber Bicecci takes the commas, semicolons, and periods in Villoro's printed book as points of departure for a visual imagination of "the mechanisms of silence, the negative language that builds itself in an invisible manner throughout the text."[50] As Nona Fernández discusses, evoking Josefina Ludmer's theorizations, Gerber Bicecci's work is that of a postautonomous artist for whom it is irrelevant if their interventions could be considered literature or not; what is important is that "they construct the present [as] that is precisely their objective."[51] It is just this casual attitude toward generic artistic conventions and her avowed mission to extract hidden meaning from objects that allows Gerber Bicecci to intervene in the construction of a body politic beyond the restrictive identitarian rhetoric of nationalism. Her strategy has the potential to circumvent any possible ascription of meaning to the objects she works with.

A similar postdiscursive move informs the realization and reception of *Concierto para fotógrafos* (Concert for photographers; 1989), an artistic video-performance collaboration between Spanish poet and photographer Ángel Cosmos (b. 1949) and Mexican composer Arturo Márquez. This experiment is "a manifesto against the conventions of music and the stereotypes of photo-realism" that illustrates a specific moment in the history of conceptual and performance art in Mexico while helping to proleptically reconfigure the artistic character and legacy of a composer-turned-celebrity like Márquez.[52] For anyone unfamiliar with Márquez's biography and his experimental output from the 1970s and 1980s, it is particularly startling to stumble on his name at CCAMM, especially given the neo-nationalist-like cult status that composing Danzón No. 2 (1994) and the popular music–inspired works that followed it has bestowed on him. However, the retrospective look at *Concierto para fotógrafos* within the montage of "Emanations of the Body" facilitates a rhetorical removal of Márquez's legacy from the ideological representations that co-opt him for an oxymoronic neo-nationalist project of sorts. It is a strategic move that hints at making him whole again; it prevents his conflation into a reduced body of his work and reclaims for him a posthuman place beyond the people as body politic.

Placed along an ecopolitical audiovisual installation by Ariel Guzik that focuses on "the sophisticated acoustic transmissions emitted by the biggest sea mammal, the gray whale," Murcof's out-of-the-body spirituality, Gerber Bicecci's postautonomous visuo-writing interventions, and Cosmos and Márquez's aesthetic experience out of the body politic, Prieto Acevedo develops a montage that resignifies the posthistorical to delineate the posthuman as a body without organs.[53] Deleuze and Guattari define the body without organs as having "nothing whatsoever to do with body itself, or with an image of the body. It is the body without an image [that] belongs to the realm of antiproduction."[54] As such, the montage in Emanations of the Body speaks of a posthuman body that is not productive in the sense that it no longer reproduces itself as the quintessential body politic of twentieth-century Mexican history and is, thus, posthistorical. This reconfiguration of the archive reimagines the amorphous mass of the multitude identified in the monstrous episode within the coordinates and emanations of a posthuman body, a body without organs.

Listening Through the Noise I: Spatiality, Leakage, Montage, and Schizzes at CCAMM

Prieto Acevedo states that his intention in designing CCAMM was "to make a museographic performance, as if these objects were relics. Which they are really not—collectors keep them in their homes. But putting them all together in the same place creates a kind of scenography, an installation. . . . [I]t is like a theatrical play without actors but with sounds and music."[55] This dramatic and performative character guides Prieto Acevedo's uses of space as well as the audience's experience of traversing the exhibit. It is a way to "dissolve" the exhibit's objects and turn them into "physical extension[s], tincturing a space[; this is a type of dissolution] that can only take place through manipulating the space in-between objects and in-between objects

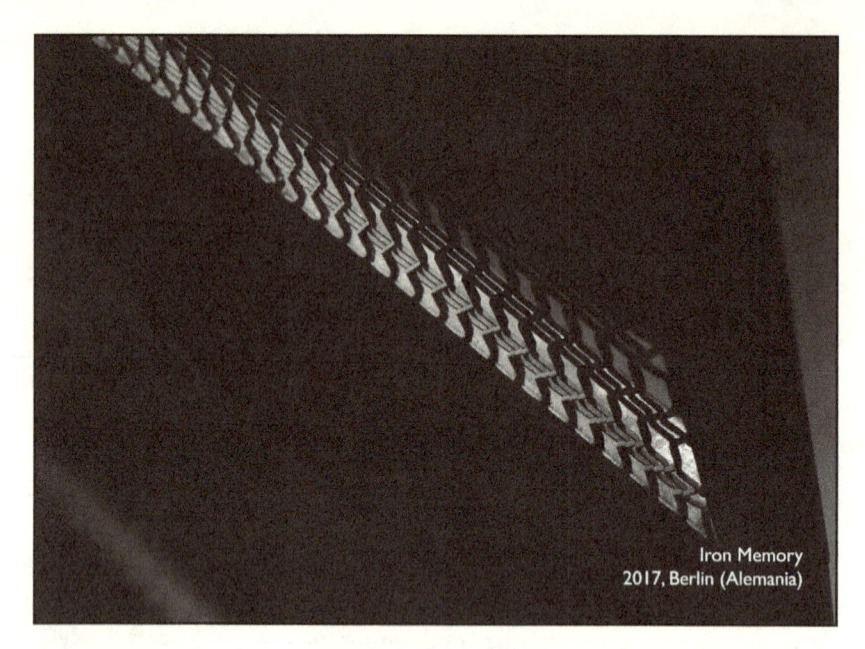

FIGURE 3.4. Félix Blume's *Memoria del hierro* (Iron memory), installed in 2017 at the entrance of the CCAMM exhibit at Kunstraum Kreuzberg/Bethanien. Photo courtesy of Félix Blume.

and audiences; i.e., through staging atmosphere."[56] Andrea Bohlman's review of CCAMM opens with a description of the exhibit's uses of space that highlights how Prieto Acevedo's curatorial labor creates such an atmosphere. There, Bohlman focuses on the exhibit's spatial setting, paying special attention to how *Memoria del hierro* (Iron memory; 2017), by French sound artist Félix Blume (b. 1984), a sound sculpture in the shape of a movable police barricade that vibrates to the recorded sounds of a 2013 protest in Mexico City, placed at the entrance to the exhibit in Berlin, "sets the tone for the acoustic experience [while] it also shapes [the] acoustic encounter with CCAMM" (figure 3.4).[57]

The CCAMM exhibition opened at Kunstraum Kreuzberg/Bethanien twelve days after Seiffarth's *Entre Límites/Zwischen Grenzen* closed there on January 15, 2017. Regardless of the fact that Prieto Acevedo's exhibit was thoroughly predesigned thematically, he pragmatically adapted its space layout to Seiffarth's floor plan for *Entre Límites/Zwischen Grenzen*.[58] Thus, CCAMM was organized in twelve rooms within a larger structure, a section of a former nineteenth-century hospital transformed into a cultural center in the Kreuzberg district in central Berlin, which was

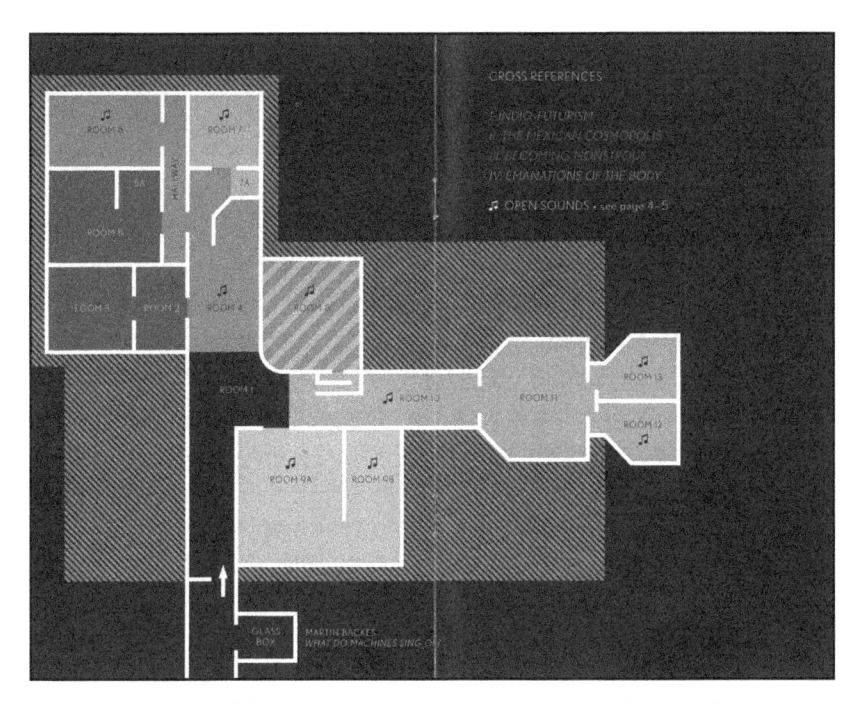

FIGURE 3.5. Map of the CCAMM exhibit at Kunstraum Kreuzberg/Bethanien.

itself divided into two blocks connected by a large lobby (which featured Blume's sound sculpture). Some of the rooms were themselves subdivided into smaller exhibition areas. Figure 3.5 shows a map of the exhibition and its spatial organization according to the four leitmotifs described earlier. Prieto Acevedo explains his strategy regarding the placing of sound in the exhibit: "There are many crossed visions about how you should feature sound in an exhibit, and I was very transgressive. I decided to have some things in headphones and keep others sounding openly [in the room]; and I tried to synchronize them. So, there is a kind of noise in the exhibit which, by areas, allows you to experience certain things although you are always hearing the other ones. There are other pieces that are confined to project rooms, where you can enter and look at them, while others are featured only on headphones."[59] Bohlman addresses these sonic leakages in her review, stating that "museums often struggle with sound: tinny speakers, no regard for bleed through. And broken technology left alone. This exhibit solves such persistent problems with its agenda.... The cacophony of the gallery resonates with the poetics of modernity and the techno-scientific imaginary.... The noisiness of

Critical Constellations of the Audio-Machine in Mexico 109

the space, as I hear it, is the post-utopia of the present and the failure of the cosmopolitan dreams that curator Prieto Acevedo aims to present in this 'proliferation of audio-topias.'"[60]

Indeed, the sonic leakages—a corruption of the purity and autonomy of the work of art pragmatically built into the design of the exhibit—work as metaphors, symptoms, and signposts of the exhibit's mission and the essentialism it reacts against, while guiding the audience through its chaos. These leakages generate a myriad of effects, thus "setting in motion a second process, project, or concern."[61] This corruption of the work of art's autonomy, its dissolution, is the curator's way to force the audience to find aesthetic and political meaning beyond the objects themselves and the archives they emanate from. This construction of new meaning is triggered precisely by the exhibit's style, its structural nature as an audiovisual montage in real time and space. Just as in a filmic montage the careful editing of a visual collage produces in the eye of the spectator a third meaning beyond the individual meanings of the short scenes that constitute it, the dissimilar objects juxtaposed in CCAMM elicit a process of estrangement of the exhibit's archive. This type of estrangement, as in Bertolt Brecht's adoption of the term, sidesteps the field of automatized perception in order to ensure that the audience remains aesthetically engaged while at a critical distance that allows them to reimagine the new potentialities of the archive.[62]

An important aspect of the montage's production of a third meaning was generated with the use of visual material (photos as well as musical scores) along with the sounds playing in specific rooms. As a case in point, the juxtaposition of Lavista's *Contrapunto* and Iturbide's pictures of the Avándaro Festival puts in evidence a new Mexican body politic and a skeptical relation with the state's messianic rhetoric in the 1970s. The transformative power of the montage is also evident in the use of scores as theatrical props. In room 8, a space shared by indio-futurism and the monstrous, Prieto Acevedo creates the illusion of a large, four-thousand-page musical score by Carlos Chávez sitting on top of an office desk while the composer's music plays in the background. The effect is a sarcastic commentary on the delirious identity of an artist who was both a bureaucrat and a composer, putting a humorous spin on an otherwise serious and intense reflection about the intersection of indio-futurism and the monstrous.

If the materiality of the archive in its actual spatialization is fundamental to understanding the estranging potential entailed in the curator's labor, the objects that did not make it into the archive are also important windows into libidinal landscapes that inform of the larger deconstructive project. In the case of CCAMM, Prieto Acevedo confesses that he had the idea of including a hall "with an archive of things that have not yet happened or did not exist. Fake books, LPs with fake center labels about collaborations that could have happened—such as Robert Moog and Héctor Quintanar or pre-Hispanic music made by Artificial Intelligence. That was the idea, which was more an installation than an archive . . . although the idea of the archive remains because it is still about 'historical' objects."[63] Although these imaginary objects did not make it into the final version of the exhibit, the revelation of their potential existence makes them into a phantasmatic presence that puts in evidence the carefree fanzine-like attitude at the core of CCAMM. Prieto Acevedo explains that he thought about that possibility because "the importance of curators is that they can make these [archival] objects more [discursively] productive by taking them out of the artists' narcissistic spheres [and questioning] how [the artists] want to see them, how they want to present them, the limits they want to place on their readings of these objects. So, curators should be able to make these archives say something different."[64] Within these dynamics, montage functions as the operation that allows for estrangement to take place while the leakages between the rubbing parts of that montage are the source of real and metaphoric noise that produces the deterritorializing schizzes that can destabilize the status quo and its narcissistic, self-validating discourses.

An Itinerant Archive: CCAMM at the Museo Morelense de Arte Contemporáneo Juan Soriano

From October 2018 to March 2019, CCAMM was presented at the Museo Morelense de Arte Contemporáneo (MMAC) Juan Soriano in Cuernavaca, Mexico, as *Constelaciones de la Audio-Máquina en México*. The concept and content of the exhibit presented at Kunstraum Kreuzberg/Bethanien remained largely the same. However, on this occasion, the layout of the exhibit was designed in collaboration with Peruvian architect Giacomo Castagnola and Erik López from Germen Estudio, an architectural and design firm originally founded in Tijuana/San Diego by Castagnola with the goal of exploring novel ways of exhibiting art archives and material culture in an attempt to conceptualize museums as public spaces.[65] The studio's website explains that the main challenge of this exhibit was to present and contain sonic materials in the museum's open gallery, for which they "designed

FIGURE 3.6. Use of space at the CCAMM exhibit at the Museo Morelense de Arte Contemporáneo (MMAC) Juan Soriano in Cuernavaca, October 2018–March 2019. Photo courtesy of Erik Eduardo López Rodríguez.

spaces and appliances that helped for a better interaction between the audiences and the works; from cylindric loudspeaker-like rooms that could completely contain sound inside, to sonic-shadow-like spaces where one could listen by getting close or reclining onto a wall, to small sound niches where one could sit inside to listen while still being able to continue looking at the work on the walls."[66] This type of carefully designed reiteration of CCAMM provided Prieto Acevedo's concept with a kinetic ability to transit through different physical as well as cultural and political spaces. Figure 3.6 shows the organization of space in the exhibit as seen from the upper entrance to the museum.

The audiences engaged by the Berlin and Cuernavaca exhibits were very different and came to these galleries with vastly dissimilar knowledges, clichés, and experiences regarding Mexican history, culture, and music, as well as sound art in general. If everyday visitors to the Berlin exhibition were generally unfamiliar with the ideologies behind narratives about Mexican history and music production, they may have been more conversant with cutting-edge sound art and its technical and aesthetic practice. On the other hand, Mexican visitors to MMAC may not have been as knowledgeable about the latest trends in sound art, but they had the cultural background to better understand the noise and feedback that Prieto Acevedo's curatorial efforts introduced into the archive. While Andrea Bohlman's review of the Berlin exhibit highlights the power of its anarchic sounds to unveil the "instability of history," Nick Herman's review of the Cuernavaca exhibit is more poignant in signaling that the audience left with an unmistakable "diagnosis that the nationalistic myth of modern Mexico is an incurable ideal."[67]

Prieto Acevedo states that one of the advantages of having presented CCAMM in Berlin first was that he did not have to deal with Mexican artists attending the exhibit. This gave him the freedom of "not protracting the artists' narcissism" and focusing instead on a larger critical engagement with Mexican cultural and historical narratives.[68] For that reason, he argues that producing CCAMM in Germany was very easy: The infrastructure was available, he had complete freedom of action, and the visitors—more than seventeen thousand in the three months the exhibit was open to the public—were informed and willing to deeply engage with the works and the ideas informing the display.[69] However, because the exhibit was developed within the framework of an electronic dance music festival and not within the logistics of a museum display, Prieto Acevedo's German collaborators lacked the expertise and experience that seasoned museographic experts have.

On the other hand, the exhibit at MMAC was produced by a team of museographic experts that skillfully translated and adapted Prieto Acevedo's ideas into a more malleable and receptive, although still problematic architectural geography. Nevertheless, the curator acknowledges that the final output of an exhibit is always deeply connected to the types of audiences it seeks to engage. In the case of Mexico, Prieto Acevedo argues that audiences, sensitive to the specific cultural overtones of the exhibit, were "more open to let themselves be affected by new possible combinations and ways of saying something, [which made the exhibit into a] very attractive and rich [experience]."[70] Nevertheless, when it came to the artists themselves, Prieto Acevedo suggests that although most of those who visited the exhibit said they liked it, many of them lacked a critical engagement with the ideas behind the display, often simply stating that they would like him to include their work in a possible future iteration.[71] So, although the noise and feedback Prieto Acevedo introduced into the archive made it attractive to a general audience, the artists' professional aspirations and their desire to be part of the exhibit may have trumped a more productive intellectual engagement with it.

Against Nihilism: Machines, Schizzes, and Anti-Oedipus

In its hopeless and pessimistic outlook on the archive that delusionally refuses to acknowledge its unproductivity and sterility, Rothery's *Archive* features the nihilistic attitude that characterizes much contemporary science fiction under neoliberalism. If change is impossible or insignificant, and if a vacuous discourse is reproduced simply because it is easier than attempting to make new sense of a series of worn-out signifiers in which nobody believes anymore, the logical responses are either hypocritical or cynical nihilism. Hypocritical nihilism describes those who understand that the system is problematic but prefer to maintain the status quo for their personal benefit. Cynical nihilism describes those who conclude that there is no other option than the status quo. Against these nihilistic attitudes, Prieto Acevedo's archiving labor shows us the estranging potential of introducing chaos and noise into the system to make it work against itself. This strategy resonates with the words in Jonathan Nolan's "Memento Mori"; engaging with the canon in this way allows us to transcend the sense that it may be doomed, thereby avoiding the nihilistic responses this "lost cause" might provoke.

Nevertheless, like all constellations, Prieto Acevedo's is unstable. Like all archival narratives, his also has the potential of becoming obsolete and sterile. However, the strength of Prieto Acevedo's labor is that "at least for a short time," as Polina Barskova suggests, he is able to "transport the souls . . . from a folder where no one will ever hear, to one where someone will."[72] His labor assembles a constellation that allows the listener to hear these old sounds in newly meaningful and productive ways. Prieto Acevedo's curatorial labor is a political statement that redraws "maps of the visible, trajectories between the visible and the sayable, relationships between modes of being, modes of saying, and modes of doing and making."[73] Thus, Prieto Acevedo's curatorial schizzes act as effective redistributors of the sensible in Rancière's terms; they are the "apportionment of parts and positions . . . based on a distribution of spaces, times, and forms of activity that determines the very manner in which something in common lends itself to participation and in what way various individuals have a part in this distribution."[74]

The image used on the exhibit's program cover as well as on the promotional leaflet is a computer-processed version of a picture in Félix Blume's photo series *Mientras escucho* (While I listen), itself part of an exhibition

FIGURE 3.7. Image from Félix Blume's photography series *Mientras escucho* (While I listen), 2016. Photo courtesy of Félix Blume.

called *Realidad amplificada* (Amplified reality), cocurated with French video artist Jérôme Fino and presented in Mexico City's Casa de Francia from May 5 to August 15, 2016. The original photo shows the face of a woman wearing a set of headphones and deeply engaged in the experience of listening to something *inaudito*, something we are literally not permitted to hear. It is a close-up that gives us full access to an extraordinary amount of facial expression detail and makes us accomplices in a moment of deep intimacy while keeping us at an aural distance (figure 3.7). On the other hand, the image on the booklet's cover makes those facial features into the laminated, expressionless mask of an uncanny cyborg that simultaneously attracts and repulses the viewer (see figure 3.1). The transformation from photograph into supernatural posthuman body is a perfect illustration of the redistribution of the sensible afforded by Prieto Acevedo's curatorial labor as well as the types of processes that inform and are signaled by the leitmotifs of CCAMM: estrangement, abjection, deterritorialization, the arrival of the body without organs.

It is no coincidence that Prieto Acevedo invokes the notion of machines, a quintessential concept in Deleuze and Guattari's conceptualization of the anti-Oedipus, as a central aspect of his curatorial endeavor. His project is anti-Oedipal as it recurs to anarchy and chaos to counter repression. If every machine "has a sort of code built into it, stored up inside," CCAMM identifies the code within the nationalist/modernist Mexican archive through schizzes that render it audible.[75] Prieto Acevedo's curatorial efforts identify the mechanisms and the logic embedded within the archive, the codes built into it to guide and allow its reproduction that also have the potential to destroy it. Indeed, I have argued that Prieto Acevedo's archiving/archival labor generates a libidinal economy that, in tune with Deleuze and Guattari's theorization about schizophrenia, may provide the epistemic conditions for new socio-personal orders. Prieto Acevedo's critique of the archives of nationalism and modernization of the Mexican Lettered and institutional Aural Cities identifies a series of new constellations made possible due to the introduction of chaos into the archival system. The CCAMM exhibition offers options to move beyond the system's ossified narratives to listen to *lo inaudito*; in doing that, it advances a way out of the labyrinth of nihilism that sterile archival formations may engender. However, this possibility also entails the audiences' labor. Only the audience's active participation can make this a productive process. The libidinal labor of those who walk through the sounds of the exhibit resides in listening with open ears and allowing the noise to performatively seduce them.

Things, Sound Objects, and Legacy at the Berliner Phonogramm-Archiv's Konrad T. Preuss Collection

Amo las cosas que nunca tuve
con las otras que ya no tengo.
(I love the things I have never had / with those I no longer have.)
—Gabriela Mistral, "Cosas" (1938)

The Humboldt Forum in Berlin, a museum that incorporates two institutions devoted to non-European cultures and traditions, the Ethnologisches Museum Berlin (Ethnological Museum of Berlin) and the Museum für Asiatische Kunst (Museum of Asian Art), opened its eastern wing on September 16, 2022. The ceremony was amply publicized in German media as the last step in a long process that sought to consolidate the non-European collections of the Staatliche Museen zu Berlin (Berlin State Museums), a group of museums, research institutes, libraries, and archives overseen by the Stiftung Preußischer Kulturbesitz (Prussian Cultural Heritage Foundation), to be housed by the Humboldt Forum in the newly reconstructed Berlin Palace. This fifteenth-century structure had been demolished in 1950 and cost around EUR 644 million to rebuild from scratch.[1] During the opening ceremony, Claudia Roth, the German minister of culture, offered a speech that acknowledged the debates about the legacy of colonialism at the core of the Stiftung Preußischer Kulturbesitz's mission and activities: "I know that not all of these collections have a colonial history. However, the fact that these exhibits are at the center of so much attention is not an unfortunate circumstance. It is a legacy of our past. It is important and necessary. It challenges us. What we are going to see in the future in this [forum] forces us to relate to this heritage, to our history. All the more because we are not only called upon to confront our colonial past but also its present."[2] This preoccupation with Germany's continuous colonial legacy made visible by the objects and materials overseen by the Stiftung Preußischer Kulturbesitz has been central to the debates about the Humboldt Forum since the German government first announced it in 2002.

Two performances that took place simultaneously during the forum's opening ceremony reveal the tensions at the core of the institution's seemingly contradictory mission—per the forum's website, to provide a space for the critical examination of Germany's colonial past while highlighting and exhibiting the objects acquired by German museums and archives during the country's moment of colonial and imperial expansion.[3] At the same time that a group of members of the Orquesta Experimental de Instrumentos Nativos (OEIN)—a Bolivian ensemble devoted to musics that use Indigenous Andean instruments in traditional as well as contemporary compositional settings—performed within the walls of the Berlin Palace as part of the official ceremony program (figure 4.1), members of the local Coalition of Cultural Workers Against the Humboldt Forum (CCWAHF) and a group of Latin American activists held a protest outside of the building in opposition to what they saw as the forum's celebration of the looting and genocide at the root of European imperial wealth.[4] Having been informed earlier in the day that they were not welcome inside the building with their flyers, slogans, drums, and panpipes, they had to stage their protest beyond the walls of the magnificently rebuilt Berlin Palace (figure 4.2). If the presence of the performing bodies of people of Indigenous background playing Indigenous instruments as part of the official program of the event could be read as an attempt to use those bodies and their musical objects to validate the forum's avowed mission of cross-cultural reconciliation, the presence of other individuals (those whose Indigenous bodies and voices were not welcome) outside the museum's building raises the question of who is able to speak and what are they allowed to say in the political context that informs the Humboldt Forum.

I take this vignette as a point of entry into invoking the Konrad T. Preuss Collection, a historical set of wax cylinders that, as part of the Berliner Phonogramm-Archiv (BPA) in Berlin's Ethnologisches Museum, is now housed by the Humboldt Forum (figure 4.3). The Preuss Collection

FIGURE 4.1. A group of musicians from Bolivia's Orquesta Experimental de Instrumentos Nativos performing at the Humboldt Forum, Berlin, September 16, 2022. Photo courtesy of Ekaterina Pirozhenko.

includes materials gathered and recorded by Prussian ethnologist and linguist Konrad Theodor Preuss (1869–1938) in Mexico and Colombia between 1905 and 1915. Here, I focus on the materials he recorded during his 1905–7 expedition to Náayeri (Cora) and Wixárika (Huichol) territories in Mexico's Sierra Madre Occidental. Preuss's field trip was conducted precisely within the historical context of the German colonial and empire-building projects that color contemporary debates about the Humboldt Forum. I am interested in the implications of looking at this archive as a collection of sound recordings that triggers a fetishizing fantasy about the past and its presence in the present while concealing the simulacrum-like character of the mediation processes that inform its production. Branching out of this, I also examine the uses of the archive in a variety of historically circumscribed cultural, social, and ideological settings.

Archives and the information they contain are designed, structured, and organized according to narratives that shape the type of knowledge their users can retrieve from them. The objects and documents in an archive tend to tell and retell very specific stories that in turn performatively reproduce the ideological frames that inform the dynamics between things,

FIGURE 4.2. Protesters outside of the Humboldt Forum, September 16, 2022. Photo by the author.

objects, documents, representations, and users. The central question informing this chapter is whether it is possible to make an archive tell us stories that deviate from the types of narratives they were originally designed to tell. To address this, I follow on Mexican ethnologist Margarita Valdovinos's assertion that Preuss's recording method "changed the materials he recorded" and that, for that reason, "regardless of his interest in symbolism and other ceremonial gestures, [he] was unable to get close enough to the logic of the ritualistic action to use it and understand the chants."[5] I suggest not only that the ways in which the sound objects in the Preuss wax cylinder collection were created respond to the type of questions he wanted to ask about the Náayeri and Wixáritari (Wixárika people), rather than to how these folks conceptualized their music/ritual practices, but also that the episteme behind the setting up of his archive tells more about him than about the communities he visited. Listening to the archive in detail renders audible the imperial entitlement of the archival project, to use Frederick Cooper's characterization, as well as the German colonial experience.[6] Considering how it was created, structured, and meant to have information retrieved from it, the archive tells us the story that Preuss wanted it to tell

FIGURE 4.3. Phonograph and wax cylinders from the Konrad T. Preuss Collection in the Berliner Phonogramm-Archiv exhibit at the Humboldt Forum, November 25, 2022. Photo by the author.

us and nothing else. Nevertheless, the goal of this chapter is not to reveal the imperial entitlement of the Preuss collection; instead, I am interested in taking this imperial entitlement as a point of entry into an exploration of how users can engage the archive in ways that circumvent its shortcomings and enable its objects to tell their own story.

I conduct this exploration in the context of the circulation of archival knowledge connected or ascribed to the Preuss Collection as it has been used within specific historical moments of empire building and as it has been made meaningful by contemporary scholars within the rise of a Mexican Aural City at the turn of the twenty-first century. In doing that, I keep in mind that, as Thomas Richards has argued in the case of the British Empire, imperial formations are kept together not only through acts and spectacles of violence but also through information.[7] Furthermore, I suggest that empires are also united by the ways in which such information circulates and is made meaningful beyond the political boundaries of empire and beyond the formal boundaries of its imperial archive. Thus, in the spirit of the need to find strategies to estrange historical sound archives to more productively engage the information they store and account for the voices they silence, I pay special attention to the contemporary work of Valdovinos herself. Her long, deep, and critical engagement with the Preuss Collection has resulted in effective ways to not only "give a voice again" to the objects in the archive, as she originally set out to do, but also endow these objects with an agency that allows them to say what the archive's structure, design, and the episteme behind it have prevented them from saying since their inception as archive.[8] In that sense, this chapter is as much about Preuss and his archive as it is about Valdovinos and her articulation of that archive. Valdovinos's listening in detail to the Preuss Collection and the trajectory of the transnational mobility of its sound objects recognizes and introduces these materials into a new understanding beyond the logics of the Berlin archive. This exemplifies the type of performative labor that characterizes the Aural City.

Since this chapter focuses on exploring the uses of the Preuss Collection as part of diverse institutional colonial and decolonial projects, its narrative traces the collection's history while being attentive to the ideologies that have informed its production, maintenance, and circulation. Thus, the text is divided into four large sections. The first one provides a detailed exploration of the unfolding of Preuss's expedition, highlighting its extractivist character—in terms of the mining and gathering of empirical data as well as the materials that came to conform the collection. This story shows that regardless of Preuss's better intentions, the expedition ended up reproducing the colonialist logic of the German Kaiserreich. The second section pays attention to the historical trajectory of the collection itself as part of the BPA in order to contextualize the influence of Preuss's legacy within the twentieth century's changing German and European cultural and political landscapes. The last two sections focus on the nature, mission, and potential of the objects in the BPA as well as the theoretical implications of the formation, circulation, and eventual repatriation of Preuss's archive. The third section studies the repercussions of the archive in relation to the sound objects created to conform the collection itself but also as a conduit for the discussion of ideas about difference and identity prevalent in early twentieth-century Germany. Finally, the fourth section discusses the Preuss Collection, the BPA, and the Stiftung Preußischer Kulturbesitz in the context of debates about cultural appropriation and the adoption of new patrimony and cultural heritage policies in Germany and the European Union at the beginning of the twenty-first century. This is concerned specifically with the creation of new objects connected to Valdovinos's repatriating project in an effort to explore the impact her endeavor had on the local Náayeri and Wixárika communities where the sounds in this collection were originally extracted from.

Listening to the Other in Mexico's Sierra Madre: An Expedition to the Gran Nayar

On August 17, 1905, Konrad Theodor Preuss sent a letter to the administration of the Königliches Museum für Völkerkunde (KMV; Royal Ethnology Museum), where he served as assistant director in the American division, requesting a leave of absence in order to spend a year conducting research in Mexico. His agenda was overly ambitious. He proposed to conduct ethnographic and archaeological research among several *Indianerstämme* (Indian tribes) of northern Mexico, including the Náayeri, Wixárika, O'dam (Tepecanos and Tepehuanos), and Rarámuri (Tarahumaras).[9] The idea of the expedition originated with Eduard Seler (1849–1922), Preuss's direct supervisor at the museum's American department and an expert on pre-Hispanic Mexican cultures. Seler had planned to conduct this field trip along with Franz Boas (1858–1942), an anthropologist and curator at the American Museum of Natural History (AMNH), whom he had met in 1885, when the latter worked as a curator for the KMV. However, Boas quit his job at the AMNH in 1905, forcing them to abandon the idea of a cosponsored expedition. In the end, Seler reluctantly chose Preuss to take over the project.[10] Seler seems to have been hesitant about Preuss not only because the younger scholar lacked fieldwork experience but also because Preuss often preferred to dwell on theoretical and abstract issues rather than "focusing on concrete matters."[11]

Nevertheless, although the field trip was originally Seler's idea, Preuss still had to provide a justification for the funding to be granted. In his proposal, Preuss argued three main points: (1) that the museum had almost no archaeological or ethnographic materials from these communities in its collections; (2) that given the linguistic and cultural connections between them and the people of the pre-Hispanic Aztec civilization, their contemporary religious rituals would shed light on the latter's ancient mythological and religious pictographs; and (3) that these communities continued to live under similar conditions to those of their ancient relatives in central Mexico, with little to no contact with people of European descent.[12]

The proposal's rationale aligned perfectly with Seler's original mission for the field trip, which in turn reflected the vision Adolf Bastian (1826–1905) had for the KMV when he convinced Kaiser Wilhelm I to create it in 1873. Bastian believed that the museum's main goal should be to gather as much material culture as possible from the *Naturvölker* (primitive peoples) of the world—as opposed to *Kulturvölker* (civilized, European people)—and to record as much of their belief systems before colonialism and the expansion of Western civilization drove them to their inexorable extinction.[13] Seler and Preuss's plan—to gather objects from a wide variety of Mexican Indigenous communities through archaeological excavations and to analyze their religious practices using comparative methodologies—reverberated with both the scholarly goals of the museum and the colonial practices of the German Kaiserreich, its de facto sponsor.[14]

Preuss was born in Eylau, at the heart of the Kingdom of Prussia, in 1869, only two years before the unification of Germany under Kaiser Wilhelm I. He grew up and was educated in Königsberg, the kingdom's capital during the consolidation of the German Empire, in a social, cultural, and academic environment where ideas such as social Darwinism, messianism, and racialism were becoming entangled with the new empire's fervent nationalist rhetoric. At the Albertus-Universität zu Königsberg, Preuss studied history and geography and completed his doctoral studies in 1894 with a dissertation entitled "Die Begräbnisarten der Amerikaner und Nordostasiaten." (The Burial Patterns of Americans and Northeast Asians). Here, Preuss's main argument, that "the diversity of burial patterns among these people made them particularly suitable to stand as representatives of all humanity," points toward the type of comparative methodology that late nineteenth-century German ethnologists and folklorists favored in an attempt to access the "mind and soul" of humankind and the nature of the world (or the world of Nature).[15] This type of comparative methodology and linguistic relativism, which followed on the long-standing relativist philosophical and linguistic traditions of Johann Gottfried Herder (1744–1803) and Wilhelm von Humboldt (1767–1835), would be central to the larger intellectual project behind Preuss's future ethnographic work in Mexico's Gran Nayar.

In 1885, after studying geography and *Völkerkunde* (ethnology/ethnography) at the Friedrich-Wilhelms-Universität in Berlin for a year, Preuss started working as a volunteer at the KMV's Africa and Oceania department. In 1900 he became assistant director at the museum's American department, which was headed by Seler. At that time, Bastian was still the museum's director. Under his leadership, this institution and the scholars it housed were fundamental in shaping the field of *Völkerkunde*. Bastian's original encyclopedic, empiricist intellectual project emphasized the study and classification of *Naturvölker* on the strict basis of fieldwork and the gathering and collecting of material culture. For him, as Andre Gingrich summarizes, this "exotic people had little or no culture and no history, and thus they could reveal the true nature of humans."[16] Although a younger generation of moderate positivist scholars sought to reconcile the nature/culture and anthropology/history dichotomies that characterized Bastian's original project and the race science that dominated the field at the time, the ideas of Bastian and Seler were still a strong influence on Preuss's early work and his proposal for the field trip to Mexico.[17] His essay "Der Einfluss der Natur auf die Religion in Mexiko und den Vereinigten Staaten" (The Influence of Nature on Religion in Mexico and the United States; 1905) illustrates that although he used a more holistic, less racist, and more interpretative approach than his superiors, the epistemic and theoretical framework informing Preuss's approach to the Nayarit expedition was still intensely influenced by their positivist ideas.

Under Seler's influence at the museum, Preuss refocused his research interests on pre-Hispanic Mexica civilization, religion, and art. Between 1900 and 1904, Preuss studied classic Nahuatl and published several articles dealing with Mexica codexes, hieroglyphs, and cosmogony. However, it was "Der Ursprung der Religion und Kunst" (The Origin of Religion and Art; 1904–5), an extended article in which he argued that by studying "demons and gods as magical natural objects with their transformations . . . one comes to understand all the religious facts of the primitives," that cemented his reputation in anthropological circles around the world.[18] This article and the essay "Der Einfluss der Natur auf die Religion in Mexiko und den Vereinigten Staaten" illustrate the epistemic and theoretical framework that informed Preuss's approach to the Nayarit expedition and the ways he listened to the people he encountered in the field and understood the objects he found.

The novelty of the Gran Nayar as a site for the exploration of the Other in the early twentieth century was largely the result of two pioneering figures who had undertaken field trips among the Wixáritari and Náayeri in the late 1890s, the French naturalist Léon Diguet (1859–1926) and the Norwegian ethnographer Carl Lumholtz (1851–1922). After his field trips in 1896 and 1898, Diguet provided the first general descriptions of the area's geography and of Wixárika culture. He brought back to France a large collection of photographs as well as botanical and archaeological materials. Particularly important were the gathering and transcription of fragments of myths from the *mara'acame* (wizard or singer) sacred chant and *xaweri* songs related to deer hunting, food offerings to the gods, and the *yaawi* (coyote) dance.[19] Diguet was not a musician; therefore, in order to produce these musical transcriptions, he required the assistance of Félix Bernardelli (1866–1908), a Brazil-born Mexican painter and violinist whom he met in Guadalajara.[20] Nevertheless, his documentation of the myth of origin of the *xaweri* (an instrument similar to the *rabel* or folk-fiddle from Spain), his speculation about how the Wixáritari adopted the instrument, and his thorough discussion of the instrument's physical characteristics and performance technique are a unique path of musico/cultural inquiry that neither Lumholtz nor Preuss pursued even though they were interested in engaging with Wixárika music practice.[21]

Lumholtz organized a total of six expeditions to Mexico: in 1890–91, 1891–93, 1894–97, 1898, 1905, and 1909–10.[22] All his trips, apart from the last one, were at least partially sponsored by the AMNH. The goal of these expeditions was to collect material culture from the north of Mexico and to study possible relations between the Indigenous peoples of the American Southwest and the pre-Hispanic civilization of the Aztec Empire in central Mexico. It was only during his fourth expedition, in 1898, planned as a trip to complete the gathering of ethnographic and anthropological objects he started collecting during the three previous expeditions, that Lumholtz brought a phonograph to record traditional music (for *xaweri* and *kanari* [guitar]) and ritualistic chants.[23] By the end of the expedition, Lumholtz had recorded seventy wax cylinders, thirty of them among the Wixáritari. Other than three transcriptions made by Edwin S. Tracy that were published in Lumholtz's *Unknown Mexico* (1903), and one more transcription (of a Rarámuri rain chant) made by Alice C. Fletcher and published in her book *Indian Story and Song from North America* (1900), Lumholtz did not really study the music stored in those cylinders.[24] His project was largely a collecting effort, not an interpretative one.

It was precisely Lumholtz's lack of intellectual reflection on the materials that prompted Seler and Boas to plot their own expedition to the Gran Nayar. Aware of Diguet's and Lumholtz's work, Seler and Boas decided to pay systematic attention to the Indigenous festivities, their chants and prayers, and the languages spoken in these communities to remediate what they believed were their predecessors' most important shortcomings. For Seler and Boas, it was only through a deep understanding of the uses of language within the religious rituals of these communities that they could unlock the secrets of their cultures and their ties to the pre-Hispanic past.

Thus, questions about linguistics became central to the project.[25] These issues shaped the planning of the expedition when Preuss inherited the plan from Seler in 1905. On his way to Mexico, Preuss visited Diguet in Paris as well as Boas and Lumholtz in New York to discuss their previous experiences in the area and the feasibility of his upcoming endeavor. Margarita Valdovinos argues that it may have been only after his visit to Lumholtz in New York that Preuss decided to follow the Norwegian ethnographer's lead and also record the chants of these communities. Since it was only after he arrived in Mexico City that Preuss acquired a phonograph and a set of wax cylinders, he failed to receive the technical training that the KMV usually provided to those they sent on these kinds of field trips.[26] Nevertheless, the history of European contact and engagement with the Gran Nayar area and the ideas about difference and Otherness that dominated ethnology in the context of the Kaiserreich were fundamental in determining Preuss's ability to listen in the field.

It took Preuss only a few months after he started his fieldwork in Mexico to realize the magnitude of the task he had proposed to the KMV. On December 24, 1905, he wrote his first letter from the field to Seler to notify him that his trip to the Sierra Madre was scheduled to start the following day and also to inform him that on arriving in Mexico City, he had "realized how difficult it would be to bring antiquities out of Mexico."[27] On March 14, 1906, after his initial stay in the Náayeri village of Jesús María, Preuss wrote to Seler again, asking to extend the expedition by an extra year since it was impossible to accomplish what he had promised to do in just one year. He also proposed to limit himself to working with just some of the communities he had originally proposed to visit (the Náayeri, Wixáritari, and Mexicaneros), letting Seler know that up to that moment he had only done research among the Náayeri.[28]

Preuss's trip to Sierra Madre began with a six-day journey from Tepic, the capital of the state of Nayarit, to the village of Jesús María, which he took as his headquarters. There, Preuss was assisted by the Catholic priest of the village, who not only gave him lodging but also introduced him to Francisco Molina, a Náayeri who helped him with learning the Cora language and with establishing relations with other members of the community.[29] This was extremely important for Preuss, who needed to develop a sense of trust among people who were extremely wary of his presence in their community, tended to avoid him, and in some cases even threatened him.[30] Preuss arrived in Jesús María during the dry season and witnessed the Náayeri's intense communal life that characterizes that time of the year.

He was especially impressed with the *pachitas* carnival celebrations and the mitote rituals and was particularly interested in the lyrics of prayers and *cantadas* (chants) from the latter, which he believed encompassed the essence of pre-Hispanic Mexica religious life.[31] Thus, the recording and translation of mitote ritual texts became central in Preuss's project (although he also recorded local mythology and legends, from Molina himself, as well as *pachitas* songs). After several months in the area, Preuss secured the assistance of three men from the nearby village of San Francisco as well as Jesús María (Leocadio Enríquez, Santiago Altamirano, and Ascensión Díaz), to whom he paid substantial fees (*hohen Lohnes*) and often gave liquor (*mit Hilfe von Whiskey*) to encourage them to sing the mitote chants into his phonograph's recording horn.[32]

The term *mitote* comes from *itotía*, the Nahuatl word for the verb "to dance," and refers to the fact that dancing is a central aspect of these celebrations.[33] In Jesús María and San Francisco, where Preuss conducted his fieldwork, the mitotes are a group of three annual ceremonies connected to the seasonal cycles and the agricultural cycle of maize. Thus, the mitote of planting is celebrated in June, the mitote of harvesting is celebrated in October, and the mitote of grain storage is celebrated in January.[34] As a group of ceremonies, mitotes feature a cosmovision in which the secular and the religious, the divine and the mundane, the Catholic and the Náayeri, are ontologically interwoven and are inseparable. Mitote celebrations can last from three to five days. The climax of the ritual arrives at night, when the ceremonial singer begins the performance of ten mitote chants that could last for over twelve hours. Singers accompany their chants with the *túunami*, a single-string musical bow played percussively with two wooden sticks, which provides a monotonous but stable rhythmic foundation on which the single vocal line unfolds. Dancers and other participants join the performance in very codified ways that respond to the actions described in the singer's singing. Thus, mitotes are characterized by a type of chanting that describes the ritualistic actions that are taking place as they take place.

Preuss's determination to record these chants was triggered by the linguistic interest at the core of the expedition's mission but also by the fact that witnessing a man sing through the whole night during the mitote celebration for roasted corn seemed to confirm his presupposition that these chants were part of an ancient Indigenous tradition that predated European contact.[35] Valdovinos's work shows that Preuss's interest in the lyrics and the Cora language determined the methodology of his textual transcriptions, his approach to the recording technology he had access to, and the process of recording these chants.[36] Preuss chose to meet with his interlocutors in sessions that often lasted for a whole day. The working method consisted of asking the singer to speak the chant text in Cora and dictate it to Preuss first. It was only when the text was entirely transcribed that Preuss and his interpreter attempted a preliminary translation. The purpose of this translation sketch was to double-check with the singer any specific questions about the chant's lyrical content. At the end of the session, once the text had been transcribed in its entirety, Preuss would proceed to record the song with his phonograph.[37] This must have been a truly bizarre experience for Enríquez, Altamirano, and Díaz since the exercise required them to sing the chants into the horn of an alien machine, without musical instruments, and in a complete performatic and ritualistic vacuum.

The sonic result captured in Preuss's cylinders is equally strange for someone familiar with the rituals described above. For example, cylinder number 13 starts with the voice of one of Preuss's singers identifying the chant as "Cantada de los dioses en la que se representa la ceremonia del venado y baile del mismo sobre el hombro de los capitanes, en Jesús María, por Leocadio Enríquez en la fiesta de la siembra" (Chant of the gods, which represents the ceremony of the deer and its dance on the shoulder of the captains, in Jesús María, by Leocadio Enríquez during the celebration of sowing). After that, Enríquez proceeds to sing a prayer expressed in a melodic fragment. It is composed of four segments of equal duration and analogous rhythmic pattern. This basic melodic sequence is repeated six times with slight variations in the lyrics before fading out in the middle of the last repetition. The structure of the lyrics follows a very clear pattern: Each repetition changes the addressee of the praying and the type of offering that is being made for them but retains the overall geographic setting (a *cerro* [hill]) where the praying and ritual action are supposed to be taking place. The only other regular sound one hears throughout the recording is the rhythmic noise of the winding up of the phonograph with a pulsing regularity. There is no *túunami* supporting the monophonic chanting nor any noise referring to the stomping and dancing that should accompany the unfolding of the lyrics' narrative, and, obviously, no trace of any of the performatic actions and use of objects that are central to the mitote ritualistic context. All of that was lost in the process of reducing a rich ritualistic ceremony into a dislocated chanting utterance.[38]

By mid-March 1906, Preuss had already recorded and translated twenty-three mitote chants as well as prayers in the Cora language and had collected over four hundred Náayeri objects. Having already become familiar with the language and mitote ceremonial tradition, he concluded that to trace the origins of this ritual, he needed to collect and transcribe the songs of the other Nahua communities in the area.[39] Thus, in June he moved to San Isidro, in Wixárika territory, in the northeastern part of the Sierra Madre. There, Preuss received support from José María Carrillo, a Wixáritari man who helped him with the Huichol language and introduced him to members of the Wixárika communities familiar with the *neixa* ritual festivities. After some time in San Isidro, Preuss identified these festivities and their dances and chants as the Wixárika equivalent to the Náayeri mitote chants.

The *neixas* are ceremonies centered on maize and indicate the beginning and end of the rainy season. These are the *hikuri neixa* (peyote celebration) and the *namawita neixa* (planting celebration)—which take place in May and June, at the end of the dry season—and the *tatéi neixa* (celebration of Our Mother), which happens in late October, at the end of the rainy season. The *hikuri neixa* marks the end of the dry season and the arrival of the first rains. The *namawita neixa* celebrates the triumph over fire of the goddess Takutsi Nakawe (the grandmother goddess of creation, growth, and vegetation). The *tatéi neixa*, also known as *wimaxkawa* or *fiesta del tambor* (drum celebration), is connected to the harvest cycle; this is the main Wixárika festivity and marks the beginning of the pilgrimage to the five cardinal points, one of the main community activities during the dry season.[40]

Preuss spent the rainy season in San Isidro, where he conducted most of his research about the *neixas* and transcribed forty myths. He sent a shipment of collected materials, including over five hundred objects and thirty-six wax cylinders of Náayeri and Wixárika chants, to Berlin in early December 1906.[41] Through the beginning of 1907, Preuss moved back and forth between San Isidro and Santa Catarina, where he continued working on transcribing Wixárika texts, recording chants, and documenting the New Year festivities. Although he prolonged his fieldwork in the Gran Nayar region—mostly among Mexicaneros but also among Náayeri—until the summer, he did not record any additional chants or myths. The bulk of his cylinder collection was completed by the end of April 1907. Preuss's working method among the Wixáritari was the same one he had implemented with the Náayeri: He would spend most of the day having his singers and narrators (in this case, José Fernando, Centavo, Marcos, José Antonio, and Ramón) dictate him the lyrics of the chants before proceeding to record them into wax cylinders at the end of the day. Thus, the shortcomings of Preuss's methodology and his determination to understand the lyrics of the *neixa* chants also had a detrimental impact on how these chants were delivered and presented to him.

Preuss's working method puts in evidence that the final sonic result recorded into the wax cylinders was mediated by two factors; one was epistemic, the other material. On the one hand, there was Preuss's goal of understanding and faithfully transcribing the chants' lyrical content (which generated serious misunderstandings of what the chants and their tradition were, as becomes clear below) and, on the other, the material shortcomings of the recording technology at his disposal. Reflecting on the type of epistemic logics behind the work of early sound ethnographers such as Preuss, Erika Brady argues that "the introduction of a 'mechanical presence' embodied in the recording both determined the form in which information was preserved and significantly altered the balance of the entire fieldwork interaction."[42] Indeed, to say that the sounds captured in these cylinders are in fact the Náayeri and Wixárika ceremonial chants or that they somehow represent the ritualistic experience they are an intrinsic part of would be not merely disingenuous; it would also be a colossal misunderstanding of what the mitote and *neixa* traditions are and what they are about.

Valdovinos has noted that the main problem with Preuss's working method is that it took the chants out of their ceremonial-ritualistic-performatic context. In his desire to understand and transcribe the lyrical content of the rituals, Preuss asked the singers to speak rather than sing the chants. This action meant that the chants were severed from the ceremonial spaces and the performatic and musical actions they are not simply meant to be in dialogue with but are a part of.[43] As Valdovinos argues, these actions are ontologically connected to the chants and work as mnemonic devices that aid the singer in the process of verbalizing their descriptive/narrative content.[44] Thus, in dictating the texts to Preuss, in the absence of the normal ceremonial-performatic context, the singers had to reimagine those ritualistic actions in order to enable themselves to produce something akin to what Preuss was expecting.[45] This was a very challenging process given that these reimaginings had to occur within the largely peculiar and artificial conditions of Preuss's working space. The process of dictating and transcribing the chants' texts shows that in transforming chants into speech, the ethnologist's choices did violence to these Náayeri and Wixáritari practices and objectified them metaphorically as well as materially. As such, the sounds stored in Preuss's wax cylinders are also a clear indication of how this process of objectification unfolded. On the other hand, the recordings did precisely what Preuss wanted them to do; they worked as aids in the process of transcribing the chant's lyrics. In doing "something out of something else," as Barbara Titus has stated, Preuss's wax cylinders are a patent indication of the blatant "exertion of power" that characterizes extractivism.[46]

When one listens to the digitized versions of Preuss's wax cylinder recordings available at the BPA, one is struck by several issues. The most salient are the short duration of each of the chants and their apparent fragmentation; the absence of instruments (an important difference between Preuss's cylinders and those recorded by Lumholtz in 1898); the level of noise, which sometimes makes it very difficult to understand both pitches and lyrics; and, for me the most telling, the singers' evident hesitations. Some of these problems refer to specific technological shortcomings. For example, the cylinders were only able to record up to three minutes of music, which forced Preuss and the singers to divide each chant into several fragments. However, even when the chants were fragmented in such a way, it was impossible to register the chants in their entirety since each of them could last for hours during their ceremonial performance.[47] This is an important factor to keep in mind because it forced the singers to come up with shortened versions of the chants that, as Valdovinos states, "are only the result of the narrators' adaptation to the recording context" and have very little to do with the Náayeri's and Wixáritari's tradition of verbal art that characterizes their normal performance.[48] However, the singers had to do that to provide recognizable renditions of texts that would normally unfold over very long periods of time. Preuss's interest in primarily making sense of the text also determined how he chose to record the chants. For him, the wax cylinders were largely meant as sonic verifications of the texts he had already transcribed. He was interested in producing them to have a type of control over the chants when working on the final transcriptions and translations back in Berlin. In the interest of making the texts sonically intelligible, Preuss got rid of the *túunami* accompaniment, which would have been very difficult to capture with a type of technology that required the singers to sing with their heads practically inside the phonogram's recording horn (although not an impossible mission, as Lumholtz's earlier recordings show). Interestingly, Preuss's reduction of chant to speech through the use of the phonograph reverberates with Thomas Edison's original identification of this invention "as a textual device, primarily for taking dictation."[49]

These were the alien circumstances in which Preuss's singers had to remember and record the chants. These circumstances may explain the singers' hesitations and false entries during the recording process. Rather than mistakes or evidence that the singers did not know the chants well enough, these moments of hesitation show these men trying to "shorten the length of the chants for dictation and phonographic recording [in order to] adjust their narration and performance to the reduced and decontextualized format imposed by the context of [Preuss's] research."[50] In fact, their hesitations are evidence that these chants are not fixed songs or entities to be reproduced the same way every time they are sung, as one would expect for songs or pieces of music in the learned Western music tradition. Instead, the nature of chants in Náayeri mitotes and Wixárika *neixas* is to change within well-defined structures every time they are performed. As Valdovinos has argued, against conventional Western conceptions of performance, in the mitote and *neixa* celebrations, chants and actions exist in close performative relation. The lyrics of the chants do not simply describe actions being performed; they are, along with those actions, verbal art that is essential in the very configuration and structuration of the mitote as a ritual and the reality that surrounds and gives meaning to it.[51] Chant and action in tandem are the type of reiterative enunciation that discursively does what it says and thus creates knowledge. In other words, Náayeri and Wixárika musico-descriptive and kinetic doing is communal knowledge that cannot be separated and fragmented into discrete individual actions or activities as Preuss requested his singers to do.[52]

In Preuss's goal to fix into wax cylinders a cultural practice whose specific representation is always in flux, his methodology became a de facto process by which Náayeri and Wixárika ritualistic traditions were made into objects. By highlighting certain aspects of the practice, the act of recording became a process of objectification that transduced ritual actions into texts (both written and sonic) that in turn translated them into Preuss's epistemic paradigm and made them manageable within the limitations of the technological template available to him.[53] In brief, in his encounter with the people of the Gran Nayar, Preuss understood what his episteme allowed him to understand; heard what he was trained to hear, in the way he was trained to hear it; and was able to capture only what his technology allowed him to capture. These shortcomings define the processes of mediation that frame the collections he was able to gather and the archives that came to store them. Thus, regardless of Preuss's good intentions of truly getting at the core of Náayeri and Wixárika culture and cosmovision, his listening was performative in the very sense that Dylan Robinson describes with the notion of "hungry listening." Preuss's listening "bolster[ed] an intransigent system of presentation guided by an interest in ... Indigenous content, but not Indigenous structure."[54] And as such, his archive, organized within this epistemic logic, is also inoculated with the performative potential to continue reproducing this representation of Náayeri and Wixárika culture and to guide the type of questions scholars could ask about the materials that constitute it.

"Habe ich etwa 100 Walzen mit religiösen Gesängen der Cora- und Huicholindianer heimgebracht": A Collection's History

The Preuss Collection—which includes his wax cylinders and the objects from the Nayarit expedition, as well as his papers and the materials from his later Colombian expedition—is formally kept at Berlin's Ethnologisches Museum (as the KMV was renamed after World War I). The BPA, which belongs to the Ethnologisches Museum, officially guards Preuss's sound archive. However, since the Humboldt Forum was created to house the non-European collections of the Staatliche Museen zu Berlin (which itself includes the Ethnologisches Museum), the Preuss Collection was moved there. Understanding the journey of Preuss's wax cylinders from the Gran Nayar to the Humboldt Forum, and grasping their significance in broader discussions about world heritage and the German and Prussian legacy, is crucial for comprehending the potential of this archive.

Preuss's Gran Nayar adventure finished in the summer of 1907. He was back in Berlin that fall and needed to listen to the chants in his wax cylinders in order to edit, consolidate, and finalize the German translation of the Náayeri and Wixárika ritual texts gathered in Mexico. However, he was aware that due to the nature of the technology, the original wax cylinders would irremediably deteriorate every time he played them, and he needed to find a way to listen to the recorded sounds repeatedly without damaging the cylinders. On October 4, 1907, on the suggestion of the KMV's new director, Felix von Luschan (1854–1924), Preuss sent a letter to the Friedrich-Wilhelms-Universität's Psychologisches Institut. The letter informed the addressee that Preuss was in possession of one hundred wax cylinders with religious chants from the "Cora and Huichol Indians" ("Habe ich etwa 100 Walzen mit religiösen Gesängen der Cora- und Huicholindianer heimgebracht") from Mexico's Sierra Madre and requested the expertise

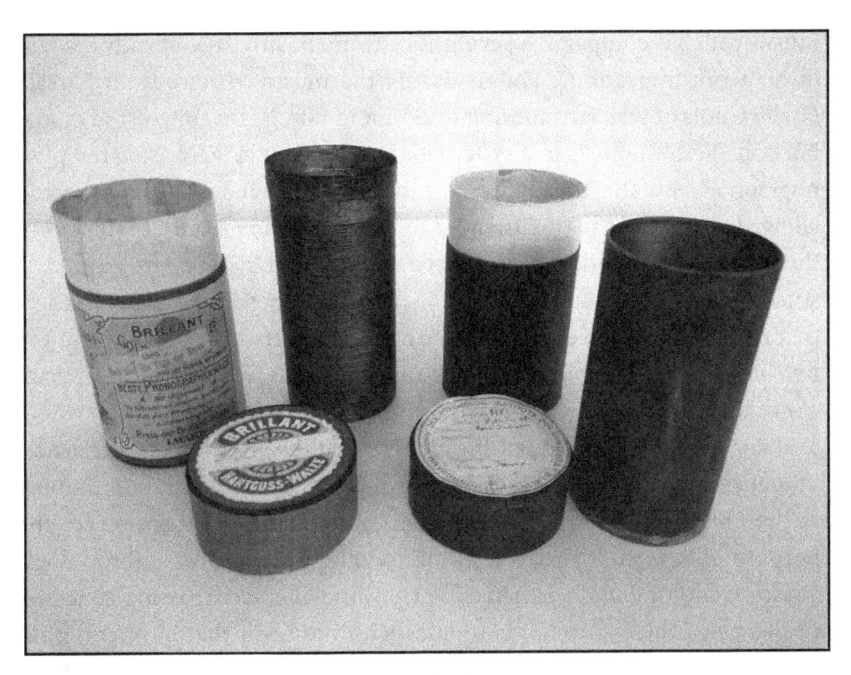

FIGURE 4.4. Copper negative ("galvano") copies of wax cylinders, ca. 1907–8, in the Preuss Collection at the Humboldt Forum. Photo by the author.

of one of the institute's professionals to make "galvanos" (negative copper copies) of these wax cylinders.[55] The galvanos were necessary in order to make working copies of the wax cylinders that Preuss could play repeatedly without fear of destroying the originals in the process (figure 4.4). Although the letter does not include the name of the addressee, one could argue that it was sent to Erich von Hornbostel (1877–1935), who was the director of the BPA—housed by the Friedrich-Wilhelms-Universität's Psychologisches Institut—and himself an expert on the difficult process of making galvano negative copies out of wax cylinders.

The BPA was created by Carl Stumpf (1848–1936) in 1900, but it was not formally founded until 1904 by him and Hornbostel, who became its director in 1905. Stumpf was a philosopher interested in the study of perception and sensation of sound and music in relation to language, human emotions, and mental states. *Tonpsychologie* (1883–90) is Stumpf's most important contribution to this field, a work that foreshadows both Gestalt psychology and philosophical phenomenology. Here, Stumpf understands sounds and pitches as physical phenomena to be measured in order to better study how they are perceived sensorially. His approach to the study of music, which is comparative, evolutionist, and positivist, is already evident in his work transcribing and analyzing the music of a group of Nuxalk (Bella Coola) musicians touring in Berlin in 1885.[56] This interest in studying non-Western music in a systematic way led him to start using the phonograph when a Thai theater troupe visited Berlin in 1900. The phonograph allowed Stumpf and his assistant, Otto Abraham (1872–1926), to transform the immateriality and impermanence of the music practice he wanted to study into a more measurable and manageable object. The wax cylinders that came out of this encounter became the first twenty cylinders in the BPA inventory. The collection grew quickly, as Hornbostel and Abraham continued recording non-European musics.[57] In fact, as Hornbostel would explain years later, the goal of the archive would be to "collect the musical utterances of all the peoples of the Earth, which were swiftly succumbing to the universal leveling effects of civilization, and offer them up for the purpose of comparative studies in musicology, ethnology, anthropology, ethnopsychology, and aesthetics."[58] Thus, the all-encompassing character of the BPA project strongly resonates with the spirit that informed both the colonialist outlook of Kaiserreich imperialism and the comparativist-evolutionist perspective of turn-of-the-century German anthropology under the leadership of first Bastian and later Luschan.[59]

Stumpf's original interest in the phonograph as technology in service of sound measurement within larger projects to understand human emotions and psychic states explains the seemingly arbitrary choice to originally house the BPA in the Psychologisches Institut. This also provides a link to Preuss's motivation behind his intellectual project, the comparative and transhistorical study of mentalities, or what he described as the "Seelenleben der Naturvölker" (life of the soul of primitive people).[60] But for Preuss to continue with this project, he needed to have safe and reliable access to the objects that would make his final measurements and analysis possible.

The process of making, first, negative galvano copies and then new working wax cylinder copies out of those galvanos was an intricate one, and in the case of Preuss's collection, a long one owing to its large size.[61] It took almost a year for Preuss to get working wax copies, and although he was not completely satisfied with the result—he claimed from the very beginning of the process that the copies were not as clear as the originals—he was finally able to continue with his analysis of the Náayeri and Wixárika texts.[62] As would become the standard BPA policy, the archive kept the galvanos in its collection and sent Preuss the newly produced wax working copies.[63]

In 1922 Stumpf retired from his position at the Psychologisches Institut, and a year later the Prussian state acquired the BPA, transferring it to the Staatliche Akademische Hochschule für Musik. Although Hornbostel continued as head of the archive, the project was overseen by the Hochschule's director, Georg Schünemann (1884–1945), a musicologist who had worked with Stumpf at the Königlich Preußische Phonographische Kommission (Royal Prussian phonographic commission), a committee established in 1915 to coordinate the ethically problematic recording of the music and speech of war prisoners in World War I German camps.[64] This affiliation with the Hochschule für Musik was considered beneficial for both institutions as it made the archive readily available to musicologists and music students and not just ethnologists.[65] In 1933 the Nazi Party removed Hornbostel from his post at the BPA and forced him into exile. The following year the BPA was reassigned to the Ethnologisches Museum, where the Preuss Collection remained until 1944–45, when the archive was evacuated due to the Soviet occupation of Berlin at the end of World War II. Eventually, most of the archive's collections, including Preuss's cylinders, were confiscated by Soviet authorities and taken to Leningrad, where they were kept at the Phonogram Archive of the Institute for Russian Literature (Pushkin House). In 1959 most of the archive's collections were repatriated to the German Democratic Republic (GDR). However, since the Ethnologisches Museum was in West Berlin, in the Federal Republic of Germany, the collections remained in the GDR's Academy of Science until 1991, when, after the reunification of the country, the archive finally made its way back to the Ethnologisches Museum.

What unfolded once the wax cylinders were back at the museum was a process of making the collections accessible that led them to also become officially recognized as heritage. The immediate task on receiving the archive back was to identify, collate, and catalog the collections according to the documentation about them that had been kept at the museum.[66] In 1998, financed by the Stiftung Preußischer Kulturbesitz, the BPA began a process of digitalization of their historical wax cylinders. The goal was to make new casts of every recording in the archive—based on the surviving copies (originals, galvanos, or wax copies) that featured the best sound quality—and then produce digital copies.[67] In 2000, two years after the digitalization process started, UNESCO recognized the uniqueness and historical value of the BPA wax cylinder collections by inscribing them in the Memory of the World Register (figure 4.5). The UNESCO nomination of the BPA emphasized the digitalization process as a way to make the collections

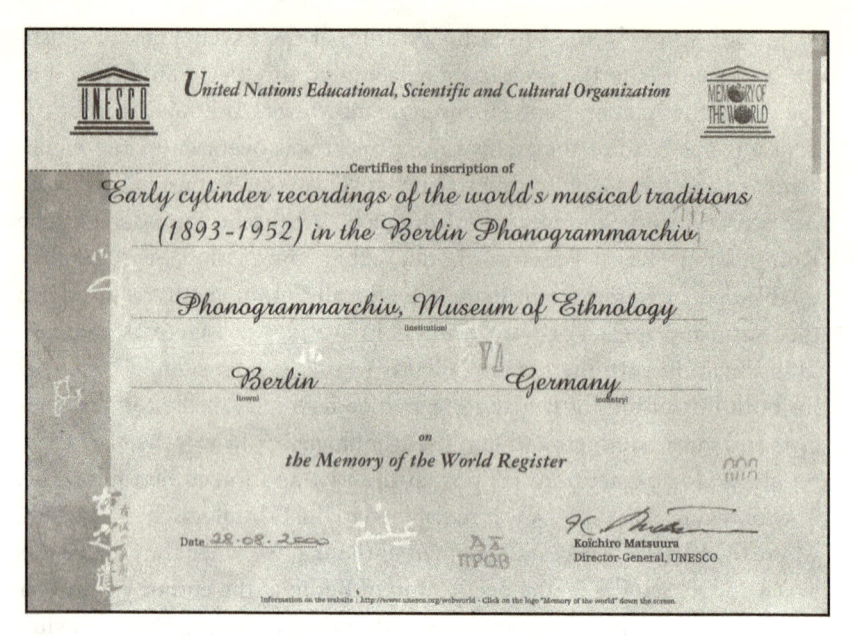

FIGURE 4.5. The Berliner Phonogramm-Archiv as Memory of the World. UNESCO certificate, August 28, 2000. Courtesy of Staatliche Museen zu Berlin, Ethnologisches Museum.

accessible so that "recordings of the past [could be] compared with present performance practices" and serve "as a basis for a planned revival of more or less obsolete performances [and] as source materials for extinct music cultures or even peoples."[68] The Preuss Collection was one of the first to be digitized. It was made available to selected specialists as early as 1999.[69]

The Loud Voice of a Silent Sound Archive

Preuss used the wax copies of his recordings to work on what he envisioned to be the final document of his Gran Nayar adventure, a multivolume comparative study of ritual and mythology among Indigenous groups in western Mexico. The first volume was to focus on Náayeri myths, the second was intended for Wixárika mythology, the third was meant to focus on Mexicanero mythology, and the fourth was supposed to be a comparative analysis of the languages of the three communities and the ethnographic objects Preuss collected during his field trip.[70] In the end, Preuss was only able to publish the first volume of his monumental study, under the title *Die Nayarit-Expedition: Textaufnahmen und Beobachtungen unter mexi-*

kanischen Indianern, vol. 1, *Die Religion der Cora-Indianer in Texten nebst Wörterbuch* (1912). The book includes a description of the project's methodology; an explanation of the basic features of Náayeri religion, deities, and ceremonies; transcriptions of mitote chant texts in Cora; an interlineal German translation of these texts; a basic Cora-German dictionary; and an appendix with Hornbostel's musicological study of two of the chants. Evidently, the title of the book indicates that Preuss was still planning to write the remaining volumes when the first one was published. However, unforeseen circumstances prevented him from continuing to work on this project. In 1914, the beginning of World War I found Preuss conducting fieldwork among the Murui-Muinan+ from the Colombian Amazon rain forest.[71] He was unable to return to Germany until 1919, and on his return, he focused on writing and publishing about the Colombian expedition, which was fresher in his mind. It was not until 1932 that he was able to return to the Nayarit project, when he published, with Franz Boas's support, his article "Grammatik der Cora-Sprache," a linguistic study of Cora language based on his fieldwork from almost thirty years earlier. The article was well received, and Boas encouraged Preuss to go back to work on the remaining volumes of *Die Nayarit-Expedition*. Gerdt Kutscher states that Preuss was working on the second volume of this project when he died in 1938 and that the two existing copies of the manuscript were destroyed in the 1943–45 bombings of Berlin and Leipzig at the end of World War II.[72] Fragments of the manuscript of the third volume did survive and were later edited and published in three parts by Elsa Ziehm in 1968–76.[73] Even though only the first volume of *Die Nayarit-Expedition* was published during Preuss's lifetime, its success in European academic circles—the book received the 1916 Graf-Loubat Award from the Königlich Preußische Akademie der Wissenschaften zu Berlin (Berlin Royal Prussian Academy of Sciences)—immediately cemented Preuss's international scholarly reputation.

Preuss's systematic, analytic, and comparative approach to mythology was important in Europe as a harbinger of structuralist anthropology—particularly the Leiden tradition and its search for deep common structures among different cultures of the world. His emphasis on a comprehensive study of history, expressive culture, and language that takes verbal art as a central aspect of analysis was especially influential. In that sense, as Valdovinos argues, Preuss's transdisciplinary ethnolinguistic approach foreshadowed the work of many contemporary anthropologists and the cross-disciplinary goal of the humanities at the beginning of the twenty-first century.[74] However, Preuss's influence on Mexican anthropology remained oblique and marginal at best. Before his death, Preuss had published only two articles in Spanish that were available in Mexico: a general explanation of the Nayarit expedition and an interpretative essay about the Aztec calendar.[75] A couple of his comparative articles about Indigenous cosmogony were published in Mexico later, in the 1950s.[76] Nevertheless, none of them provided Mexican scholars with any in-depth account of his working method or even the deep intellectual questions that guided his research. Besides sporadic and partial mentions of Preuss in early works about Náayeri religion, it was not until the 1990s that a real interest in his scholarship materialized when Jesús Jáuregui published translations of some of his writings about mitote.[77] This interest in Preuss's work in the 1990s led to the initiative to translate *Die Nayarit-Expedition* into Spanish. This ended up being a long project that proved complex and contradictory in the context of growing concerns about the colonial overtones of many of the collections in German and Prussian museums and archives, and the role of non-European scholars in exposing the source of those concerns.

It should be noted that when Preuss's work was finally discovered and taken seriously in Mexico, his collection of recordings was still largely unavailable to Mexican scholars. They knew about it, but for them, the wax cylinders remained a phantasmatic and elusive object of desire. Undoubtedly, the wax cylinder collection of Náayeri chants was central in Preuss's writing of the first volume of *Die Nayarit-Expedition* and in the academic validation of his scholarly work. However, it could be realistically argued that only a handful of persons had listened to it in detail before it left the Ethnologisches Museum in the 1940s. Given the historical trajectory of the BPA and the vicissitudes it experienced throughout the twentieth century, it is not hard to imagine Hornbostel—who published his analysis and transcription of two Náayeri chants from the Preuss Collection as an appendix to Preuss's *Die Nayarit-Expedition*—and Preuss himself as the sole listeners of this collection before it was confiscated by Soviet authorities after World War II.[78] Indeed, before it became an object of desire among Latin American scholars at the beginning of the twenty-first century, the Preuss Collection was paradoxically a rather silent sound archive. Nonetheless, even in the quietest of its incarnations, this archive speaks volumes that transcend the unique and impressive patrimonial character of its holdings. It speaks of the ideological, perceptual, and colonialist spirit and intellectual expectations that conditioned the way Preuss was able to listen in and to the field. As Miguel A. García suggests regarding the work of other European ethnographers who recorded Indigenous communities throughout the Americas in the early twentieth century, Preuss's sound archive features a "listening attitude or perspective that operates vertically and that is validated by and validates a cartographic project structured on the basis of a center expanding its power towards the periphery."[79] I would add that this kind of listening also produces the periphery as well as the Other as ontologically different from the Imperial Self in the violently dialogic performance of Empire.

The strict split between the study of *Naturvölker* and the study of *Kulturvölker* had informed the separation between anthropology as a natural scientific field and history as a humanist endeavor through the end of the nineteenth century. This division was also at the core of the fantasy that validated the Kaiserreich's imperial and cultural projects and their political attempt to homologize German identity with civilization. In that context, the scientific gathering and classification of the Other was clearly a project about self-identity. The understanding and symbolic taming of the Other through science became an excuse for the validation of the self. However, at the turn of the century, a younger generation of ethnologists—among whom one could count Preuss—was eager to question that split.[80] At the KMV, Karl von den Steinen (1855–1929) promoted a more comprehensive type of scholarship when he became director of the museum for a short period in 1905. Thus, it is not a surprise that he approved and supported Preuss's proposal for the Nayarit field trip, even when the latter's holistic approach and philosophical concerns, which somehow merged the goals and questions of the social sciences and the humanities, moved away from Seler's pragmatic positivistic idea behind the original planning of the expedition. Nevertheless, regardless of the generational, methodological, and political gaps that separated Seler's taxonomic approach from Preuss's theoretical and comparative concerns, their intellectual projects shared the common extractivist spirit and hungry listening behind the Kaiserreich's imperial and colonialist project. In that sense, despite his good intentions in understanding the Other from a progressive emic perspective, ontologically trapped in the ideological web of the Kaiserreich and his own ideological assumptions, Preuss's intellectual endeavor reflected a type of imperial entitlement that continued to discursively produce and reproduce the Other and mediate its representation within the logic of empire.

The "objectivist" character of Preuss's project is evident not only in the assumed objectivity of his scientific gaze, in the way in which he transformed expressive culture into objects, in the way in which he reduced chant to speech, and in the way his listening regime relegated and objectivized the

Other into timeless *Naturvölker* outside of history (even when trying to be critical of such ideas) but also in the morally and ethically problematic extractivist character of its mission. Preuss's correspondence with Seler and other members of the KMV team makes it clear that a central aspect of the Nayarit expedition was the gathering of objects. In a letter written to the administration of the museum on September 4, 1905, Preuss stated that he was aware that "the Mexican government has forbidden taking antiquities out of the country."[81] Most likely, Preuss's statement was informed by the Ley sobre Exploraciones Arqueológicas (Law about Archaeological Expeditions) and the Ley sobre Monumentos Arqueológicos (Law about Archaeological Monuments), from 1896 and 1897 respectively. Both pieces of legislation sought to stop the unabashed way in which foreign travelers, scholars, institutions, and self-appointed explorers took archaeological objects out of the country through the end of the nineteenth century.[82] However, awareness of these new laws did not persuade the administration of the museum to steer away from trying to gather objects for their collections. In a letter to Seler written on December 24, 1905, Preuss states, "Only when I arrived in the capital [Mexico City] did it become clear how difficult it would be to take objects out of the country."[83] Such a statement puts in evidence that even though everyone involved in the expedition was already aware that it was now illegal to take archaeological objects out of Mexico, they continued trying to find ways to get away with it. Nevertheless, more alarming is Preuss's acknowledgment, in a later letter to Seler, that "the Cora do not know that I took [some of these materials]."[84] The objects Preuss was referring to were ceremonial feathers, arrows, belts, woven bags, and several mitote altar items, for which he paid certain individuals in the community. In the end, besides photographs and the wax cylinder collection, Preuss was able to send to the museum a collection of well over two thousand ethnographic objects from Wixárika, Náayeri, and Mexicanero communities. The unethical overtones of Preuss's collecting endeavor as well as the imperial entitlement behind the project are not surprising given the ways in which many museums and archives in Europe and the United States at the turn of the twentieth century were fighting over the acquisition of objects for their collections. This is especially relevant given the questionable ways in which many of the KMV collections ended up there in the first place. In the context of the uneven power relations framing these encounters and exchanges, the use of the term *extractivism* to characterize the production, gathering, and relocation of objects for display or archiving at the KMV is not metaphoric. The removal of these objects from their original cultural context, their reification, and their accumulation as archive or museum collections that eventually fuel an economy of colonial and imperial prestige drove the extraction of local resources by foreign entities in order to make them into global consumption goods. Emily Hansell Clark proposes that "not just substances and labor, but forms of creative practice and knowledge," could be subjects of extractivism.[85] In that sense, the process of producing, gathering, and relocating objects for display or storage in European institutions is literally that, extractivism. It is an extractivism that aims at the production of not only cultural capital but also, in many cases, financial capital—as the economy generated just by rebuilding the Berlin Palace to host the Humboldt Forum makes evident. Dylan Robinson suggests that this type of "extractivist drive to accumulate is embedded within settler modes of listening, looking, and gathering [and that] this gaze, this listening is felt regardless of the decolonizing self-work an ally may have undertaken."[86] One could also argue that this is precisely the logic that characterizes imperialist modes of listening, looking, and gathering, and as such, Preuss's project, even when trying to actively move away from Seler's ideas about the *Naturvolk*, continues to make us feel this gaze and this aurality. Preuss's listening is an example of what Ronald Radano and Tejumola Olaniyan explore under the concept "audible empire"; it is a performative action that makes Wixárika and Náayeri chants "discernible" within European logics and reveals itself as "part and parcel of the regimes of knowledge, understanding, and subjectivity key to the constitution of both belonging and unevenness in the making of the modern."[87] As such, Preuss's wax cylinders are the inscriptions of his act of listening and the epistemic regimes that informed it; they encapsulate the powerful voice and agency of an archive that, ironically, remained silent for almost a century.

Making and Repatriating Other Objects

If making objects was a crucial aspect of the Preuss expedition as well as the constitution of the Preuss Collection in the BPA, it has also been a central activity in the engagement of the Preuss Collection by Mexican scholars. As mentioned earlier, the attention to Preuss's legacy among Mexican scholars in the 1990s led to an interest in translating *Die Nayarit-Expedition* into Spanish. While working on this project, Margarita Valdovinos garnered the support of Barbara Göbel, director of the Ibero-Amerikanisches Institut zu Berlin, to work on a second collaborative initiative, the publication

FIGURE 4.6. Cover of the CD *Walzenaufnahmen der Cora und Huichol aus Mexiko 1905–1907/Grabaciones en cilindros de cera de los coras y los huicholes de México* (Recordings on wax cylinder of the Cora and Huichol of Mexico; 2013). The recordings are drawn from the Preuss Collection.

of a CD with selected recordings from the Preuss Collection at the BPA. Published in 2013, the CD, entitled *Walzenaufnahmen der Cora und Huichol aus Mexiko 1905–1907/Grabaciones en cilindros de cera de los coras y los huicholes de México* (Recordings on wax cylinder of the Cora and Huichol of Mexico; 2013) (figure 4.6), was the first of two objects produced by engaging an archive that had been dormant for many decades; the second was the book *La expedición al Nayarit*, released in 2020 (figure 4.7). The process of making these objects provides a productive blueprint for retrieving from, engaging with, and interrogating archives to allow them to say something different from what they were originally meant to say.

The project to translate Preuss's book began in the mid-1990s. German ethnologist Ingrid Geist Rosenhagen, a professor at Mexico's Escuela Nacional de Antropología e Historia (ENAH), was put in charge of translating the text. However, the editorial team quickly faced a major conundrum:

FIGURE 4.7. Cover of *La expedición al Nayarit*, vol. 1 (2020), edited by Margarita Valdovinos, which translates Konrad Theodor Preuss's *Die Nayarit-Expedition* (1912).

Did it make sense to translate the Cora text to Spanish from Preuss's German translation rather than from Cora itself? In other words, did it make sense to subject the original religious Náayeri texts to a double process of translation/mediation? Jesús Jáuregui, coordinator of the project, proposed that it seemed more sensible to translate the texts directly into Spanish from the Cora language as printed in Preuss's book, without having to retranslate from the German translation. Thus, in 1998 Jáuregui invited Margarita Valdovinos, then an ethnology undergraduate student at ENAH, to join the project as editorial coordinator.

Valdovinos began working on the project by tracing back Preuss's steps. She moved to Jesús María, Nayarit, where she first experienced the mitote rituals. However, not being an Indigenous individual nor an expert on Náayeri traditions at the outset of the project, she soon realized that the cultural specificity of the texts required her to find local collaborators who were familiar with these rituals. Thus, she enlisted the efforts of

Bolívar Celestino Celestino, the teenage grandson of Florentino Celestino Zeferino, who was the traditional governor of Jesús María, a Náayeri chant singer, and a specialist in the mitote ritual, and Antonio Gutiérrez Rafael, a middle-aged man from San Francisco, Nayarit. Together, the team worked on transcribing the Cora texts in *Die Nayarit-Expedition*. While working on the translation, Celestino Celestino and Gutiérrez Rafael suggested numerous changes to Preuss's Cora text to make it more idiomatic and in tune with contemporary Cora language practice. As Valdovinos explains, the final result was a "text inspired by Preuss's text, and thus the [Spanish] translation would have to be more linked to this new [Cora] text and not so much to Preuss's work."[88] Celestino Celestino and Gutiérrez Rafael realized that Preuss's book did not showcase their communities' ceremonial chants but rather unorganized fragments of them.[89] Reassessing what to do regarding the new Cora translation and Celestino Celestino and Gutiérrez Rafael's findings delayed the project considerably. Around this time Valdovinos also became aware of Preuss's recordings at the BPA when a German ethnologist she met in the field in Nayarit—Viola Hörbst—played her a tape with some of the Náayeri chants. Valdovinos recalls that this encounter was a bit strange because the German ethnologist did not allow her to make copies of the tape and was very adamant about not playing the recording to any members of the community.[90] It would take Valdovinos five years before she was able to listen to the digitized versions of Preuss's wax cylinders at the BPA. Listening to these recordings allowed Valdovinos to corroborate Celestino Celestino and Gutiérrez Rafael's claims; she realized that "those recordings did not feature the chants but rather fragments [of them]."[91] However, in the meantime, the translation team's findings encouraged her to begin a doctoral project herself, focusing on the chants of the Náayeri mitote rituals. This project led Valdovinos to make her own in situ recordings of the Náayeri's ceremonial chants.

Having access to state-of-the-art digital technology allowed Valdovinos to record the Náayeri mitote chants in their entire length and within their ritualistic performance context. In highlighting this, I do not intend to put forward a techno-utopian argument since evidently, despite their avowed digital fidelity and their length, these recordings are also mediated representations unable to fully capture the ceremonial rituals and their performatic nuances. Nevertheless, the new recordings did eventually allow Valdovinos to thoroughly compare the contemporary iterations of these chants to the fragmented versions in Preuss's cylinders. Valdovinos's goal in doing this comparative work was not to find out whether the chants had or had not changed over time. She understood that impermanence is an integral aspect of the tradition but also knew that there were important performatic structures that were continued and shared transhistorically that guided how these chants changed from performance to performance while allowing them to retain the essential arc of their mythological narrative. Rather than a claim to authenticity, Valdovinos's work intended to identify the key performatic elements in the tradition that Preuss was unable to recognize as ontological to the chants, and how this misrecognition led to the messy presentation of the texts in his book. Thus, Valdovinos was no longer simply following in Preuss's footsteps; she was now retracing them to reassess his work, account for his shortcomings, and understand the reasons behind the problems with the chant texts pointed out by Celestino Celestino and Gutiérrez Rafael.

Analyzing the mitote chants as part of the ceremonies they belong in allowed Valdovinos to understand that they are

> part of a complex system of interactions. [The system is] established based on the actions of three types of participants—singers or ritual experts, apprentices, and common participants—and is organized around the enunciation of the chants. The mitotes' interaction structure presupposes in the first place the presence of a ritual specialist who, more than knowing the chants by heart, knows how to connect their enunciation with the ceremonial context. The components of the ritual action constitute a sequence of events that work as mnemonic devices; in other words, they offer the singers external support to guide them through the development of their enunciation and facilitate keeping the thread of [the chant's] long narrative. At the same time, the ritual events follow the sequence of actions described in the songs.[92]

By figuring this out, Valdovinos recognized that Preuss failed to account for the most important feature of the Náayeri mitote rituals, the performatic and performative nature of the internal interactions that make them possible. This ontological attribute sets the mitote chants, their texts, and their rituals completely apart from the cultural practices that Preuss was familiar with, which proved to be an inadequate frame of reference for him to truly understand the chants.

Following this critical assessment, Valdovinos's translation of Preuss's *Die Nayarit-Expedition* goes well beyond the line of duty. Rather than simply and uncritically allowing Preuss's archive to tell the story it was designed to convey, the 2020 publication of the three-volume book project *La expedición al Nayarit* is the result of inciting and allowing the archive to tell a different story. Valdovinos's edition presents extensive critical essays that contextualize Preuss and his work historically while providing a much more nuanced explanation of the book's subject matter (the mitote chants and their religious context) that sets the record straight. Most important, she also challenges the problematic center-periphery dichotomic type of power relations that a figure like Preuss—as a pioneering anthropologist working within the ideological frame of German empire building—tends to animate. *La expedición al Nayarit* as a book is an object that, unlike the objects Preuss produced and gathered during his fieldwork in Mexico and described in his own book, tells a story against the grain. For that story to fully emerge, it is necessary to also consider the kind of work done by the making and circulating of the CD *Walzenaufnahmen der Cora und Huichol aus Mexiko 1905–1907/Grabaciones en cilindros de cera de los coras y los huicholes de México*.

Although originally conceived as a side project to the book's translation, the CD came to fruition seven years before *La expedición al Nayarit*. Cocoordinated entirely by Valdovinos and Barbara Göbel, the CD was issued as an item in the BPA series of historical sound recordings and was cosponsored by Berlin's Ethnologisches Museum and the Ibero-Amerikanisches Institut. Later, Valdovinos was able to secure funds from Mexico's Instituto Nacional de Lenguas Indígenas (INALI) to enable the production of an extra lot of CDs to be freely distributed in Náayeri and Wixárika communities. The many challenges faced during the CD production process help us better understand Preuss's book and legacy within the larger political context that informs current concerns about German heritage in relation to the work of the Stiftung Preußischer Kulturbesitz and one of its most important contemporary public outlets, the Humboldt Forum.

The obstacles to producing the CD were political, bureaucratic, financial, and ideological. Soon after listening to the digitized versions of the recordings at the BPA, Valdovinos suggested to the administration of the archive the possibility of bringing back those sounds to their Náayeri and Wixárika communities of origin. It was at this moment that the reason behind the secretive behavior of the German ethnologist she had met in the Sierra Madre years earlier became clear. Obtaining the permit to bring the recordings back to Mexico was a complicated matter because of UNESCO's recent elevation of the BPA's historical wax cylinder collection to the category of world heritage, which placed the materials in a gray area regarding ownership for export or emigration rights. The cautious restrictions of the BPA administration regarding the circulation of these recordings also came as a response to unresolved tensions between the new UNESCO status bestowed on the wax cylinders, long-standing German cultural heritage legislation, and new UNESCO and European Union conventions regarding the repatriation of objects and materials held at European museums and archives that may have been obtained as part of questionable colonial exchanges—these were international agreements and conventions that the German government had ratified.

In 1955 the Federal Republic of Germany had passed the Kulturgutschutzgesetz (Cultural Property Protection Act) to protect valuable works of art, cultural goods, and historical documents and materials held at German museums, galleries, or archives from being sold and sent abroad. The law specified that "the archives, archival collections, assets, and collections of letters that are essential for German political, cultural, and economic history should be included in the directory of nationally valuable archives in whichever country they are when this law comes into force. . . . The archival materials included in this law are not only all types of documents but also maps, designs, seals, as well as artistic, film, and sound materials."[93]

The law allowed individual federal states to decide which materials would be inscribed in the directory but highlighted that the final decision to license their sale or export lay with the Federal Ministry of the Interior. The Kulturgutschutzgesetz of 1955 remained the law of the land regarding cultural heritage in Germany until it was replaced by the 2016 Kulturgutschutzgesetz in response to the 1993 European Union regulation on the return of cultural goods and the 2007 German ratification of UNESCO's 1970 Convention on the Means of Prohibiting and Preventing the Illicit Import, Export, and Transfer of Ownership of Cultural Property.[94] The 2016 Kulturgutschutzgesetz was drafted after an evaluation commissioned by the German government in 2013 determined that, among other things, the 2007 amendments to the law were impractical regarding the repatriation of cultural property.[95]

Evidently, issues regarding the return and restitution of material culture were very important in the national discussion about the protection of German cultural heritage when Valdovinos proposed the repatriation of Preuss's recordings to the Náayeri and Wixáritari.[96] Barbara Göbel states that "there was a lot of ambivalence [among the German institutions] because it was not just a matter of world heritage but also of cultural patrimony. There was a legal issue regarding making accessible objects that are the common good of a country [Germany]. So, when I first approached [the institutions] about it, the answer was always the same: 'It cannot be done because it is cultural patrimony and there are rights over these objects.'"[97] One of the main concerns with making the recordings available in CD format was that they would quickly begin circulating as pirated copies and the German institutions would lose control over their distribution and regulation. Nevertheless, the public discussion about accessibility and transparency of information and museum collections in Europe and Germany in the 2010s led to a radical change of attitude. Thus, after several years of negotiation, the Stiftung Preußischer Kulturbesitz approved the production of the CD. In fact, the production of a CD that would be part of the BPA and Ethnologisches Museum's Historical Sound Documents series was a prerequisite before the Stiftung Preußischer Kulturbesitz authorized the repatriation of these materials.[98] However, this decision led to a pragmatic problem during the processes of making decisions about and producing the CD.

Valdovinos and the BPA–Ethnologisches Museum did not share the same goal in making these recordings available. For Valdovinos, producing the CD was the first step toward repatriating the chants back to their communities of origin; for the BPA team, including the Preuss Collection as part of a historical CD series whose avowed mission is to "publish and circulate the wax [cylinder] recordings that up to this day are little known or not known at all" was a matter of featuring and celebrating their institutional assets.[99] For that reason, the BPA staff wanted an ethnomusicologist to analyze the music and assess the recordings and were hesitant about having an ethnologist like Valdovinos take on the project. However, for Valdovinos, it was essential to have someone who could contextualize the living rituals of the Náayeri and Wixáritari in relation to the moment when the wax cylinders were recorded but also in relation to how those practices are kept alive today. For that reason, she thought that her presence as an anthropologist who was knowledgeable about these traditions was crucial for what she understood to be the main mission of the project: to bring back the materials to their living communities in a culturally respectful and meaningful way.[100] Barbara Göbel truly believed that the materials in the Preuss Collection needed to be made openly available to move beyond the logic of accumulation that often characterizes archival endeavors.[101] Thus, Göbel's unequivocal support and that of the Ibero-Amerikanisches Institut were essential in easing these tensions and mediating these disagreements.

Nevertheless, the clash of visions became very evident when it came to discussing the details of the recordings' edition and mastering process for the CD.

The BPA staff's focus on the objects as depositories of historical meaning and Valdovinos's understanding of the chants as points of entry into a meaningful intercultural conversation were also a point of contention when determining how to present the digitized versions of the chants. Albrecht Wiedmann, BPA curator and sound technician in charge of the CD project, recalls, "My approach was to keep as much original sound as possible and to remove as much noise [as possible] before you would hear a difference with the original.... Margarita said, 'I want to present them to the Huicholes there and to the Cora in Mexico, and they want to understand clearly ... the text.' So, we struggled with how to deal with equalization or removal of noise or clicks."[102] Wiedmann continues, "In the end we produced two CDs, one for the European market and one for the Mexican market."[103] Valdovinos explains, "The recordings had a lot of noise, and it was very difficult to listen to them.... [But after a difficult negotiation], for this CD, they allowed me to filter out many of the sounds and noises that were there and that were not part of the chants. So, what we can listen to in the CD [for the Mexican market] is a polished version of what was there [in Preuss's wax cylinders] originally."[104] In sum, the mix of the CDs for the Mexican market not only removed external noises and clicks but was also further filtered, cleaned, and enhanced to make the music and lyrics more understandable. The CDs for the European market maintained some of the noisy quality of the originals in an attempt to stay faithful to their assumed authenticity. Thus, the sound quality differences between these CD productions are vocal witnesses to the clash of epistemes behind the apparently innocuous decision to make the BPA objects "audible again" and available beyond the boundaries of the archive.[105]

As mentioned earlier, the INALI agreed to cosponsor the production of an extra batch of CDs with the understanding that they were to be distributed free of charge in Náayeri and Wixárika communities. However, before this could happen, Valdovinos had long conversations with local indigenous mitote specialists who considered that the fragmented recordings did not truly represent their traditions. These specialists were also concerned with the dissemination of the chants via public media as they considered that they should only be heard within their ceremonial context. Once the local experts were on board and their request to accompany the recordings with a critical explanation of the problems they feature was met, Valdovinos and Göbel finally presented the CD to the traditional Náayeri authorities of Jesús María in 2014.[106] The event was a space for the encounter of several community constituencies that do not often engage in productive dialogue. On the one hand, there were local media outlets and teachers who had rarely attempted to engage with and understand Indigenous traditional rituals, as they considered them "a thing from the past"; on the other, there were Indigenous ritual specialists whose expertise was validated beyond their communities by the national and international institutions that made the project possible. The space also allowed local specialists to critique the recordings and the way they were recorded as well as to start conversations with other local specialists whose understanding and interpretation of the rituals and their cultural significance may be different from theirs.[107] In the end, as Valdovinos argues, the process of repatriation may allow for a dual development in which local experts "maintain a critical stance [toward the repatriated objects], while less informed folks may end up creating new mythologies based on old [reified] evidence of their ceremonial knowledge."[108]

What this process of repatriation shows is that regardless of the agendas, goals, and expectations of those pushing for the return of these materials, the receiving communities would make use of them in a wide variety of ways that are useful to them according to their own specific social and historical circumstances. In other words, the receiving communities have the final word on how these archival objects will sound and re-sound again, how they will be listened to, and how they will be made significant, or not, in their everyday lives. Theirs is a project that will make the information in the archive speak again but this time, against the logic of the archive's original design and production, to benefit them in their own terms.

The colonial and imperial logic of the archive is embedded in the archive itself, in the documents it features, in the historical routes it highlights—which are codified in the materials themselves, their provenance, and the histories behind their extraction, gathering, and collection—and in the paths it allows us to research and transit.[109] Valdovinos's timely intervention in the Preuss archive, at a moment when it was being reawakened after decades of silent slumber and when contradictory conversations about world memory and German heritage rendered its problematic colonial legacy very visible, serves as a model of how one may listen for *lo inaudito* in an archive meant to stubbornly repeat the same story. Her productive interrogation of the archive and the objects she was able to extract from it are the result of a performative listening that leads to new intellectual routes and nuanced decolonial political agendas. Nevertheless, Valdovinos's labor avoids recalling the objects in the Preuss Collection in order to restitute them into national cultural property, as past cultural heritage repatriation claims from the Mexican state have argued.[110] In fact, one of the most significant aesthetic/political consequences of the way Valdovinos moved Náayeri and Wixárika objects out of the Preuss Collection is that it freed them from the perverse logics of patrimony in both Germany and Mexico, instead endowing them with an agency to become something else and generate local discussions about identity that bypass the essentialist claims of nationalistic rhetoric. This is particularly important because the move also breaks with the state's tendency to, as Ronda Brulotte argues, relegate producers of cultural patrimony "to the margins of Mexico's national project."[111] Valdovinos's intervention benefited from the Mexican state institutions' interest in the production of new objects but bypassed its gatekeeping patrimonial gaze by directly engaging Náayeri and Wixárika communities in deciding how to frame and what to do with these objects. Thus, the book *La expedición al Nayarit* and the CD *Walzenaufnahmen der Cora und Huichol aus Mexiko 1905–1907/Grabaciones en cilindros de cera de los coras y los huicholes de México* are objects that stand out as very loud examples of the kind of questioning to which one may need to subject a sound archive in order to extract songs different from those it was meant to let us hear.

Listening Through the Noise II: Voices, Objects, and the Immaterial in the Sound Archive

Like Preuss and Valdovinos before me, I found listening to the recordings in the BPA's Preuss Collection a difficult task due to the amount of noise they contain. However, I also believe that listening in detail to and through the noise and extra sounds that accompany the mitote and *pachitas* chants in these recordings provides a point of entry into figuring out the invisible materials and information hidden in the archive, the stuff that the archive does not blatantly showcase but that is nevertheless there waiting to be awakened or uncovered from beneath layers of other kinds of epistemological noise.[112] One of those extra sounds is the voice of Preuss himself, which is heard at the beginning of many of the recordings announcing the chant the local singer is about to sing. However, the scratchy sound of Preuss's voice in these recordings is not just an informative index that simply and transparently tells us what it is that we are about to listen to. It is also an index of the dynamics of mediation informing the production of the wax cylinders storing these chants. When we pay attention to the presence of Preuss's voice and listen to it in detail, it morphs into an indexical sign right in front of our ears; it becomes the smoking gun that denotes the performativity and mediation behind the sonic representation stored in the wax cylinder; it becomes a marker that speaks of the performativity of his body and his actions in the field. Against the assumed objectivity of the archive, the voice of Preuss in the cylinders speaks of the subjectivity and position-ness that the ethnologist's body brings to the data. Thus, his voice is not simply telling us the name of a chant; his recorded speech is an interpellation that makes the chant into a sonic reality subject of aural contemplation in Preuss's terms. His voice prepares us to listen to the Náayeri and Wixárika chants in the way he understood them, as objects of contemplation and eventual scrutiny. Regardless of their lack of circulation, the influence of Preuss's recordings lies in the fact that they are constitutive elements of the authority of the archive. While the archive validates itself by storing unique objects that provide it with a sense of authenticity, it also brings an aura of authoritative importance to the objects it stores. This circular economy of archival desire is central in understanding the allure of archives and their users' craving to engage them as windows into an assumed past or distant reality otherwise inaccessible to them. The performative power of Preuss's voice lies in the fact that, for the archive's purposes, it transforms the vocal fragments in these wax cylinders into the authentic Náayeri and Wixárika chants. Preuss's voice bestows these mediated sounds with the sense of authenticity they need in order to enter this economy of archival desire.

In his discussion about the ontology of "original" sounds within processes of technological reproduction, Jonathan Sterne argues that "without the technology of reproduction, the copies do not exist, but, then, neither would the originals."[113] Following on that assertion, one can understand Preuss's wax cylinders as inserting his fragmented versions of Náayeri and Wixárika chants into that logic of sound reproduction. Thus, Preuss's interpellation of the fragmented chants at the beginning of these recordings creates an "original," for the purposes of this logic, that actually has no reference in Náayeri and Wixárika ritual reality. This is the difference between Preuss's spoken announcements in these wax cylinders and those also commonly featured in most early phonograph commercial recordings. While the latter simply state a song's name that is recognized as such by those who are performing it, Preuss's voice makes the fragments that he needs to corroborate the validity of his transcriptions into ritual entities that are not recognized as such by the local practitioners who keep those traditions alive in their communities. Preuss's voice bypasses the local practitioners' knowledge and infuses his objects with the sense of authenticity of the archive. Thus, listening to Preuss's voice in detail should make us aware of the perverse dynamics of mediation and authority in and of the archive. His voice in the wax cylinders puts in evidence ethnographic authority as an essential ingredient for the validation of any of these Eurocentric anthropological/extractivist projects.

Preuss's voice in the wax cylinders is also the sonification of an invisible presence that disrupts what the archival reification of the sound object presents to us as an authentic expression of the Náayeri and Wixáritari. His voice as a presence should work against that authentication since it is the evidence that a body external to the ritual is mediating the information that the authority of the archive makes us believe is transparent. In that context, listening through the noise in detail gives us the opportunity to discover the seed of the destruction of the archive's logic at the very core of the archive, in the objects that give it its raison d'être. Thus, Preuss's voice in the archive, while assumed to be the sonification of an objective and rational presence unobtrusively documenting reality, is in fact a performative utterance, an index of the performative listening that guided the researcher's interactions with the practices he was recording and produced the reality he sought to find for the ends he needed to meet.

The colonial power dynamics made evident by Preuss's voice in the archive are precisely the types of undercurrents that problematically inform the Humboldt Forum. As an institution, the Humboldt Forum is in essence a space for the exhibition of cultural materials that were made into objects of contemplation expressly within the colonialist logic of empire building and through the intervention of the colonizer's gaze. However, conversations about the restitution and repatriation of colonial objects in the past fifteen years have led the forum to embrace a revisionist decolonial rhetoric and make it part of its avowed mission. This is evident in the curatorial design behind the museum's exhibits, as was clear in Claudia Roth's speech at the opening of the forum's eastern wing on September 16, 2022. This obscene contradiction, intrinsic to the forum's mission—expiating its original colonial sins while insisting on exhibiting the materials that its sinful concupiscence made into objects of contemplation at a palace costing over EUR 600 million to reconstruct—was noticed by the protesters outside of the German Palace that day. Their critique and claims—that the institution's "restitution achievements are baseless as they have yet to begin" and that Indigenous communities want to be heard instead of being considered objects of study or contemplation—made apparent an issue central to Dylan Robinson's conceptualization of the type of hungry listening that informs the transhistorical unfolding of a colonial archive such as the Humboldt Forum: that its extractivist colonial drive is "felt regardless of [its] decolonizing self-work."[114] This realization leads one to wonder who speaks, for whom, and what that voice says in an archive. In his critique of the Humboldt Forum's colonial and extractivist dynamics, Bonaventure Soh Bejeng Ndikung suggests that the only way out of the project's ontological contradiction would be through "listening to other voices. Listening to the whispers in the corners. Listening to the voices that do not occupy the epicentre. Dismantling the epicentre as a whole."[115] I would also argue that a different way of listening to the epicenter may also work as a strategy to help us hear the voices that the archive renders silent. This is one of the reasons why, beyond the patrimonial rhetoric, Preuss's BPA wax cylinders are so important today: They allow for a retracing of the performative listening that informed their recording process and that continues to inform much of their contemporary circulation. This exercise provides a point of entry into this timely decolonial project. It also raises important questions regarding the ontology of the sound objects kept in the archive, especially in relation to the cultural heritage and national legacy claims that framed the repatriation process started on Margarita Valdovinos's request.

In colonial and neocolonial contexts, patrimonialism and empire building go hand in hand. Patrimonialism in this instance is a mechanism that validates the extractivist character of colonial expansion. However, extending the reach of patrimonialism to moments of decolonial effort creates a problem. The obstacles to the production of the CD of Preuss's Gran Nayar recordings made this evident. The declaration of the BPA wax recordings as German legacy via the Kulturgutschutzgesetz and as world heritage via UNESCO's inscription into the Memory of the World Register reified a collection of objects that archived traditional sounds and raised institutional concerns about their repatriation. The problem lay in an institutional interpretation that homologized the physical container with the sounds it contains. By refusing the circulation of the sounds in the wax recordings even in the transduced digital versions that bypassed and made irrelevant (for circulation purposes) the original containers, this interpretation of the legislation merged the object and the sounds it contains into a single entity. This led to an attempt to control the circulation of the sounds themselves and not just the movement of the reified original containers. This perspective makes the problem self-evident: Although the wax cylinders may be German heritage, the sounds they contain are not. As ritual utterances, these sounds belong to the communities from whom they were extracted to produce archival colonial objects. They are these communities' heritage regardless of where they may have been stored.

This goes into the reasons why UNESCO declared the BPA wax recordings to be Memory of the World; it did so because the archive's sound objects provided their communities of origin with the opportunity of remembering traditions that may be in the process of being forgotten. Nevertheless, as Valdovinos's repatriation efforts made clear, when these recordings are confronted with practitioners who have not forgotten the practices they contain, their shortcomings as mnemonic devices and the colonial logic behind their inscription are rendered visible and audible.

This affair also provides food for thought regarding the nature of sound objects kept at conventional analog archives. Brian Kane has shown that in conceptualizing the notion of the sound object (*objet sonore*), Pierre Schaeffer fluctuated between identifying it as the sounding object from which the sound "*qua* vibration" emanates and identifying it as the intentional representation through which we perceive the sound "*qua* percept."[116] Schaeffer eventually settled for the latter on the basis that the transformation of sound into object requires "a chartable consistency enabled only by repetition: something to master or, at least, temper its temporal flux and ephemerality."[117] In other words, although Schaeffer originally considered sound per se as the sound object, he finally defined it in terms of the mediation that allows for a more permanent aural relation with the sound per se. This conceptualization of the sound object makes it relational as it comes into being once it is listened to. Paradoxically, this ontological dynamic also makes the sound object into a unit of perceptual interaction whose meaning is never fixed; it changes every time it is listened to according to the acoustic, physiological, psychological, cultural, and historical conditions informing each of those listening experiences. Based on the logic that "sound is lost and refers to many," Zeynep Bulut argues that sound becomes problematic in the conventional archive because "it can be read as a threat against order [as it] enlivens the fear of loss." As Bulut concludes, such dynamics force us to reconceptualize the sound archive as "a place where sonic experience is encouraged, a place where personal knowledge can be constructed," rather than a place where one seeks to encounter some sort of objective, unmediated knowledge.[118]

This is relevant for an assessment of the BPA wax cylinder collection in general—and the Preuss recordings in particular—in relation to questions of German heritage and world patrimony. Evidently, a declaration of these collections as national heritage or world patrimony should be concerned only with the physical materials kept at the archive. It should refer to the status of the sound objects (the sound qua percept) but not to the status of the chants themselves, not even the fragmented versions of the chants stored in the wax cylinders; those belong to someone else.[119] The particularly complex nature of the sound object in a sound archive led to confusion for the administrative teams of the BPA and the Stiftung Preußischer Kulturbesitz. Regardless, the German institutions' attempt to control the circulation of the sounds themselves—in versions that transduced the chants into digital information (zeros and ones in this case) that further removed them from the original wax cylinders—should not have been an issue given that the chants, as sound "*qua* vibration," are not subject to German legislation; they belong to the Náayeri and Wixáritari. Temporarily preventing the circulation and repatriation of these chants based on legislation foreign to the communities from which these cultural practices were extracted shows how patrimonialism and colonialism may act in tandem and may bypass restitution efforts intended to alleviate historical affronts. In this context, patrimonialism becomes a strategy that conceals the hungry and cannibalistic character of the colonial encounter. As such a strategy, patrimonialism is asserted to protect the archival institution as well as the status of the things and objects it stores.

Things, Subjectivity, and the Listening Other

Gabriela Mistral opens her 1938 poem "Cosas" with a seemingly nostalgic call, "I love the things I never had / with those I no longer have."[120] The lines evoke impermanence as a condition permeating the triggering of our desire for objects. We long for that which we cannot capture or that which we have lost; and that libidinal relation shapes our sense of our selves; it triggers our search for identity. It is through coming to terms with that which is no longer within our reach and the void it leaves in us that we come to be who we think we are. This connection between desire and identity also necessarily informs our relationship with the objects that we determine to be the embodiment of Otherness. Reading the opening verses of Mistral's poem that way is illuminating because it reveals the close relationship among loss, desire, nostalgia, and hunger, which informs our relationship with things. As Remo Bodei suggests, "We invest objects intellectually and emotionally, we give them sentimental meanings and qualities, we place them in coffers of desire or envelop them in repellent coverings, we situate them within systems of relationships, we insert them into stories that tell about ourselves or others."[121] Thus, our relationship with the things we love and the things we lose is not about anything ontological about them; it is about ourselves, about the ideas about belonging and identity we project onto them and the affect that goes into those projections. One of the actions informing this libidinal economy is mobility/circulation since, as Silvia Spitta's study of the transhistorical and transcultural agency of misplaced objects reminds us, "when things move, things change," and thus our relationship with them also changes.[122] In a context in which the meaning of things is "inscribed in their forms, their uses, their trajectories," as Arjun Appadurai argues, "it is only through the analysis of these trajectories that we can interpret the human transactions and calculations that enliven things."[123] This is particularly telling in the case of sound objects like those housed at the BPA's Preuss Collection because of the changes implied in the mediated transduction of a ritualistic/sound practice into a sound object or in the moving of those sound objects from the Gran Nayar mountains to a Berlin archive and back—with the losses and gains implied in these mediations. But it is also important because the very perceptual and relational "ontology" of the sound object requires a listening other that is also in perpetual flux and whose understanding of the object is also in a state of impermanence. In these processes something always changes. The mobility and circulation of sound objects highlights a discrepancy between ontology and perception that is a central principle of thing theory, the fact that, as Bill Brown argues, "however materially stable objects may seem, they are, let us say, different things in different scenes."[124] Nevertheless, although change is an intrinsic condition of our relationship with sound objects, their storage in an archive is a move against this very nature. The logic of the archive is to deprive objects of their mobility; it attempts to stabilize their meaning according to the archive's design, structure, and ideological coordinates. The archive and its organization speak of a doomed effort to keep something from changing or being lost. In response to that, I suggest that an active engagement with the archive should seek a way to provide archival objects with a sense of agency that enables their performative transformation while circumventing their fetishization in a commodity-based economy. This is just what Margarita Valdovinos's critical work with the Preuss Collection did. Not only was she able to "give [the objects in the archive] a voice again," as she set out to do at the outset of her project, but her intervention did much more than that.[125] Her listening is precisely the type of strategy suggested by Ndikung to counter the Humboldt Forum's colonial overtones. Valdovinos's labor endowed Preuss's sound objects with an agency to say something different from what they were meant to say while rendering loud *lo inaudito*, the voices that the archive stored but that its episteme had rendered systematically mute since its creation. She was able to achieve this while effectively avoiding both commodity and patrimonial forms of fetishization. This was accomplished by circumventing what Sandra Rozental refers to as the "possessive individualism" inherent in nationalist views of cultural property.[126] Indeed, Valdovinos's work shows not only that this epistemic reversal is possible but also that the potential for that turnaround is often hidden within the configuration of the archive itself. And as one listens to Preuss's voice whispering in the archive, one is reminded that the source of archival estrangement is right there. One only needs to learn how to listen.

Mexican Rarities, *Disco pirata*, and the Promise of a Sound Archive of Postnational Memory

Hay que "tocar" a los documentos, parafraseo ahora,
como si fueran las teclas de un piano.

(You have to "touch/play" the documents, I paraphrase now, as if they were the keys of a piano.)

—Cristina Rivera Garza, *Los muertos indóciles* (2013)

After looking for him for several hours, Luz María finds Memo in an alley outside La Estrella, one of Tijuana's most iconic working-class dance halls. Memo was robbed during his first night in the city and is now without money to pay for the node job he needs in order to join the transnational workforce that would allow him to send money to his family in Santa Ana del Río, Oaxaca. *Nodes* are high-tech gadgets that, implanted in their bodies, enable individuals to directly connect their brains to a series of global computer networks with the intention to perform virtual labor and have access to sources of technological entertainment available only to members of this global village. Luz María, a *coyotek* who helps workers access cheap hardware plugs, takes Memo to the Tijuana Node Bar, a place where he can get clandestinely connected.[1] Aesthetically, the bar epitomizes the dystopian character of the futuristic world Memo is now a part of: red, green, and blue neon lights pouring over bodies in an otherwise darkened room with walls covered by mirrors; live node girls; a tired worker getting virtual shots of tequila; and a *cyberfichera* dancing with a client to the hypnotic snare drum beat, accordion loops, and hallucinating trumpet melody of Bostich's "Norteña del sur," which loudly fills the bar.[2] When his node job is complete, Memo reflects on his new cyborg condition: "Finally, I could connect my nervous system to the other system—the global economy." He would soon work as a tele-migrant, operating construction machinery in a high-rise somewhere in the Global North from a virtual sweatshop in a Tijuana shantytown.[3]

It is not surprising that when Alex Rivera, the director of *Sleep Dealer* (2008), imagined a soundtrack for his examination of migrant life in a not-too-distant dystopic future, he borrowed the sounds of Bostich and the Nortec Collective, with their modernist reinvention of tradition and their way of making the often-disdained sounds of working-class music from the north of Mexico (banda and norteña) into symbols of an oddly cosmopolitan hipster coolness.[4] In the 2000s, and not without reason, this music came to epitomize Tijuana, the border between two apparently opposed financial, social, and cultural worlds. In the late 1980s, Néstor García Canclini infamously referred to this region as a "laboratory of postmodernity," before rectifying that and calling it a "laboratory of the social and political disintegration of Mexico."[5] Its sudden international success in the early 2000s also made Nortec into the soundtrack of the city that Antonio Navalón guilelessly and uncritically commemorated with his magniloquent and problematic binational art project *Tijuana, Tercera Nación* in 2005.[6] Indeed, for those who are open to listen in detail, beyond the celebratory rhetoric of naive scholars, politicians, and cultural brokers, the sounds of the Nortec Collective have also come to embody the symbolic and real violence generated by the friction caused by the asymmetries between the so-called developed and developing worlds. The global success of these sounds not only put these asymmetries on the dance floor for everyone to hear but also provided a new point of entry into making sense of the Mexican experience without having to pass through Mexico City, thus subverting the nationalist discourse that represents the US-Mexico border as the margins of the nation-state. Instead, Nortec put Tijuana in the spotlight and relegated Mexico City to the background. In that sense, the sounds of Nortec became an archive for the construction of a postnational memory. Here, the idea of the postnational refers to "a point of view beyond the nation-state as a frame of reference," and a way to engage the experience of those who have to negotiate the contradictions of living in a multi-ideological geographic and cultural matrix on an everyday basis.[7]

As Nadim Khoury suggests, postnational memory refers to the development of new identities and narratives that build on "resources within national identity—suppressed voices and counternarratives—to disclose alternative futures.... Postnational memory is a way of thinking about the past in this possible and more just future."[8] Thus, postnational memory implies a way of remembering that evades the epistemic placeholders of nationalism and remains critical of the exclusions generated by its discourse of difference. In describing postnationalist memory, Nigel Young also highlights the need to look into the nationalist archive to "break a wide range of silences" and eventually "transcend a national framing of the past."[9] In other words, postnational memory indicates a reassessment of past narratives that listens to the voices that nationalist histories have kept silent in order to reimagine not only the past but also a future that includes them. Nonetheless, the violence behind these histories of silence and suppression often implies a new dystopic relationship with the narratives and frameworks from the past.

In *Sleep Dealer*, the Nortec Collective's sounds represent the dreams and nightmares about modernity and the collapse of the nationalist fantasy informing Rivera's cinematic premonition. These aspirations and desires speak of a geography where past and present meet, engendering all sorts of imaginatively informed fantasies about the future. Thus, the fantastic utopian and dystopian imaginaries displayed in Rivera's *Sleep Dealer* put in evidence the central shortcomings and delusions of the Mexican projects of modernity and nation building. They reveal a postnational condition in which seemingly desperate individual actions respond to unfulfilled needs, desires, and aspirations; a condition in dialogue with the promise of global mobility and interconnectedness that the political and economic project of the nation-state is unable to deliver even when its materialization has become essential for the everyday survival of these individuals. These are the conditions that give Alex Rivera's dystopian futurist fantasy its eerie sense of reality and imminence. In that sense, *Sleep Dealer* works as an archive of postnational memory that not only stores evidence of the cultural crisis of the Mexican nation-state and uses it to imagine an alternative from the bottom up but also stores documentation of how individuals have already started developing cultural practices and strategies that deal with this crisis and highlights new notions of identity and cosmopolitan—if dystopic—belonging. As an archive, the film speaks of both the imaginary and the real, of what could happen and what has already happened in a country where the desires expressed in national symbols remain while the system that should give everyday cultural and political valence to those symbols collapses.

The benefits and shortcomings of Memo's new connectivity inform the film's plot thoroughly. One of these problems is that the nodes allow Luz María to access Memo's memories, creatively edit them to give them the patina of stylistic authenticity expected by a global north market hungry for "real stories," and sell them as commodities through an online memory market platform called TruNode. Although this situation is only tangentially touched on in the film in order to dramatize Memo and Luz María's romantic relationship, I want to dwell on it because, as Kristy Ulibarri argues, it speaks of "traces of memory" that act as the building blocks of a "speculative realism [that] parses both the visible and the invisible processes that produce these violent entanglements of the virtual and the real."[10] However, rather than focusing on the slow death of the worker that Ulibarri highlights in her analysis of *Sleep Dealer*, I want to pay attention to the potential of these traces of memory and the testimonies they inspire as what Cristina Rivera Garza calls *noriginales* (nonoriginals). Rivera Garza coined this neologism to talk about how, rather than fetishized loci of authenticity, archival documents are produced through collaborative negotiations that continuously assess, reassess, and rewrite the past affectively in the present to invoke and confirm that very present.[11] For her, the potential of these *noriginales* lies in their questioning of the authority of the archive and the assumed originality of its documents while keeping in mind that the mediating processes that actually make them into documents in the first place are fueled by individual and collective efforts to listen to the silences in the past in order to understand the present in new ways. In other words, as Rivera Garza poetically puts it, it is all about activating "in the present a past that is always about to happen."[12]

Chapters 1 through 3 discuss archival formations that shed light on the development of the Mexican national fantasy in relation to an imaginary past and an anticipated future, the types of knowledges that the reproduction of this fantasy enables, and a series of strategies to either escape the patrimonial gaze embedded in these formations or turn it upside down. These chapters deal with what a postnational revamping of the information stored within these archival formations allows one to know, see, or listen to. In this chapter, I move away from the logic of the institutional archive and the types of rhetorical actions it supports, or the efforts to subvert it, to focus instead on the logics of the individual collector as seen in two very different but oddly similar mid-2010s archival ventures: Mexican Rarities and *Disco pirata*. Like Rivera in *Sleep Dealer*, I do not intend to estrange a nationalist archive but rather explore the labor and potential of putting together an archive that generates postnational memories and subverts the values and logic of the national archive model from the outset.

Mexican Rarities: An Oddity Among Mexican Sound Archives

On February 13, 2021, in the middle of the COVID-19 pandemic, music collector Arturo Castillo (b. 1963), digital artist Alfredo Martínez (b. 1986), underground musician and visual artist Víctor Garay (b. 1985), and sound artist Juan Pablo Villegas (b. 1986) launched Mexican Rarities via an online event. Presented as an "archive, label, channel for distribution of content, and a platform for underground events," the project's avowed mission is to store, preserve, and recirculate Mexican music that by virtue of its oddity remains hidden "in different layers of the Mexican subsoil."[13] The event was the formal presentation of the project's web page—the internet interface of an analog archive—which was accompanied by the launching of the project's first LP, *Inframundo* (Underworld), by Esteban Aldrete (aka Nicolas), and the reissue of a limited edition of *Cometa 1973/Cromometrofonía No. 1*, a cult LP recorded in the early 1970s by Óscar Vargas Leal and David Espejo, Julián Carrillo's last Sonido 13 disciples. The title of Aldrete's album and the underground character of Vargas Leal and Espejo's LP perfectly highlight Mexican Rarities' task of unearthing little-known music and sound projects that escape the commercial logic of the Mexican music industry and giving them a chance to circulate or recirculate among a community of folks eager to listen to and engage unconventional musical practices.

The virtual launching of Mexican Rarities was the crystallization of an idea born more than twelve years earlier, when Martínez and Villegas, then teenagers, met Castillo and became fascinated by his huge collection of alternative music. Back in the 1980s, Castillo was already well known in the underground Mexico City music scene as someone who, in a pre-internet and pre-NAFTA (North American Free Trade Agreement) world, had access to an unprecedented number of avant-garde musical materials and projects from around the world that were largely unavailable in Mexican music stores and media. At El Chopo, the legendary Mexico City alternative music flea market, where one could find bootlegs of everything from Nueva Trova to Krautrock, Castillo was known as the person to go to for

Rock in Opposition and Wax Trax.[14] He had spent most of his life amassing an impressive collection of international avant-garde and experimental music that had very little to no circulation in Mexican commercial music networks. But he was not interested in simply storing these piles of LPs at home; he wanted to share the music with anyone curious and interested. In the 1980s El Chopo provided the perfect subcultural venue for that; it gave Castillo a space for musical exchange and buying and selling, as well as a community where his erudition was valued and his expertise sought after. In 1989 Castillo's extensive collection and growing reputation as an expert in international avant-garde musics led to an invitation by performance artist Roberto Escobar to collaborate in *La Mecánica del Concepto*, a celebrated radio show devoted to the dissemination of international experimental music for the pioneering Mexico City radio station Rock 101. In a way, this project brought Castillo out of the underground shadows and into the mainstream. In the 1990s he opened the first of several music stores he would manage through the 2010s and ventured into organizing concerts in Mexico for some of the European music projects he was interested in. At the same time, he continued to accumulate a vast collection of LPs, CDs, and cassettes that, by the turn of the twenty-first century, took up most of the space in his Roma neighborhood apartment.

It was at this time that Castillo met Martínez and Villegas. The two youngsters were budding music collectors interested in the type of international experimental music Castillo was a known champion of. Since Castillo's apartment had become an informal point of encounter for many collectors, musicians, and people interested in these musics, it was only a matter of time before Martínez and Villegas found their way to his place. Martínez recalls his first visit to Castillo's apartment: "At the time one did not see much vinyl [in Arturo's place]; the LPs were in boxes in an out-of-sight room. Everything was CDs. It was very impressive. The space was large, but there was no furniture. All you could see were towers and towers and towers and towers of CDs. I talked to him and explained my musical tastes, and he selected six CDs, gave me a beer, and sent me into a listening room. I listened to everything he recommended. After that day, I started going there every single weekend."[15] These open-house weekends attracted well-established artists and young folks alike. While Castillo became a mentor for Martínez and Villegas, the presence of important figures from the Mexico City art scene also allowed them to articulate new artistic and intellectual networks.[16] Meeting Castillo not only encouraged Martínez's and Villegas's collecting efforts but also provided them with a solid aesthetic, artistic, and historical orientation for their future careers as visual and sound artists.

Although Castillo was known for his interest in circulating European and US avant-garde music, all along he kept collecting Mexican music that he considered the counterpart to the foreign alternative projects he advocated for. This included bands like Nazca, La Banda Elástica, and La Nopalera; blues singer Guillermo Briceño; Indigenous musics; microtonal maverick Julián Carrillo; and electronic music pioneers like Antonio Russek, Roberto Morales, and Vicente Rojo Cama, among many others. Regardless of their initial interest in European and US alternative musics, when Martínez and Villegas approached him, Castillo was adamant about instilling in them an awareness of the sound of these Mexican projects. Villegas remembers:

> Arturo opened the doors to certain materials for me. I was like: "What am I doing?" There was an impressive number of artists, musicians, and projects that were not being valued, researched, or analyzed and that came from [Mexico]. I did make a value judgment and said, "This is amazing! Sonically it is an impressive thing. But people do not know it. I do not know it. Why is this happening?" So, something that became clear to me was that there was a lot of carelessness on the part of institutions, of the artists, of the labels, about generating a memory or about taking responsibility for [making sure people knew] that this is part of our culture, that these discs are not only commodities or that they are inserted in a market but that they are also part of an identity and that in part that music and those sounds make us. They make us who we are and are part of our history.[17]

Villegas acknowledges that meeting Castillo was fundamental for him and Martínez since it sent them into a frenzy to find these strange, largely marginal Mexican musical projects. In the 2000s, under Castillo's mentorship, Martínez and Villegas became avid collectors of alternative experimental Mexican music while pursuing their degrees in visual arts—Martínez at the Escuela Nacional de Pintura, Escultura y Grabado "La Esmeralda" and Villegas at the Centro de Diseño, Cine y Televisión, later pursuing graduate studies at Le Fresnoy Studio National des Arts Contemporains in Tourcoing, France.

One evening at his store, in the early 2010s, Castillo was looking through his collection of music and began asking himself, "It's been thirty years of advocating for these [foreign] avant-garde musics... but I am Mexican. What's up with the Mexican stuff?"[18] This reflection led him to the idea of creating Mexican Rarities as an archive devoted to those initiatives. He mulled over the idea for a couple of years and finally, in 2015, invited Martínez, who had already stockpiled his own substantial collection of CDs and vinyl of experimental Mexican music, to work together with him on developing the archive. Castillo explains that the original idea was "to rescue things [from oblivion]. To rescue recordings of artists and bands who were almost unknown or whose existence went by largely unnoticed.... The idea was to reissue a few unknown old titles and then to issue new stuff."[19] Two years later, Castillo and Martínez presented the project to the community of vinyl collectors at the 2017 Expo Vinylo Oaxaca (EVO). This presentation laid out the mission of Mexican Rarities as a project devoted to the preservation and organization of materials from Mexican subcultural projects. A day after the presentation, Castillo and Martínez, advertised as Los Eclipses, also presented a DJ set of Latin, folk, and psychedelic rock based on materials from the archive. Although this incarnation of Mexican Rarities still looked quite different from its final configuration, as formalized two years later, the Oaxaca performance of Los Eclipses made it evident that bringing the archive to life through the artistic and performatic reactivation of its documents was already one of the most important missions of the project.

Between 2017 and 2021, Mexican Rarities was consolidated with the incorporation of Garay in 2020 and Villegas in 2021. When it was finally launched on February 13, 2021, its legal status as a Sociedad de Acciones Simplificadas (SAS) had been established, and the organization chart and division of labor were in place: Castillo contributed his expertise on the materials of the collection as well as his knowledge of exchange networks; Martínez designed, programmed, and maintained the project's internet page; Villegas was in charge of the reactivation of the archive via the coordination of events and parties; and Garay was the video editor and the person in charge of managing the project's connection to artists.[20] Later, they were able to hire Daniel Kobelkowsky to work on transferring the digitized information and data to the website. By this time, the iconography of the website—especially the finalized version of the project's symbol, a Quetzalcoatl circling around an LP (which came out of a dream Martínez had) rather than around a treble clef as in the earlier version presented at the EVO—was well established. Its layout and the multiple identity of Mexican Rarities as an archive, label, store, and regulator of the socialization of the

FIGURE 5.1. Home page of the Mexican Rarities website, https://mexicanrarities.com. Screenshot, November 6, 2024.

archive's holdings through the organization of various events to guarantee their recirculation were in place (figure 5.1).

Although its main window to the world is its website, Mexican Rarities is actually not a virtual or electronic archive. It is an archive of analog documents and recordings whose content was put together by combining the personal LP, CD, and cassette collections of Castillo, Martínez, and Villegas. In fact, the intention was never for the website to be an archive but rather for it to work as a partial window into the actual archival repository and its analog holdings, with the idea that interested folks could get in touch with members of the project if they wanted to access any particular item, whether for further research, for acquisition purposes, or for developing other types of creative projects.

Mexican Rarities as an Archival Project

As an archive, Mexican Rarities stores more than twenty thousand items. However, by the beginning of 2024, only about one thousand had been documented and digitized, and only a small portion of those digital documents had been uploaded to the website. Castillo and Martínez decided that the more practical and organized way to go about documenting these materials was to organize them, digitize them, and upload them according to series or collections. The process started with the digitalization of

Mexican Rarities, *Disco pirata*, and Postnational Memory   169 the vinyl records related to Julián Carrillo (which included the series of records he self-produced in the late 1950s and early 1960s, as well as other materials related to him, his students, or his Sonido 13 project). After that, they incorporated the LPs of Ángel Cosmos's *Colección Hispano-Mexicana de Música Contemporánea* (Hispanic American collection of contemporary Mexican music); *Testimonio Musical de México* (Musical testimony of Mexico), by the Instituto Nacional de Antropología e Historia (INAH), a long-standing collection that bears witness to the country's Indigenous and ethnic diversity; the Instituto Nacional Indigenista series documenting its ethnographic and audiovisual archive; a series of recordings connected to the Grupo Tribu's seminal ethnomusicological work in the 1970s; the *Voz Viva* (Live voice) series of the Universidad Nacional Autónoma de México (UNAM), which features authors reading from their own works; the series of *Encuentros de Música Tradicional Indígena* (Encounters of traditional Indigenous music) from the Fondo Nacional para Actividades Sociales (FONAPAS); and the Mexican recordings from the Smithsonian Folkways series; as well as a few other collections of independent labels. Mexican Rarities features these collections in agreement with the institutions or individuals who first produced them.

According to Martínez, the process of documentation implies uploading all the information of every disc, including pictures of the cover and back cover, pictures of the A and B sides of the actual record, transcriptions of the booklets and credits, transcriptions of the track lists, and transcriptions of the notes. Besides this information, every item's entry on the website also includes links to other websites that may contain information about those items, information about the highest and average price paid for that particular record, information about and links to the original publishing label, and links to similar musical materials. Martínez acknowledges that the Mexican Rarities internet database is inspired by the Discogs free marketplace model, which encourages the preservation of information as well as the circulation of sonic materials.[21] However, since the Mexican Rarities staff is well aware of possible copyright infringement issues, the website very rarely provides actual sound files of the music on these records. However, if existing recordings of a particular item are available on the internet, the database entry would contain information about them and even links to them.[22] Finally, since the archive's internet interface is meant to be bilingual, once all the information has been gathered and digitized, it is translated into English. Nevertheless, the materials are uploaded only once an entire series has been completely digitized, formalized, and systematized.

It is clear that Mexican Rarities is an archive of established series and collections that have received little or no commercial distribution or that have slipped through the cracks of the Mexican music market. As such, its holdings are not original or unique documents that may not exist anywhere else. Instead, they are the type of *noriginales* that Rivera Garza refers to. They are significant and meaningful not for their presumed originality but rather because of the interpersonal dynamics and the collective meaning they generate in the present. As Rivera Garza would put it, whether these documents are originals or not is not important; what is crucial is that they are appropriated and revamped "locally and in an everyday reality to 'generate present,' that is precisely their meaning."[23] Thus, the value of their rareness is not connected to their exceptionality as documents as is the case for materials in conventional archives. Instead, their value refers to how their oddness is reevaluated in the present precisely because it challenges the placeholders of traditional narratives where these documents are either marginalized or rendered invisible, forcing us to reassess the very ideological framework that gives meaning to those narratives in the first place. In that sense, Mexican Rarities breaks away from the logic of traditional archives by refusing to adhere to the idea of the repository as a site of memory and authority. Instead, it follows the logic of the collector, highlighting the circulation of materials and the construction of memory as a collective endeavor. This is precisely the reason the other identities of Mexican Rarities—as a store, label, and platform for regulating the recirculation of its holdings—are fundamental in understanding the archive's promise of a postnational memory.

Mexican Rarities as a Store, Label, and Platform

If the logic of the collector is already in evidence in Mexican Rarities' archiving/archival labor, it is in its work as a label, store, and regulating platform that this logic and its postnational potential become more apparent. The idea of linking the archive to a label and a store was an answer to Castillo and Martínez's long-standing collecting endeavor, which one could trace back to Castillo's work at El Chopo in the 1980s and his desire to circulate certain materials among a small elite of interested audiophiles regardless of their purchasing power. Forty years later, the economic situation, technological access, availability of materials and information, and circulation networks at Castillo's, Martínez's, Villegas's, and Garay's disposal are all very different. However, the desire to share music is still what motivates their labor in Mexican Rarities. Thus, following on Castillo's experience with his music stores as a central aspect in his collecting efforts, the idea of pairing the archive to a system that would facilitate the circulation of some of its materials started with the very conceptualization of the archive. Moreover, creating a label and a store, and imagining a system to manage the public performatic recirculation of the archival materials via DJ sets and through the organization of talks and workshops, was at once a way to reactivate the documents in the archive (in a manner that reverberates with the goals of the Fonoteca Nacional and some of its projects, although following different regulation strategies), a way to educate and inform people about these materials, and a pragmatic way to help finance the project. These workshops and DJ set events allow the members of the Mexican Rarities team to sonically feature the archive's materials for an interested audience, while the label gives the project the possibility of producing objects that would eventually be sold at their internet store.

This financial model is not without shortcomings. One of them is precisely that the advent of the internet has changed the way most audiophiles relate to music and the material objects used to store it. Evidently, the internet is an instrument that facilitates the circulation of materials that used to be almost impossible to access for folks who were not part of international networks of people who knew about these underground projects and had access to their recordings. However, the massive availability of musical materials in digital formats has also helped shape a new generation of listeners who are content with streaming or owning digital versions of their favorite music and are not interested in buying LPs or CDs. This would seem to be an offer-demand trend working against the Mexican Rarities model. Nevertheless, Castillo states, "Throughout time I have generated an interest in collecting Mexican things.... Regardless of the digital boom, at least in Mexico there is an interest in collecting.... Fortunately, there is a worldwide vinyl boom [that helps]; so, I do see that young people in Mexico want to learn and have more information. Even though the younger generations are not used to handling [these types of formats], there is a growing interest."[24] As Castillo describes, Mexican Rarities has been able to circumvent the apparent lack of a market for their product resulting from the development of streaming systems and other internet servers by appealing to the collector's logic and the way it generates a type of aura for their product, a cultural capital that is appealing to a very particular community of connoisseurs. Although this is not a strategy that aims at engaging a large mainstream market, its articulation of the desires of a small aural elite—undoubtedly a segment of the Aural City, as I explain below—has been enough for the project to be successfully self-sufficient.

Martínez explains the business model and the funding of Mexican Rarities as follows:

> At the beginning, Mexican Rarities was financially funded by me personally. There is no governmental or private support. At some point we looked into that, but the reality is that institutions in Mexico are very much neglected; [they] have very little support. In a way, doing it with our own means was an anarchivist way of generating this archive. So, early on, I personally paid for the cost of the server, the platform, and the person who helps us. But now the project has started crawling by itself thanks to the label and the different [LPs] we have released.... [Publishing] the work of bands that have a large underground following has helped maintain the economy of the label.... The idea is to [alternate the] release of an experimental art LP of which we only make a few difficult-to-sell copies, and then an underground LP with a larger fanbase that is easier to sell, interspersing between them reissues [of historical material].[25]

It is not coincidental that, invoking Jacques Derrida, Martínez refers to their DIY strategy as an anarchivist project since avoiding state support and achieving financial independence is precisely a first step toward emancipation from "the archon in the archive," a first step toward overcoming the archive's embedded sense of authority.[26] This move affords Mexican Rarities a freedom of action and epistemological agency that is often missing in the type of top-down hierarchies that institutional archives validate. Escaping the logic of validation that characterizes institutional archives also provides a foundation for the reimagination of the archive's authority as a dialogic enterprise at the nexus of archiving labor, circulating labor, retrieving labor, and consumption. This plan of action has allowed Mexican Rarities as a label to strategically combine its mission to reissue old and unavailable recording productions and to issue recordings by new marginal and alternative experimental musicians while staying true to its original creed of making *lo inaudito* heard. The desires informing these types of labor in relation to the musics they engage are also fundamental in this process of collective archival reevaluation.

Castillo, Martínez, and Villegas agree that one of the foundational projects for Mexican Rarities was the documentation of Julián Carrillo's work. Martínez confesses that his passion to collect music was triggered by his obsession with "finding hard copies of Sonido 13 records, the discs produced by Sonido 13 as well as other labels related to Julián Carrillo."[27] Villegas further explains this fascination with Carrillo: "When I discovered the immensity of Carrillo's universe, it just overwhelmed me. I became passionate about his personal story and everything related to the manufacture of his [microtonal] instruments... because I myself make machines... so, all his work about developing and designing the instruments, the musical notation, the philosophy behind it, the theory; it was very exciting for me."[28] Under the spell of Carrillo and his Sonido 13's aura, which has enchanted many citizens of the Mexican Aural City, the team focused the initial efforts of Mexican Rarities on making accessible the collection of recordings produced by Carrillo in Paris between 1960 and 1963, and on reissuing the even more mystic *Cometa 1973/Cromometrofonía No. 1*, a real rare cult item among collectors of alternative experimental music.[29] Castillo and Villegas had decided to reissue this LP even before the initial presentation of the Mexican Rarities project at the 2017 EVO. Castillo says, "I discovered this disc many years ago, and I just fell in love with it.... I heard the recording and found it fascinating."[30] Villegas further explains, "It was very sad for us that [David and Óscar] were lost to memory, that there was no recognition [of their work], that their [microtonal] harps were in the state they were, that [Óscar's] cabin and the scores were all decaying. That's when I said, 'My God! If the Fonoteca is not doing it, if the art spaces are not doing it... What's up? Someone has to do it. We have to do something!' So, Arturo told me, 'Let's the two of us do it if we can.' And that's how it happened."[31] The testimonies of the Mexican Rarities team speak of a believer's crusade informing the early labor toward the formal establishment of the archive. This type of work responds to their own fascination with a very particular musical world, an *inaudito* world that was both astonishing and unheard, as well as a response to what they felt were the shortcomings of state institutions such as the Fonoteca Nacional. Documenting Carrillo's work and reissuing *Cometa 1973/Cromometrofonía No. 1* not only exposes the logic of the collector behind Mexican Rarities but also reveals the face and labor of a subset of the Mexican Aural City, one that, as mentioned earlier, I had identified when attending Sonido 13 concerts in Mexico City during the fieldwork that informed the writing of my book about Julián Carrillo. Indeed, the labor and individual profiles of the members of the Mexican Rarities team could be taken as perfect descriptions of what the Aural City may be and do in reaction to what they perceive as the inaction of national institutions and in a move to separate themselves from these institutions' patrimonial mission and the excluding nature of their nationalist outlook.

Mexican Rarities, Sediments, and *Noriginales*

Mexican Rarities announces itself as an archive "of music found in different layers of the Mexican subsoil" (see figure 5.1). There are at least three interpretations of this language.[32] On the one hand, it refers to the fact that the project's mission is to unearth musics that are literally unheard due to having been lost or misplaced in the catacombs of Mexican memory. It speaks of the underground character of the records and musical projects that the Mexican Rarities curatorial team is interested in. And, finally, it also refers to the massive character of the Mexican Rarities physical archive as well as the piles of vinyl they had to go through in flea markets and record stores in the hopes of finding a unique musical gem. Villegas elaborates on this practice in great detail:

> We talked about that a lot. To me, it is very beautiful to think about what is beneath the surface of the ground. For example, everything about tubers, potatoes and everything that grows underground, which in the end is what nourishes society in times of crisis.... When above the ground everything is desolate, there is still life to be found underground. They are the sources that can suddenly bring back energy and vitality to a society that is in decadence. And on the other hand, it is also about the idea of strata. There are these superheavy layers of concrete, which are these gigantic [transnational] labels that generate stunning quantities of petroleum; they distribute it everywhere and literally generate a crust of information that often crushes these local micro-manifestations. So, it also has to do with that; there is a homogenization that forms a thick stratum that seems to be all there is, but below that there are other realities, other movements, other fluxes.... And we also thought about it in relation to the piles of materials in the archive, the physical part of the archive that also accumulates in strata. Often when you look at the archive, you do not see these layers because the LPs are organized horizontally. But if you take a pile of discs and spread them around, you generate a series of layers that are like rocks sedimented on top of each other. They become literal strata of petroleum. You see? The stuff that materializes this memory comes from the underground... and we store information in it.[33]

Villegas's eloquent and poetic description of what the idea of the underground means for the Mexican Rarities curatorial team begs to be read through the lens of Cristina Rivera Garza's notion of *escrituras geológicas* (geological writings). For her, an *escritura geológica* is a strategy to dig out sediments that "reveal not only the persistence of the past, its agglomeration in futures that begin with us right now, but also the arduous, and often joyful, process of research ... as a form of imagination and care."[34] Indeed, Villegas's description of the piles of LPs in the archive reverberates with Martínez's memory of the "towers and towers and towers and towers of CDs" in Castillo's apartment (figure 5.2). These visualizations are an invitation to reimagine the concept of "digging in the crates"—the notoriously time-consuming practice of searching for the right music break to sample that defined classic hip-hop—in the context of looking for that underground music gem that will allow us to reimagine pasts and futures that "begin with us today."[35]

The premises of Mexican Rarities' promise of a postnational memory are evident in the project and labor leading to the documentation of the Carrillo collection and the reissuing of *Cometa 1973/Cromometrofonía No. 1*. Postnational memory in this case refers to a movement away from the patrimonial logic of institutional archives to focus on *noriginales* that acquire meaning in the collector's logic of circulation. It also speaks about Khoury's idea of postnational memory in terms of the desedimentation of voices that the nationalist archive rendered silent or kept hidden. If the nationalist rhetoric is currently empty and its archive unproductive, these forgotten underground practices are the nourishment needed for us to engage a past that is "always about to happen" as we imagine a new anticipated future.[36]

On the other hand, the premises of that postnational promise appear to be more ambiguous in the efforts to document the INAH collection, the second curatorial project of Mexican Rarities. How does a postnational project engage a collection whose very inception was the result of a nationalist project that sought to document the Indigenous sounds of the nation-state by making them into patrimony? A point of entry into answering this question is a text also found on the Mexican Rarities website; its title is "¿De qué hablamos cuando hablamos de México?" (What do we speak about when we speak about Mexico?).[37] Written by Rolando Hernández, a former collaborator of Mexican Rarities, the piece sheds light on how the project's understanding of this collection may differ from the ways in which the INAH and its sound archive, the Fonoteca del

FIGURE 5.2. Partial views of the Mexican Rarities physical archive at the apartment of Arturo Castillo, Mexico City. Photos courtesy of Arturo Castillo.

INAH, have ascribed meaning to this set in the loci of production and storage. Hernández's argument is that there is a paradox in the rhetoric that informs nationalist institutions and their relationship to Indigenous cultures in Mexico. While, on the one hand, this rhetoric celebrates these musics and their communities in order to validate its own claims to national authenticity through their maintained autochthonous purity, on the other, the homogenizing nature of the nationalistic rhetoric actually makes these communities into patrimonial icons whose representation features them largely as essentialized cultures frozen in the past. As Hernández succinctly explains, the paradox lies in the fact that in this rhetoric

Mexican Rarities, *Disco pirata,* and Postnational Memory 177 the idea of the Mexican is a straitjacket for the multiplicity of cosmogonies and cultural manifestations of the [country's] sonosphere, which, paradoxically, have been taken over by the national project without even granting them due recognition.... This premise leads us to paradoxical situations where the people who exercise violence and take on racist and classist stances toward people from towns in resistance [Indigenous populations] are the same ones whose chest swells with pride when talking about the wonders of their last trip to Oaxaca, the importance of the preservation of traditions, and how many of them should remain intact.[38]

In her research about the invention of the category of the Indigenous in Mexico as part of larger processes of nation building during the first half of the twentieth century, Marina Alonso Bolaños articulates these concerns. There, in relation to the INAH series, she explains that regardless of the fact that the collection "clarifies that Indigenous music features a multiplicity of styles, purposes, and historical traditions, the notions of integration of pre-Hispanic and European elements, syncretism and sacredness, the importance of the collective over the individual and other 'markers of Indianness' always appear as premises to present the pieces contained in the phonograms."[39] Alonso Bolaños recognizes the importance of this collection in generating interest in Indigenous communities that have been marginalized throughout the history of the country. Nevertheless, she is also quick to point out that the patrimonial character of the collection—a natural response to the homogenizing nationalist policies of the modernizing Mexican state—idealizes Indigenous communities and individuals into imaginary representations that conceal the fact that in reality they remain marginalized and often denigrated. Furthermore, this patrimonial enunciation is successful because it rests on the reproduction of deeply embedded stereotypes that have rendered Indigenous cultures as relics of the premodern world. Within this episteme of hungry listening, in a move that ironically resembles the motivation behind the gathering efforts of Adolf Bastian and the Königliches Museum für Völkerkunde in Berlin, the individual items in the INAH collection arrogantly stand as witnesses of an imaginary pure tradition that must be protected before the advancement of modern national civilization inevitably corrupts it.

In its treatment of the INAH collection as *noriginales*, Mexican Rarities transcends the patrimonial model privileged by institutional archives. They evade the reification that characterizes the INAH collection because their interest in documenting this collection does not emanate from its imagined aura of authenticity or a desire to identify any type of cultural roots. Instead, their interest lies in the recognition of Indigenous diversity per se, which the very existence of the collection puts in evidence regardless of the essentializing and homogenizing national project that informed its birth. In a sense, this move takes advantage of identifying the seed of the archive's ideological self-decimation within the archive itself. This aspect of the Mexican Rarities project evokes Young's understanding of postnational promise as a reassessment of past narratives that listens for *lo inaudito* in the archival documents themselves in order to "transcend a national framing of the past."[40] In sum, the postnational promise of Mexican Rarities stands on its challenge to the patrimonial archival model, a challenge that takes *noriginales* and their circulation and activation beyond the space of the archive(s), as its emancipatory reagent. In doing that, the curatorial team of Mexican Rarities comes across as a group of informed and passionate dilettantes whose important labor and cultural capital signal the Mexican Aural City and its ties to a number of alternative musical projects.

Disco pirata as Action Piece, Performance Intervention, and Kitsch Listening

Polifonía ambulante (Ambulant polyphony), an exhibit of four sound installations by French sound artist Félix Blume (b. 1984), opened at Mexico City's Fonoteca Nacional on June 9, 2016. The event was publicized as an homage to Mexico City's *cantos* and *pregones* (street vendors' chants and cries). Indeed, the four works featured in the exhibit, *Coro informal* (Informal choir), *Coro polifónico* (Polyphonic choir), *Los gritos de México* (The cries of Mexico), and *Disco pirata* (Pirate disc), are based on Blume's own samples of cries by Mexico City street vendors, which he identified as unique sonic features of the city's soundscape. Inspired by Clément Janequin's *Les cris de Paris* (The cries of Paris, 1530), a four-part polyphonic chanson based on the cries of vendors in sixteenth-century Parisian markets, the first three sound installations in Blume's exhibit offer a musical way of listening to these Mexico City urban cries.[41] *Coro informal* is a sound installation by Blume and Daniel Godínez Nivón that features ten short street vendor chants in ten individual wooden boxes.[42] Invoking the title of Janequin's chanson, *Los gritos de México* is a twenty-nine-minute-long soundscape composed by Blume using samples of everyday sounds from

Mexico City.[43] *Coro polifónico* is a video-contrafact, a video montage of four singers singing Janequin's chanson in which the original lyrics are replaced by the words of Mexico City street vendor cries.[44] The fourth work in the program, *Disco pirata*, is a selection of one hundred sounds from the city that differs in presentation from the previous works but still argues for a somehow musical approach to the listening of these sounds.[45] In this case, the musical point of entry is the bootlegged CDs sold informally in the city's subway system. I am particularly interested in *Disco pirata* given its transformation from a sound installation and performance art piece into an open-access archive that has frequently been used and referred to by the Mexican Aural City, being particularly popular among sound designers in the local film industry. The fact that *Disco pirata* has become a source of sounds for this group of professionals when trying to sonically re-create Mexico City makes it into an excuse for a discussion about authenticity, representation, and the uses of the archive that problematize the identity claims at the core of nationalist rhetoric.

Trained as a sound engineer, Blume began his career as a technician recording sounds for films and documentaries. He traveled to Mexico City for the first time in 2009 to collaborate on a project by Belgian video artist Francis Alÿs (b. 1959). One of the first things that immediately captivated Blume about Mexico City was the abundance of street vendor sounds. He explains, "I realized that the cries and voices of Mexico City give it a strong sense of sonic identity that other Western cities have lost. I hear those voices, and for me, they are like a polyphonic choir that is part of the city's being."[46] He began recording and storing these sounds in order to document their originality and distinctiveness. In 2010 Blume decided to start sharing his recorded sounds in Freesound, a collaborative database of Creative Commons licensed sounds that allows for the free sharing of this sonic material and its eventual use to creatively build on them.[47] Among the first sound files he shared in this platform, along with sounds from Argentina, Ukraine, Mali, Italy, and France, were the sounds he recorded during his first trip to Mexico.

Although Blume originally began uploading sound files to Freesound with the intention of freely sharing them with anyone interested in using or listening to them, he eventually realized that mere availability was insufficient. To fully convey his sound experience, he needed to use these sounds to express his feelings about the places where he recorded them. This led him to start working more creatively with his sound files and begin composing sound pieces. In 2012 he used the sounds he had recorded for

Aurélien Lévêque's film *El puesto* (2010), about a man living alone in Patagonia, to compose his first sound piece, *Terre de feu: Les moutons du bout du monde* (Land of fire: The sheep from the end of the world). After composing sound pieces about the Fula people from Mali and the Venezuelan grasslands, in 2014 Blume composed *Los gritos de México* using the Mexico City urban sounds that had captivated him during his first visit to the city five years earlier. Part of the reason he composed this piece was that he was back there, living in Mexico City, and wanted to convey a listening experience of the city in which noise can morph into sound that can be aesthetically pleasing. For him, the idea was to make these quotidian sounds available for *chilangos* (people from Mexico City) to pay attention to and to show them a way of listening that bypasses their biased belief that these are just noises they have to endure. Furthermore, as Blume stated in the program notes to the piece, this soundscape is meant to celebrate "all of [Mexico City's] shouts, as they take part in the sound memory of a time that will be over sooner or later."[48] The piece was featured at art festivals in Argentina, Austria, Chile, France, Germany, Italy, Spain, Taiwan, the United States, and Uruguay. It received an honorable mention at the 2014 Bienal Internacional de Radio de México and won the Pierre Schaeffer Award at the 2015 Phonurgia Nova Festival in France. On its presentation at the 2015 Loop Festival in Barcelona, Spanish political scientist Ariadna Rissola wrote:

> The story told by the artist invites us to make two reflections. First, the homogenization of cities so often denounced by contemporary artists is also taking place in a city as unique, as genuine, as Mexico City, but not only due to the knock-on effect of globalization, but as a result of policies intended to make the elements of popular culture that are less pleasant for the consumer society disappear. Second, given that these customs tend to disappear, Félix Blume's work ends up exercising a recording function (between documentary and ethnography) that over time can become a historical archive or sound memory of a time destined to disappear.[49]

The growing interest in soundscapes, ecoacoustics, and the understanding of sound as patrimony among intellectual elites in the United States, Europe, and Latin America paved the way for the positive reception of *Los gritos de México*. Thus, Blume's project provides a semipatrimonial sonic gaze that seeks to document sound practices that may disappear due to the government of Mexico City's attempt to regulate street vendors and the city's informal economy. However, unlike institutional patrimonial projects like the Fonoteca's soundscapes, which focus on establishing an essentialized connection between sound and place in terms of nature and tradition, the semipatrimonial gaze in Blume's project focuses on sounds that are the result of more recent urban practices, with less permanent or natural attachments to place. The international success of this piece helped launch Blume's career as a sound artist and played a significant role in the organization of the 2016 exhibit *Polifonía ambulante*, a further reinterpretation of Blume's sound recordings of Mexico City's everyday life and the project that gave birth to *Disco pirata*.

Blume created *Disco pirata* as an action piece that incorporates the sounds of Mexico City he had been fascinated with into the logic of circulation that characterizes the city's informal pirate music economy. He did it by selecting one hundred of his Mexico City sound files, organizing them into five categories (cries associated with specific trades, sounds of specific public events, sounds of public activities, sounds associated with specific places, and sounds heard on public transportation), and producing a CD of them. Like bootlegs sold in the subway or the streets in downtown Mexico City, Blume's *Disco pirata* was packaged in individual poly plastic CD sleeves including a cheap paper jacket that imitated the kitschy designs that characterize pirate CDs (figure 5.3).[50] Diego Aguirre Fernández, a graphic artist with whom Blume had collaborated for a couple of years, was in charge of designing the jackets. He recalls that Blume and he analyzed several pirate CD jackets and concluded that the designs for these discs are collages characterized by an aesthetic of excess; "they try to tell you as much as they can. If there are one hundred musicians, they try to include the silhouette of one hundred musicians." Aguirre Fernández says that he tried to mimic that style: "We did the title in 3D, with the Zócalo in the background, some mariachis, the gas tanks, a bike selling tamales, the wrestlers, etc. We also noticed that every pirate CD has the signature of a producing company, so Félix and I came up with Cocodrilo Producciones.... We also replicated this comical attitude when they say that their pirate CDs are 100% original."[51] To capture as closely as possible the original pirate aesthetic in the jacket materials, Blume had them manufactured in Tepito, the working-class Mexico City neighborhood where many of these pirate products are made, by a printer in the bootlegging business. Thus, the paper and the print quality met the standards in this economy. In other words, as the seal of originality on the jacket claims, Blume's *Disco pirata* is truly an original pirate copy.

FIGURE 5.3. Front and back covers of Félix Blume's CD *Disco pirata* (2016). Design by Diego Aguirre Fernández. Courtesy of Diego Aguirre Fernández.

Once the CDs were ready, Blume had to obtain permission from the leaders of the bootleg mafia to be able to sell them in the subway without upsetting them or creating any friction with other vendors. The final step was to embody the performance style of Mexico City subway vendors and try to sell the CDs in the train cars. Greek theater scholar Despina Panagiotopoulou, who collaborated with Blume in documenting the performance side of the project, states that people in the subway had "a question in their eyes: 'What are these people doing?' Because we [looked] more white [sic], and they could understand that we were not Mexican.... First, we would look if there was someone selling their CDs. If there was somebody, we would go to the next wagon [train car]. Then [Félix would] give the CDs to people. Some people gave [him] money. Some people were laughing; they were a little bit confused. Some people were very interested. I did not see any indifference."[52] Indeed, Panagiotopoulou's video of the performance shows Blume making his way through subway cars and yelling in a conspicuous attempt to emulate the *chilango* accent and the singing style of street vendors: "Good afternoon, ladies and gentlemen. On this occasion I bring to you, for sale, the album of the sounds of Mexico City. It includes more than one hundred sounds in MP3 format. It's more than three hours long with all the sounds of Mexico City. It includes the one about the Oaxacan tamales, the one about the old iron to sell. None of the good ones are missing. It costs ten pesos; you pay ten pesos." Among the many reactions, one can see a young couple amusingly discussing the track list before paying for the CD as well as an older lady gleefully watching as Blume lists the qualities of the album to two youngsters who, laughing, tell him that they do not want to buy it.[53] Indeed, the overall impression is that of a sense of silly complicity. Most subway users recognize the absurdity of the situation—a foreign white man, wearing a nice polo-like shirt, who tries to sound like a local vendor and attempts to sell them pirate CDs featuring the sounds they hear every day of their lives—but amusingly choose to play along.

Blume's explanation of the performance piece avoids mentioning this reaction and distances him from the incongruity of the situation. Instead, he states that this action was meant "to give back to the streets what I had recorded on the streets."[54] Central to Blume's apology for the recording endeavor is his belief that the city sounds he records do not belong to him; "they belong to the city, to the vendor, to the vendor's bell, to the tree planted in the Madrid Park, which sounds when the wind blows . . . or they belong to the wind."[55] Thus, *Disco pirata* plays with the idea that Blume, like the producers and sellers of pirate CDs in Mexico City, is also circulating something that does not belong to him. However, unlike the pirate street vendors, Blume considers himself "a sort of bond between sounds and listeners."[56] In the end, Blume was not actually interested in making a profit from selling the CDs. He knew he would sell only a handful of copies. The real motivation for this symbolic action was the opportunity to temporarily access the distribution channels of this informal economy in order to give regular *chilangos* a chance to encounter their city soundscapes in an unexpected and estranged manner, and an opportunity to listen to them anew, thus erasing the boundary between sound and noise and promoting a type of listening that is aware of the connection between sound and local identity. As the back jacket states, the CD offers "exclusive MP3 pirate distribution [and a] different [way of] listening of your city" (see figure 5.3). Certainly, since the performance was limited to a very small pool of people, it worked only as a symbolic action. The real moment for *chilangos* to relisten to these sounds happened at the Fonoteca exhibit since the event was designed precisely to promote this type of listening along the lines of the institution's mission. Echoing Blume's justification of his pieces, critic Ana Cecilia Medina wrote that "stripped of their context, these calls awaken the individual and collective memory of the city. The visitors to the [Fonoteca's] garden listen and move to the imaginary and concrete spaces that are familiar. 'Polifonía Ambulante' is a reading of the city, to rediscover it by listening."[57]

There is a certain kitsch and simulacrum-like quality to Blume's project and its call to relisten to the city sounds that can be inferred from paying attention to specific moments in the production of *Disco pirata*. They are more evident in the sense of excess that characterizes the stylistic materiality of the CD as well as Blume's interactions with his subway clientele. From Aguirre Fernández's design of the CD jacket as a faithful emulation of original pirate iconography to Blume's imitation of pirate vendors' calls and their performance style, there is a patent excess between originality and representation that generates a productive cultural tension. The sense of artificial excess that characterizes the project is clear in Aguirre Fernández's playful approach to jacket design, which mischievously states, "We are the second-best brand in piracy. Here we do make badass records" (see figure 5.3); in Blume's imitation of the street vendors' calls and accent; and in their amused reception by subway passengers. There is always something a bit off in these examples, and that mismatch confers on the project a surplus that manifests in amusement and laughter. In a way, the surplus between emulation in the loci of production and performance and the project's mission to trigger a new type of semipatrimonial listening leads to a rather kitschy reevaluation of these sounds. In fact, Blume's proposed reassessment of everyday Mexico City sounds resonates with the kitsch approach to the city's popular visual culture that editorial projects like Cristina Faesler Bremer's *ABCDF* (2001) and Juan Carlos Mena's *Sensacional de diseño mexicano* (2001) made so trendy among the early twenty-first-century Mexican Lettered and Aural Cities. This kitsch attitude framed the successful reception of musical projects like the Nortec Collective or Nopal Beat.[58] If kitsch speaks of a sensibility that embraces artificiality as a type of parodic catharsis, one could read Blume's project as third-degree kitsch, as proposed by Celeste Olalquiaga, as a type of empowerment that comes from an outside appropriation of a tradition in an attempt to adapt it to new aural and expressive needs.[59]

Disco pirata as Open-Access Archive

After its initial intervention in Mexico City's informal economy, *Disco pirata* found its natural niche within the walls of the Aural City as the action piece that welcomed visitors to the Fonoteca Nacional for the *Polifonía ambulante* exhibit, and later in New York's Mexican Cultural Institute, where the piece was featured in 2018 as part of *Sonic Postcards from Mexico City*.

The cultural circuit of these sounds came full circle when Blume made the CD available free for download on his personal website and on Freesound. That move eventually made *Disco pirata* into an archive itself, a cult object that began to circulate widely since its Creative Commons license allowed people to use its sounds for free. Soon, the archive and its files became very popular among sound designers in the Mexican film industry. Sound designer Daniel Rojo states that one of the reasons why *Disco pirata* turned into a recurring resource in his field is that although it contains many of Mexico City's contemporary sonic clichés—such as the *fierro viejo pregón*, a recorded call for scrap metal that, due to its free licensing, has become a ubiquitous audio meme in the city since it was first recorded on cassette in 2005—it also provides users easy access to unique sound files.[60] In speaking about his own work as sound designer, Rojo shares that he has used selected files from Blume's archive "to sonically illustrate the city in the most subtle but convincing way."[61] Sound designer César González Cortés suggests that although it is always difficult to identify sounds from specific sound archives once the final mix of a movie is ready, using sounds that belong to the places being represented makes a film more credible and realistic. He argues that "the reason why *Disco pirata* has been so successful [among sound designers] is that it includes sounds that are endemic to Mexico City. Since there are so many movies about Mexico City, one always needs its typical sounds, and Félix [Blume] is one of the persons who has spent more time recording them."[62]

In speaking about the importance of sound in developing a sense of authenticity in film, sound recordist Isabel Muñoz Cota argues that "truth is in the sound.... For me it is unthinkable to try to tell a story without [using sound] as a narrative tool. You are telling a story with sound. You are not just illustrating something. You are going to generate a feeling [and other] things, through whatever the spectator hears."[63] On first reading these statements, one gets a sense that Muñoz Cota, Rojo, and González Cortés all emphasize the sound itself as a source of authenticity: "Truth is in the sound," says Muñoz Cota, while González Cortés and Rojo acknowledge the uniqueness of Blume's sounds and the fact that they are "endemic" to Mexico City. Although there is a sense of veracity associated with the sounds themselves, all of them are quick to point out that it is the representation created with those sounds that needs to be "convincing" if it is to "generate feelings" in the process of listening. In that sense, they all acknowledge that the listening experience is the site where that sense of authenticity arises and makes a film's diegetic sounds veritable.

Disco pirata makes evident that the information stored in the archive is a mediation of reality just as much as the sound environments created with these sounds are also assembled representations of reality. In both cases, there is a symbolic system in place that allows listeners to develop affective responses to the sounds and to the uses of those sounds. On the one hand, *Disco pirata* features the aurality of a French artist who was captivated with the sounds of Mexico City he heard while living in the city's downtown. Blume brought to his encounter with these sounds the perspective of an outsider who was able to disentangle them from the negative connotations as annoying noise they may have for local *chilangos* precisely because he was a foreigner. His aurality was informed by the patrimonialist ethos of an outsider for whom these sounds were an intrinsic aspect of an aesthetic experience of the urban space. However, while trying to safeguard these sounds for future generations, he also made them into objects of aesthetic contemplation. As such, the material result of Blume's aurality, the CD with the preserved schizophonic sounds, attained the status of kitsch since severing these sounds from their specific cultural and geographic contexts generates a surplus of affective value that may translate into moments of comical absurdity. In the case of a film's sound design, the tacit agreement between filmmaker and audience is about "selling and buying" an illusion that generates a veritable impression of reality. As an open-access archive, *Disco pirata* allows for a creative recovery of the sound objects that brings them back from the realm of kitsch into the realm of simulacrum—representation perceived as reality. Thus, throughout the *Disco pirata* process, listening—more than sound or the sound object per se—emerges as the central locus of signification. It is listening that operates the transformation of everyday sounds into schizophonic kitsch objects, and the eventual transformation of the excess of kitsch into the discursive audiovisual conformity in which these sounds can be conceived as part of a seemingly authentic cultural system again.

Beyond the Patrimonial Logic: Mexican Rarities and *Disco pirata* as Archives of Postnational Memory

When I interviewed Despina Panagiotopoulou, she expressed surprise that I was interested in talking to her about a project that she considered to be incomplete. She stated that these kinds of projects operate "like post-traumatic theater in which the dramaturgy [happens when you bring the project] back to the audience. It is in that sense that it [could be] political because in the center is the spectator or listener, and [they] can make [their] own conclusions."⁶⁴ However, unaware that Blume had uploaded *Disco pirata* as an open-access archive and that this move had generated an active engagement with sound designers and film audiences, Panagiotopoulou felt that this crucial step had not taken place and thus the project was unfinished. When I informed her of the postperformance life of *Disco pirata*, she realized that in fact it was the project's transformation into an open-access archive that allowed for that epistemic and aesthetic loop—the gap she felt made the project incomplete—to be closed. The transformation of *Disco pirata* into an open-access archive signals the potential of a postnational memory in the displacement of the archive's authority from the sound object and its possession to the affective act of listening in detail.

Indeed, listening in detail to *Disco pirata* offers an opportunity to notice how, as an example of digital archiving, this project refuses the institutional archive model while running into trouble in the way it decontextualizes the sounds of the Mexico City streets and packages them for their eventual use as samples. The notion of *noriginal* is crucial in understanding how this contradiction presents the potential of a postnational memory as it informs the simulacrum process that characterizes the production and uses of the archive at every step. From the outset, Blume brings a way of listening that, by way of his foreign sonic gaze, epistemically de-essentializes the sounds of the city; it separates them from their local understanding as noise. This denaturalization is followed by a split between sounds and their sources. Each single step in this process separates the perceived sounds and the sound objects generated by this perception from the original sounds qua vibration, along with the context that actually creates and imbues them with meaning. The sense of simulacrum is even more explicit in the development of the CD as a commodity that copied the pirate CD aesthetic in detail, while stubbornly and waggishly emphasizing its "100% originality" on its jacket, and in the performance aspect of the project, with Blume imitating street vendors in a planned-to-fail effort to pass as local.⁶⁵ Here, the references to originality are anything but excuses to playfully challenge the expectations typically associated with such an expression. Instead, they generate a kitsch surplus that further distances Aguirre Fernández and Blume's practices from any originality. Ironically, given Blume's initial idea to document these sounds in order to preserve their originality, this procession of simulacra emphasizes representation over uniqueness. There are no originals in the simulacrum; instead, there are *noriginales* that, as Rivera Garza argues, "attest to the collaborative work that shapes them" but do not spring out of "a past that appears stable or already finished [but rather] from and toward the present, in the vicinity of the present presence of the past itself, and even of the future."⁶⁶ The postnational memory that *Disco pirata* promises hides in the excess between the newly negotiated past and the newly imagined future that foreign and kitsch sonic gazes make possible. Regardless of Blume's intentions, because there is no sense of humor in nationalism, because nationalism never affords the possibility of laughing at itself, the transformation of this archive into kitsch enables it to evade the solemnity and gravitas of the patrimonial gaze.

Mexican Rarities diverges from the conventional patrimonial approach upheld by national archives by treating collections as *noriginales*. Rather than being fixated on the collections' perceived authenticity, Mexican Rarities emphasizes the location and digging out of materials neglected by the archive's nationalist discourse of difference and their reissuance and recirculation as new documents beyond those ideological coordinates. This approach shifts the power structure of the institutional archive and questions its traditional hierarchy. Rather than locating authority in the repository and its power to ideologically reproduce itself, Mexican Rarities highlights the agency of its users and their ability to collaboratively and dialogically manufacture the archive's documents in new socially meaningful and significant ways. Thus, postnational memory is produced collectively as we generate documents that, as Jacques Rancière would put it, assert themselves as "the principle behind a new distribution of the sensible [that unites] the act of manufacturing with the act of bringing to light, the act of defining a new relationship between *making* and *seeing*."⁶⁷ The collective generation of archival documents by unearthing hidden materials, making them visible or audible, and reassembling them anew is a process that not only reveals the *inaudito* but also frees us from the preoccupation with safeguarding original objects, controlling their circulation, and regulating their representation. The latter are the tasks that typically characterize the mission of nationalist institutional archives.

Alex Rivera used the music of the Nortec Collective to sonically accompany his postnational hallucination in *Sleep Dealer* because, by reinventing the traditional working-class music from the north of Mexico according to the dystopian coordinates of globalization, the cyborg sounds of the collective become the sonic mirrors of a newly estranged future. In a similar way, Mexican Rarities and *Disco pirata* offer new ways of listening to documents that were already there but were *inauditos* either because they were discursively neglected or because they were taken for granted.

The importance of these archival ventures is not that they grant us access to something uniquely original but rather that they allow us to, as Cristina Rivera Garza argues, play these documents like a pianist plays the keys of his instrument, that is, by retrieving familiar sounds and reinterpreting and reassembling them as part of new musical discourses. As such, akin to the music of the Nortec Collective, the postnational promise of these archives lies in their ability to enable us to reinterpret these documents, thereby sounding the past in ways that facilitate the imagination of new futures beyond the constraints of the nationalist archive. This is accomplished through the affective agency of those who reactivate these unearthed materials. If understanding archives as metaphoric instruments provides an avenue to make the archive's documents anew, thinking about instruments as archives may also give us an opportunity to reimagine the affective possibilities of the archive(s). Building on the concept of the open-access archive explored in this chapter, the following chapter embraces that challenge.

Aurality, Materiality, and the Carrillo Pianos as Archives

A veces, el piano que nadie toca es un piano de reflejos.
(Sometimes, the piano that nobody plays is a piano of reflections.)
—Jaime Moreno Villarreal, "El piano que nadie toca" (1995)

In his 1995 prose poem "El piano que nadie toca" (The piano nobody plays), Mexican writer Jaime Moreno Villarreal provides an intimate portrait of a piano that has lived through better times. Having once been at the center of a family's social life, the piano has remained largely silent since the passing of the woman who played it. However, in the hope that a relative may be interested in taking piano lessons in the future, the family has decided not to sell the instrument. Thus, the piano stoically stands as a silent witness to the arrival of new relatives and to the settlement of family disputes. Although its keyboard is untouched, its soundboard reverberates with—while keeping secret—the many conversations that happen around it. As time goes by, the piano that nobody plays deteriorates, and its role in the family quarters morphs. It becomes a piece of furniture for ashtrays and wine glasses to be placed on, an antique embellishing a living room wall, and eventually a dust collector and the perfect secluded space for that elusive mouse to keep its nest. Moreno Villarreal is very detailed in describing the piano's changing materiality—the type of wood it is made of, its thinning varnish, the color and aroma of the cloth covering its keyboard, the cast iron plate and its loosening tuning pegs; however, it is in the poetic imagery that recounts the affectivity ascribed to and projected onto the piano's aura that his writing shines more vividly. In invoking the structures of feeling that people bestow on the instrument, Moreno Villarreal writes that "keeping the fragrance of a world that it will take with it, [t]he piano detaches, rises and dissipates into the air."[1] He suggests that "at night, [the piano's] black lacquer lights up a drizzle, and on the board something like the calligraphy of an Italian gelateria is drawn, almost frost and red, like a Christmas showcase; the piano makes a long way under the cold, it floats in the flood, it sets off to sensations of a platform or a flashlight that is left on reverse," and concludes with an expectant and elegiac assertion, that "sometimes, the piano that nobody plays is a piano of reflections."[2]

The piano that nobody plays is many things and has been many things. Some of them are confined within the instrument's decaying materiality. Others are recorded in the memories of those for whom the instrument was an important life companion. Yet most of them are prospects stored in the postponed desires, yearnings, and hankerings that the physical instrument furtively embodies and conceals. The materiality of the piano stands for a type of archive that alludes to the palpable acoustic information contained within it and sanctioned by its design, its soundboard, the configuration of its keys, the arrangement of its strings, and the specificity of its action frame; at the same time, the instrument as an object obliquely alludes to something more immaterial while simultaneously disguising it. This immaterial surplus is an elusive archive of affectivity made of traces, echoes, and rumors that one may not be able to access through conventional archival retrieval strategies or express through standard quantitative techniques. This is an invisible archive whose data and stories one may only be able to recall, grasp, and convey through alternative means of coding and decoding, of storing and retrieving. Valeria Luiselli alludes to these potential techniques in her novel *Lost Children Archive* (2019) when she recognizes that the apparent nonsense of the voices of her main character's children as they play and reenact the bits and pieces of stories they have been hearing from her and her husband during the family's cross-country road trip that frames the book's story is "the only way to record the soundmarks, traces and echoes that [the] lost children [she has been desperately trying to locate as the journey unfolds] left behind."[3] The children's games are in fact a performative way of reordering and rearranging the information they have been listening to throughout the trip, and of activating the affective archive such data concealed. In dealing with this silent and mute affective surplus of the archive, Polina Barskova suggests that it is in the "emotional experience and possibility" of poetry that the "images, sounds, shadows, and memories" of those whom the archive has rendered invisible, and who thus exist beyond knowledge, can be recovered and made sense of.[4]

What kind of materiality does "El piano que nadie toca" focus on? What kind of archive does that materiality imply? Moreno Villarreal's prose poem is itself an archive of echoes in the sense alluded to by Luiselli and Barskova, as it poetically renders visible the affective surplus that the piano as archive keeps hidden. It does so by taking the decaying materiality of the piano as a point of contact between two ways in which an instrument can be understood as an archive: as a collection of acoustic possibilities and as an assortment of tacit aspirations, fantasies, and unforeseen promises. In Moreno Villarreal's story, these unspoken and invisible desires are envisioned or imagined in the metaphoric reflections of futurity of a piano that captures the fragrance of a world as it dissipates in the air.

This chapter takes the materiality of instruments as a point of entry into a realm of affectivity that is always at the core of the archive, even when not visible or readily expressible via everyday language or linguistic strategies. Central to my larger argument here is Alexander Rehding's contention that musical instruments are objects that carry specific sources of knowledge, or Thor Magnusson's suggestion that "instruments contain music, theory, and culture," that they extend "our senses and ideas of the world, and through them, another world emerge[s]."[5] In other words, instruments are archives whose design stores certain types of knowledge. My case study is a series of pianos that, as in Moreno Villarreal's prose poem, hardly anyone played for several decades but that, nevertheless, reflect and store a number of desires and aspirations in their design, construction, ascribed usage, and contemporary reawakening: the nineteen Carrillo Pianos designed by Mexican microtonal composer Julián Carrillo (1875–1965) and Mexican-German piano maker Federico Enrique Buschmann Kampmeier (1911–??) in the late 1940s and early 1950s.[6] The reason I have chosen the Carrillo Pianos to illustrate a discussion of instruments as archives in the context of an exploration of sound archives and the Mexican Aural City is twofold. On the one hand, as Ricardo Miranda has shown, the piano was the quintessential instrument of the nineteenth-century Mexican bourgeoisie.[7] This is not unique to Mexico, but the fact that the instrument is an icon of the Mexican enlightened elite's cosmopolitan aspirations makes it into an ideal symbol of the political and cultural civilizing project of the country's

Lettered City. On the other hand, as seen in chapter 5, a certain fascination with the almost esoteric mystique of Carrillo and his microtonal music (the so-called Sonido 13) and their aura of forbidden authenticity among young Mexican musicians, sound artists, and proponents of sound studies has made the composer and his microtonal project, including his Carrillo Pianos, into significant phantasmatic patrimonial presences in the imagination of the contemporary Mexican Aural City. It should be noted that this current fascination with the Carrillo Pianos is partly fueled by the fact that the instruments' presence in the Mexican musical world has been mostly symbolic precisely because they have remained largely unplayed. This silent presence makes them into a type of patrimonial asset, one with the potential to trigger and activate a variety of powerful underground and countercultural fantasies and desires. Hence, the figure of the piano that nobody plays materializes factually and metaphorically in the Carrillo Pianos as archives with the potential to cut across historical temporalities and illustrate continuities and discontinuities between the Mexican Lettered and Aural City projects.

In their exploration of aesthetic artifacts that operate as analog archives, Carla Maier and Holger Schulze focus on how the discursive and material attributes of instruments "become sonic affordances for instrumentalists or sound artists who play, interpret or manipulate the instrument."[8] For them, instruments are not just metaphoric archives; they are material archives since their "physical and semiotic properties create... sonic affordance[s] that need... to be actualised and appropriated in the performance of playing, disturbing, [and] manipulating [the instrument]."[9] In other words, in order for the acoustic possibilities stored in an instrument—its sonic affordances—to be realized, they need to be retrieved from the instrument's materiality through specific techniques. I take Maier and Schulze's idea of sonic affordances as "the tacit grooves stored in instruments/apparatuses," or as the "functional and relational aspects which frame, while not determining, the possibilities for agentic action in relation to an object," as Ian Hutchby conceptualizes it, to further theorize instruments as open-source archives.[10] By *open source*, I mean to look at them as repositories of information that anyone can interact with and modify freely according to their aesthetic needs and aspirations. In doing this, I am certainly interested in studying how instruments are designed and built with certain sonic affordances in mind that imply particular storage and retrieval strategies. But I am also interested in exploring how individuals may circumvent these sanctioned pathways in order to develop retrieval strategies that reveal unforeseen sonic affordances and thus produce new knowledge out of the information stored in the archive. Thus, here I think of instruments as open sources whose materiality procures the means of their own estrangement once chaos or anarchy is introduced into their original design and system. Doing that presents us with the opportunity to conceptualize and theorize the possibilities of opening archives to new ways of aesthetically and affectively reimagining and making sense of the information they store.

"Contienen ese manantial sonoro": The Carrillo Pianos as Blueprints of Desire

One of the many reasons why Julián Carrillo's brand of microtonality failed to catch on among musicians, audiences, and fellow music experimentalists during the composer's lifetime was that despite the futurist rhetoric he used to validate it, Sonido 13 was essentially conceptualized as a "closed normative system [that] precludes future particularization."[11] Magnusson asserts that "by way of ingenious design, [musical instruments] serve as surfaces of musical inscription. The musical theory of each musical culture is written into the functional body of the instrument itself. [And] the instrument is concretised music theory."[12] This is certainly the case in the design and construction of the Carrillo Pianos. Carrillo was very specific about why and how he meant microtonality to be the future of music (for example, a blind belief in teleology and the overtone series as a source of that teleological drive for musical progress), what kind of intervallic system Sonido 13 should be (for example, equal microintervallic divisions of the whole tone), and how the music written in this system should relate to the Western music tradition and its history (Carrillo originally thought of his microtonal system and himself as fully rooted in that tradition). He was adamant about this creed and accepted no other interpretations of microtonality. Therefore, in designing his Carrillo Pianos as "containing that sonic wellspring which offers the possibility of unsuspected richness that will provide spiritual food for the future," the composer and theoretician projected these desires, fantasies, possibilities, and limitations into the blueprints of the instruments.[13] Nevertheless, once the instruments were built, their materiality, their very physical existence, provides the platform to overcome Carrillo's restrictive normativity and to make the information they store meaningful in new and unpredicted ways. The case of the Carrillo Pianos corroborates that, as suggested in chapter 3, in their very design, archives have the potential to store the source for their own ideological decimation but also that of their own resurrection.

One of Carrillo's earliest references to microtonal pianos appears in an article entitled "Los pianos con cuartos de tono construidos en Alemania y Estados Unidos," (The Pianos with Quarter Tones Built in Germany and the United States), published in *El Universal* in December 1924.[14] The piece was based on a journalist's conversation with Carrillo in which the composer reacted to a note published in the French journal *Le Ménestrel* that mentioned a German piano maker who had recently made a quarter-tone piano and to an article in the *Musical Advance* that not only verified this news but also told of Moritz Stoehr's efforts at making a quarter-tone piano in New York City.[15] The bulk of Carrillo's article focuses on the interfaces of these keyboards and the type of sonic affordances they provide to the musicians playing them. In both cases, the article argues that the keyboards in these pianos are designed as extensions of the standard piano keyboard that allow for the easy production of quarter tones but either utterly prevent the performer from playing music in twelve-tone equal temperament (12-TET) or drastically hinder the possibility of moving back and forth between the quarter-tone and 12-TET systems.[16] In a style typical of Carrillo's adventurous and often hyperbolic statements at the outset of his microtonal crusade, the article closes with a statement that attempts to claim the high ground for Carrillo as a music revolutionary and instrument inventor: "The piano being designed in Mexico has great advantages over those built in Germany and New York. . . . It does not enlarge the length of the keyboard; it would play all existing music besides that of the Sonido 13 revolution. Neither the piano made in Germany nor the one made in New York go beyond the quarter tone, while Mexico's will include up to the sixteenth of a tone."[17] A year later, on February 22, 1926, Carrillo had the opportunity to play Stoehr's quarter-tone piano in New York. On studying the instrument, Carrillo seemed to confirm his earlier impression of the shortcomings of its design. He wrote in his journal, "The quarter-tone piano is a complete failure, with the aggravating circumstance that it was built after going to Germany to look at the piano built there, which is another complete failure since it is unpractical."[18] Nevertheless, Carrillo's diary entry is also revealing since the composer confessed to having been surprised that "one of the keyboards [in Stoehr's piano] is very similar to the one I invented in Coyoacán, which I placed on a piano that belonged to Antonio Gomezanda."[19] Although Carrillo's journal note does not provide any details about the keyboard he had designed in the early 1920s nor about how the piano he had in mind could play both music in 12-TET and music in microtones, it does show that his early conceptualization of a microtonal piano also relied on somehow altering the traditional interface of the instrument.

Carrillo had to wait more than twenty-five years before presenting a concert that featured one of his microtonal pianos. On September 29, 1949, as part of a cycle devoted to a retrospective of Carrillo's music at Mexico City's Anfiteatro Simón Bolívar, Dolores Carrillo, the composer's daughter, played his Prelude for Third-Tone Piano (1949). Although Carrillo had been working on the designs of his microtonal pianos for several years, the instrument used on that occasion was his own Steinway, a baby grand that Federico Buschmann modified to produce the desired intervals. Buschmann's modification consisted of replacing the Steinway's original cast iron plate with a special cast frame made specifically to hold the strings in the arrangement needed to tune them in thirds of a tone.[20] In its materiality, other than the internal modifications in relation to the strings and the frame, the rest of the piano remained unaltered; thus, its interface is that of a standard instrument: eighty-eight keys (fifty-two white keys and thirty-six black keys) arranged in the traditional keyboard pattern. However, instead of having intervals of a half tone between adjacent keys, the instrument features intervals of a third tone between them. Therefore, its eighty-eight keys cover a range of four octaves plus five whole tones instead of the standard seven octaves plus three semitones. The lower key in the instrument is an A0 that sounds like an A1, and the highest one is a C8 that sounds like a G6.[21] While the length of the keyboard remains the same, because each key presents an interval slightly smaller than the semitone in a standard piano, the instrument's actual range necessarily contracts.

Between 1953 and 1954, Carrillo and Buschmann produced and patented the plans for the nineteen Carrillo Pianos: two whole-tone pianos (one per each whole-tone scale), three third-tone pianos (two baby grands and an upright), and a piano for each subsequent partition of the whole tone, from quarter tones down to sixteenths of a tone. All the instruments are upright pianos, with the exception of the two baby grands in thirds of a tone. Each plan provides a view of the extension of the cast iron plate and the strings (some of them also show a side view of the instrument and its mechanical action). In them, Buschmann offers detailed indications regarding the materiality of each of the pianos, including the pitch for each key, the distance between the place where the hammer hits the string and the upper end of the bridge, the distance between the upper and lower

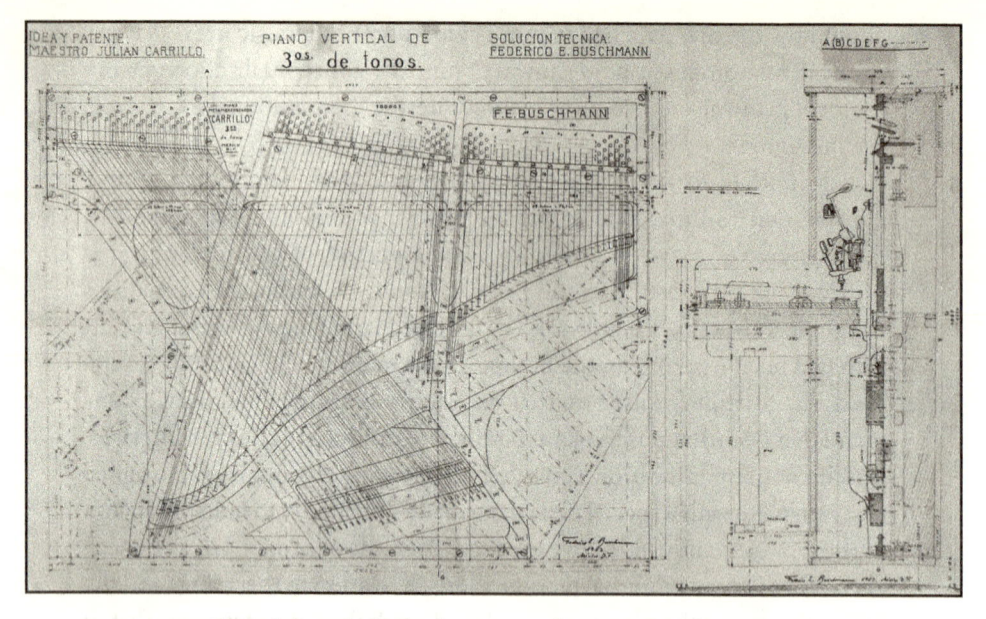

FIGURE 6.1. Plan of the upright third-tone Carrillo Piano, 1953, by Federico Buschmann following Julián Carrillo's specifications. Archivo Julián Carrillo.

ends of the bridge, the diameter and length of the strings and the distance between them, the number of strings per pitch, and so on.[22] Figure 6.1 shows the plan for the upright third-tone piano. Figure 6.2 is a picture of the instrument and its inside.

The upright Carrillo Piano in Buschmann's plan is in essence the same as the Steinway modified in 1949 with two exceptions. One, since it is not a grand piano, the inside of the instrument—the cast iron plate, the arrangement of the strings, the disposition of the piano action, and so on—is different. And two, the range of the instrument is also different; while the Steinway is tuned A1 to G6, the Buschmann upright is tuned three semitones higher, C2 to A-sharp 6.[23] Tuning the instruments this way allowed Carrillo to have access to the two whole-tone scales and thus to two sets of frequencies and pitches that are mutually exclusive in uneven partitions of the whole tone.[24] This was important for him since by that time the whole-tone scale had become one of the structural sonic frameworks within which he used microtones in his mature compositional style, and thus it makes sense that he tried to retain access to both sets of pitches.[25]

The remaining Carrillo Pianos, from quarter tones down to sixteenths of a tone, are all upright instruments that entail similar types of estrange-

FIGURE 6.2. Third-tone Carrillo Piano made by Federico Buschmann ca. 1953. Photo by the author.

ment to those enacted by the third-tone pianos. However, they also present material differences in both the instruments' interior design and their keyboard interface. The sixteenth-tone Carrillo Piano, the instrument featuring the smallest intervals between contiguous keys, provides an entry into exploring the commonalities and differences among these instruments. It also allows us to further interrogate what it is that the materiality of these instruments stores and how Carrillo meant for the information they contain to be retrieved in order to create (or reproduce) knowledge.

The keyboard extension in the sixteenth-tone Carrillo Piano is larger than in a standard piano. Since there are ninety-seven sixteenths of a tone in an octave, the keyboard had to be expanded for the instrument to cover a single octave, from C4 to C5. This is the largest keyboard extension of all the Carrillo Pianos. With the exception of the fourteenth-tone and fifteenth-tone pianos, which also have a span of a single octave, all other Carrillo Pianos have eighty-eight keys.[26] For that reason, the configuration of keys on the keyboard interfaces of these three instruments is slightly different from that of a standard piano. The lower key in the fourteenth-tone piano looks like a c1 (although it is tuned to a c4), and the lower key in the fifteenth-tone piano looks like an F-sharp 0 (although it is also tuned to a c4). Like for the fourteenth-tone piano, the keyboard of the sixteenth-tone Carrillo Piano starts with a c1 tuned to a c4 but ends in a c9 tuned to a c5.[27]

Given that the range of the sixteenth-tone Carrillo Piano is only one octave, the arrangement of the strings and the shape of the cast iron plate varies significantly from those in a standard piano and those in the Carrillo Pianos with a wider span. For example, the wider range of the third-tone piano requires strings of many different lengths and a frame shape designed for the longer strings from the lower register to cross over the shorter strings from the medium and higher registers—this is how the strings inside of a standard piano are organized. On the other hand, since they all belong to the same register, the strings in the sixteenth-tone piano are all of very similar length, which requires a very different cast iron shape to stand the more even tension of the strings. Following on this, the third-tone piano uses single strings per pitch in the lower register (from A to D two-thirds of a tone), double strings for the following octave (E to D), and triple strings for the remainder of the instrument. In the case of the sixteenth-tone piano, all the pitches are triple strings (see figures 6.1 and 6.3).

The Carrillo Pianos at the Expo 58 in Brussels: Instruments as Archives of Affectivity

It took a few years for Carrillo to find a piano manufacturer willing to make his microtonal pianos. In 1957 the Sauter Pianofortemanufaktur in Spaichingen, Germany, took on the challenge to make the instruments in time for the Exposition Universelle et Internationale de Bruxelles (Expo 58 in Brussels).[28] The fifteen Carrillo Pianos arrived at the Mexican pavilion of the expo in the summer of 1958, but the space allocated to the Mexican delegation was too small to host such a large instrument collection. At the request of Francisco del Río y Cañedo, the Mexican ambassador in Brussels, the administration of the expo made the Palais 3, a venue for temporary exhibits, available for the Carrillo Pianos from July 25 to August 17.[29] The pianos were advertised as "Les Pianos Carrillo. Pianos 'métamorphosés.' Uniques au monde. Rendant divers tons chacun. Des sons jamais entendus!" (Carrillo Pianos. "Metamorphosing" pianos. Unique in the world. Making many different sounds each. Sounds never before heard!) when the exhibit opened on July 28 (figure 6.4).

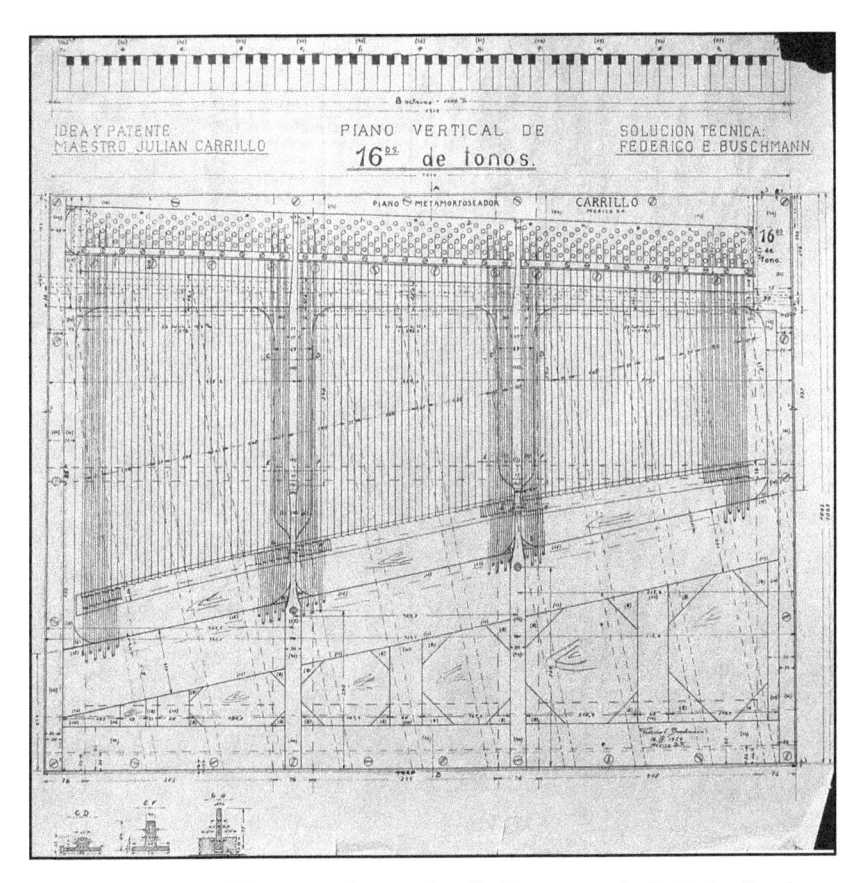

FIGURE 6.3. Plan of the sixteenth-tone Carrillo Piano, 1954, by Federico Buschmann following Julián Carrillo's specifications. Archivo Julián Carrillo.

Carrillo stayed in Mexico City while the pianos were on display at Expo 58. On August 22, after the exhibit ended, Oscar Urrutia sent him a letter informing him that the medal given to him by the General Commissary of the Expo in recognition of his participation at the expo had been mailed to him. This letter provides information about what other Sonido 13 items were featured at the expo: mainly, some of Carrillo's other microtonal instruments as well as an explanation of his numerical notation. Marcel Gaveau, manager of the French piano manufacturer Gaveau, arranged for the pianos to be sent to Paris once the Expo 58 exhibit ended, to be shown at the Salle Gaveau from October 27 to November 7. This second exhibit coincided with the meeting of UNESCO's International Music Council in Paris, which generated a lot of interest in the Carrillo Pianos.[30] The occasion

FIGURE 6.4. Announcement for the Carrillo Pianos at Expo 58 in Brussels, 1958. Archivo Julián Carrillo.

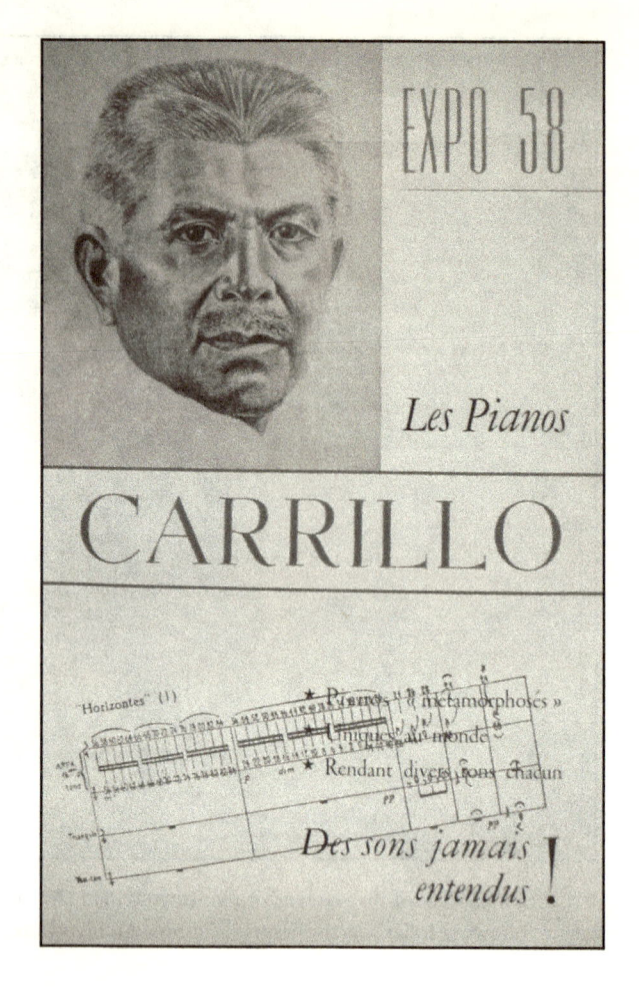

brought Alois Hába (1893–1973) and Adriaan Fokker (1887–1972) to Paris and gave Carrillo, who traveled to France for the exhibit, a chance to meet for the first time three of the most vocal proponents of microtonality in Europe, when they, along with Ivan Wyschnegradsky (1893–1979), visited the Salle Gaveau to learn about the Carrillo Pianos. Carrillo referred to this occasion in several of his writings, always highlighting the favorable opinion these microtonal pioneers had of his instruments. According to him, Wyschnegradsky stated, "We Europeans have been searching for practical keyboards which would allow us to produce quarter tones; and here, Julián Carrillo surprises us with a 'classical' keyboard capable of reproducing not only fourths, but fifths, sixths, etc. of a tone—this discovery is as miraculous as Columbus' egg"; while Hába said, "Upon my return to Prague, I shall ask my government to officially invite Julián Carrillo so we can admire his wonderful museum of fifteen metamorphosing pianos there."[31] The interest was genuine, if short-lived, and led Wyschnegradsky to compose his *Deux Pièces* Op. 44 (Two pieces, 1958) for sixth-tone piano and his *Prelude et Étude* Op. 48 (1966) for twelfth-tone piano. Jean-Étienne Marie (1917–89), who was already acquainted with Carrillo's music before Expo 58, composed his *Trois échantillons* (Three samples, 1958) for third-tone piano and *Le Tombeau de Carrillo* (1966) for standard piano, third-tone piano, and tape. As a token of gratitude for his continuous support, Carrillo entrusted Marie with doubles of the third-tone and sixteenth-tone pianos, which the French composer loaned to the Conservatoire de Paris.

Soon after the Paris exhibit ended, Carrillo traveled to Brussels for a reception at the Maison des Arts and a concert at the Palais des Beaux-Arts on November 9. The reception was hosted by Gaston Williot, editor in chief of the liberal Belgian newspaper *La Dernière Heure*, who had been at the Carrillo Pianos' showing at the expo. He was aware that Carrillo had received part of his musical training at the Koninklijk Conservatorium Gent (Royal Conservatory of Ghent) at the turn of the century and wanted to celebrate the achievements of a distinguished conservatory alumnus. The concert, financially sponsored by the Mexican government, featured the Belgian National Radio Institute Symphony Orchestra playing an all-Carrillo program under the composer's baton (figure 6.5). The program included the European premiere of *Horizontes* (Horizons, 1952), a symphonic poem for violin, cello, sixteenth-tone harp, and orchestra, with Gabrielle Devries, Reine Flachot, and Monique Rollin as soloists; the world premiere of the Concerto for Cello and Orchestra (1958), with Reine Flachot as soloist; and the world premiere of the Concerto for Third-Tone Piano and Orchestra (1958), with Dolores Carrillo as soloist. For Carrillo, a highlight of this concert was the attendance of Queen Elisabeth of Belgium, a well-known sponsor of the arts, especially music. In his autobiography Carrillo affirms that the queen offered a banquet in his honor the following day at the Palace of Laeken, the official residence of the Belgian royal family.[32]

The central place of these pianos in Carrillo's Expo 58 exhibit, the Salle Gaveau in Paris, and the Palais des Beaux-Arts concert in Brussels speaks of them as archives of cosmopolitan desire and aspiration. Besides the sounds, pitch relations, tonal and scalar patterns, and music theory they store, the pianos are containers of Carrillo's desires to belong and be recognized as not only an equal but an innovator within an international community of artists and musicians. These desires transpire from the Carrillo

FIGURE 6.5. Poster for a concert featuring Julián Carrillo as conductor at the Palais des Beaux-Arts, Brussels, on November 9, 1958.

Pianos as the locus where that respect and admiration was ignited as well as the soundboards where that appreciation and recognition could resonate and be amplified. That the instruments were made in Europe and shown there before they were presented in Mexico; that they gave Carrillo a chance to mingle and rub shoulders with members of the European royalty and bourgeoisie; that the pianos became the final excuse for Hába, Fokker, Wyschnegradsky, and Carrillo to finally meet after decades of hearing about each other's work but not shaking each other's hands; that some of these composers wrote music for his pianos; and, last, that the pianos allowed Carrillo to be ultimately recognized as a music maverick in Belgium, the country where he studied fifty-five years earlier, all point toward the Carrillo Pianos as archives of the types of "images, sounds, shadows, and memories" that transcend their materiality. These unspoken and invisible desires, aspirations, pride, and ambition go beyond the pianos' acoustic properties and speak of an archive of affectivity that literally, not only metaphorically, as we will see, "captures the fragrance of a world as it dissipates in the air."[33] This is precisely the affective surplus that the instrument as archive keeps hidden.

Of Dead Archives and Museums

In Spanish, corporations, organizations, and institutions often use the term *archivo muerto* (dead archive) to refer to files that are no longer needed in their daily operations. These types of files usually generate problems for the companies due to their lack of adequate spaces to keep them and thus are often sent to remote storage or annexes where they remain dormant and are rarely consulted. It is debatable whether archives are ever dead entities since the information they store always has the potential to be retrieved. However, in the context of the efficiency that neoliberal companies strive for, archives with such characteristics are simply unproductive, and trying to maintain or support them could not only waste resources but also lead to an unproductive environment. While conducting research in Latin American music archives, I have encountered *archivos muertos* on a few occasions. In every instance, the term was used to imply that, for different reasons, those particular archives were considered irrelevant or had been rendered useless by specific physical, ideological, or practical circumstances.

The Carrillo Pianos were advertised and presented as the highest achievement of a type of technology that, at a moment when musicians were turning to new electric and electronic gadgets, was actually becoming passé as the preferred medium to imagine the music of the future. As a powerful indication of the changing times, the pianos were presented at the marginal Palais 3 while the Philips Pavilion—designed by Le Corbusier and Iannis Xenakis (1922–2001); featuring *Poème électronique* (Electronic poem, 1958) by Edgard Varèse (1883–1965) and Xenakis's musique concrète piece *Concret PH* (Concrete PH, 1958); and intended to celebrate postwar technological advancement—occupied center stage at Expo 58 and has ever since remained central in the music historiography about the exhibit. In a way, this contingency is a painful indictment of Sonido 13 as a future meant not to be, a future where such analog technology not only was irrelevant but could be seen as an obstacle to the advancement of electric, electronic, and eventually digital technologies. That Hába referred to these instruments as Carrillo's "wonderful museum of fifteen metamorphosing pianos" is no coincidence. Under such circumstances, the Carrillo Pianos were destined to become relics of an unfulfilled prophecy and museum curiosities, essentially an *archivo muerto* that, if anything, captured "the fragrance of a world as it dissipates in the air."

Through the 1970s and 1980s, Jean-Étienne Marie organized sporadic concerts using the third-tone and sixteenth-tone pianos kept at the Conservatoire National Supérieur de Musique de Paris. However, on Marie's death, his heirs decided to sell the two instruments at auction. The third-tone piano was acquired by the Konservatorium Bern in Switzerland, while the sixteenth-tone piano ended up at the Musée de la musique in Paris. There, the pianos completely morphed from instruments to be played into objects to be seen. It was during his studies with Marie in Paris in the late 1960s that Mexican composer Mario Lavista learned about the Carrillo Pianos. On his return to Mexico City, he was able to convince Dolores Carrillo to lend him some of the pianos for performances with his improvisation group, Quanta. Nevertheless, when Dolores found out about Quanta's unorthodox ways of using the instruments—often playing inside of them, preparing the strings or directly plucking them with their fingers, or even kicking and hitting them; in other words, a usage completely outside of Carrillo's microtonal system and the theory the pianos were meant to reproduce—she forbade Lavista from ever using them again.[34]

As for the fifteen pianos featured at Expo 58, once the Salle Gaveau exhibit was over, the instruments were shipped to Mexico. The following year, Carrillo organized an exhibit at Mexico City's Palacio de Bellas Artes (figure 6.6), where the pianos were shown for several weeks before they were sent to a storage room at the Museo Nacional de Historia in the Chapultepec Castle. Carrillo argued that the instruments remained there without proper maintenance and were subject to abuse by the museum workers, who would "place cans of paint on top of them, without even covering them with newspapers."[35] The pianos were eventually moved to safer storage at the museum, where they remained for several years. A few days before his passing, Carrillo published an article in *El Universal* in which he restated his wishes to have the pianos featured in a permanent exhibit at Mexico City's Chapultepec Castle.[36] Eventually, Dolores Carrillo reclaimed the instruments and moved them into her house in Mexico City's San Angel district. There, they also remained mostly dormant for several decades until, after Dolores's passing, the family donated the whole of Carrillo's archive to the state of San Luis Potosí. Since 2011 the Carrillo Pianos have been kept there in a permanent exhibit at the Centro Julián Carrillo.

FIGURE 6.6. Cover of the booklet for the 1959 exhibit of Carrillo Pianos at Mexico City's Palacio de Bellas Artes. Archivo Julián Carrillo.

Regarding the Carrillo Pianos, permanently displaying them in museum-like settings may arguably have ensured a place for them in Mexican history as patrimony or cultural heritage. However, it also denied them a place on the concert stage as living musical artifacts. Thus, the pianos became different types of instruments. With their sonic existence largely bracketed by their status as museum objects, the pianos were freed from the dictum of performing the sounds of the future and were afforded the ability to perform Julián Carrillo himself as Mexican patrimony. Evidently, the fifteen Carrillo Pianos never really had a life as performing instruments, and in the absence of the practice of playing them, the knowledge that Carrillo embedded in them through their design failed to be transmitted. This is a critical circumstance that begs us to ponder the social and historical roles of a group of out-of-circulation instruments that become objects of display in a museum exhibit. Furthermore, if Carrillo's senses and ideas about the world were not transmitted through the pianos, it is important to explore what kind of knowledge these instruments were involved in generating. As Barbara Kirshenblatt-Gimblett argues, "Exhibitions are fundamentally theatrical, for they are how museums perform the knowledge they create."[37] By making the instruments into objects of display, Carrillo inscribed them with a type of agency that is animated by the patrimonial plot and mise-en-scène of the museum exhibit. If objects are "actors and knowledge animates them," as Kirshenblatt-Gimblett argues, placing the Carrillo Pianos on display at Mexico City's Museo Nacional de Historia, in the majestic Chapultepec Castle, activated them as part of a plot that performs national heritage and imbues it with a sense of monumentality.[38] Carrillo's desire to have the instruments on display at the museum at the end of his life was thus a move to ensure a place for himself as part of the grandiose and heroic narrative of Mexican national history. In the end, it did not work out in such a magnificent way since the pianos ended up at the regional Centro Julián Carrillo. But their permanent display there, at a research center and museum devoted to celebrating Carrillo's legacy, does play an important role in the performance of the composer and his work, including the instruments themselves, as patrimony.

The fact that Carrillo's idea of having the pianos on display at the Museo Nacional de Historia did not work out as planned may have been a blow to the composer's desire to be officially recognized as a central actor in his country's cultural history. However, their display at the family's private home for over forty years afforded them an aura of authenticity and antiestablishment that Carrillo could not have foreseen. Conspiracy theories arose that attempted to explain Carrillo's marginalization from Mexico's cultural history as a conscious attempt by individuals in governmental institutions and members of the Lettered City to render his legacy invisible.[39] These types of representations and the aura of hidden alternative knowledge that accompanies them form the allures that have made Sonido 13 and the Carrillo Pianos into objects of desire for the Mexican Aural City at the beginning of the twenty-first century. The museumification of the Carrillo Pianos may have marked their transition into *archivo muerto*, but it also meant their activation as compelling archives of affectivity.

Writing about the relevance of the information stored in *archivos muertos*, Cristina Rivera Garza wonders, "Where are the documents really going when they don't seem to move?" Her answer is that their true trajectory "is none other than eternity or oblivion. In short: the innumerable dead."[40]

Her conclusion borrows Jean Genet's idea of "the innumerable dead" to qualify "eternity" and "oblivion" as the infinite space created in the experience of every object or piece of art. For Rivera Garza, the trajectory of the documents stored in *archivos muertos* is an aesthetic experience that refuses to be historical, one that "privileges, or should privilege, discontinuity over continuity. [It] is a moment of recognition, [but it] is also and perhaps especially a moment of restitution."[41] Thus, these innumerable dead refers to the space in which the information in the *archivo muerto* is reimagined beyond its historicity as an aesthetic experience that provides a way to affectively read the archive anew transhistorically and in relation to the retrieving individual's gaze or aurality. This conceptualization of the aesthetic experience coincides with the alternative means of coding and decoding through performance and poetry that Luiselli and Barskova propose as ways to activate the affective surplus concealed in the archive's data. In what follows, I explore a number of aesthetic operations that effectively recognize and restitute the affective information stored in the Carrillo Pianos.

Scores, Bodies, Kinetic Action, and Archives of Possibility

It was almost two decades ago, in Mexico City, when the Julián Carrillo Archive was kept at the house of the composer's daughter in the San Angel neighborhood, that I first encountered a Carrillo Piano. It was a sixteenth-tone piano, and the experience was exhilarating and confusing at once. I placed my hands on the keyboard and proceeded to play the first measures of a Beethoven piano sonata I have played since I was a teenager. Although I knew that the piano would not sound like a standard piano, the sonic result was something unique and fascinating that was also completely at odds with what my body expected from the interface I was looking at and feeling through my hands and fingers. The familiarity of its interface and the disciplining of my body in relation to that interface through years of playing piano were the source of the cognitive dissonance I experienced when the sound came out of the instrument. I had to stop playing to reassess what was happening. After randomly playing through the instrument, I attempted to play a major scale. Finally, I was able to find the right pitches on the keyboard, but the visual result of what my hands and fingers needed to do in order to produce that major scale and what I was hearing were completely at odds. There was nothing in my body and its history of musical disciplining that could respond to what I was hearing, seeing, and feeling. Thinking about that moment retrospectively, I understand that what was at stake was not only an estrangement of the traditional instrument but also a need to understand the instrument itself as an archive of aurality and somatic experience. It was not only that the instrument was an archive; it was also that my body was an archive activated by my interaction with the instrument. It was almost as if the piano had an uncanny type of agency over my body.

Roger Moseley proposes to think "archaeologically" about the piano keyboard as an interface and to focus on its "rules of engagement and codes of conduct, reflecting functions of mediation that operate both analogically, by translating mechanical input into corresponding sonic consequences, and digitally, insofar as such input is typically initiated by the play of fingers as discrete entities."[42] Thinking about the Carrillo Pianos in these terms forces us to engage the fact that one of the main characteristics of these instruments is precisely that the relationship between their keyboards and the actual sounds they produce when played—their expected analog and digital correspondence—is estranged. That is one of the reasons Carrillo also advertised them as *pianos metamorfoseadores* (metamorphosing pianos), as seen in figure 6.4; because playing on them what may look and feel to the hand on the keyboard like a traditional 12-TET piece would in fact metamorphosize it into a piece with a different intervallic structure. It would subject any given piece originally written in 12-TET to a process of intervallic expansion or diminution, largely transforming it into a different, new aural experience.[43] That the visual and the aural (the somatic and the experiential) are at odds is probably one of the most striking features shared by all the Carrillo Pianos and presupposes the development of new rules of engagement and codes of conduct. In essence, this aspect of the Carrillo Pianos implies an estrangement of the traditional piano archive in terms of its repertoire and its materiality. This is the case both in relation to the instrument itself and in the relationship between the expectation in the kinetic action involved in playing it (the input initiated by the play of the fingers) and the resulting alienating aural experience.

In an article published in *El Universal* on September 2, 1965, just seven days before his passing, Carrillo described his pianos as "containing that sonic wellspring which offers the possibility of unsuspected richness that will provide spiritual food for the future."[44] The quotation is intriguing because, without conceptualizing the instruments as archives, he did refer to them as such: as objects containing an arcane type of information to be retrieved in the future. The materiality of the instruments, their design, their construction, and the compromises reached while thinking about them in relation to traditional keyboard interfaces and composition practices tell us about the type of information the Carrillo Pianos were meant to store as material archives. The pitches and intervals these pianos are able to produce—their sonic affordances—are information that points toward very specific ideas. In this case, they speak of Carrillo's aesthetic worldview (his obsession with the whole-tone scale as the tonal and harmonic foundation of a microtonal style of composition; a musical thought inescapably trapped within the boundaries of equal temperament and thus the colonial overtones of the Western art music tradition) and a type of prescriptive aurality related to Sonido 13 (the way in which Carrillo expected audiences to listen to and make sense of his microtonal music, in relation to the tonal frames and tuning systems mentioned above). In turn, the shapes and designs these instruments feature encapsulate the relationship of his futuristic project to the Western art music tradition, as an attempt to literally and metaphorically expand it based on the project's deep and immovable roots in it.

Furthermore, in their deviation from Carrillo's original ideas about what a microtonal piano should have afforded, the materiality of the Carrillo Pianos' final design is also an archive of the composer's struggle between desire and practicality. A quick look at the pianos vis-à-vis Carrillo's critique of the quarter-tone pianos built in Germany and the United States in the 1920s puts in evidence that, regardless of Wyschnegradsky's comment about the pianos being "as miraculous as Columbus' egg" in their practicality, Carrillo's early wish—to have an instrument with an interface that could be used to play traditional and new microtonal repertory and whose range and keyboard size would remain the same as in the standard piano—had in fact to be compromised when it came to designing and making the instruments. Carrillo was a practical man, and the instruments he designed always responded to practical performance matters and concerns. That is why he criticized the US and German quarter-tone pianos as impractical; he believed that their estranged interface (the multiple keyboards layered on top of each other) made them difficult for a traditionally trained musician to engage. He wanted to avoid that in his instruments and thus stuck to the familiar interface of the traditional piano. It was a nod to possible future performers. Nevertheless, as a result of his pragmatism, the instruments ended up generating a type of cognitive dissonance, a distance between the "input of the playing fingers" and the "rules of engagement and codes of conduct" needed in order to make sense of the retrieved information. Carrillo responded to this estrangement in a very practical way: When he composed for the Carrillo Pianos, he moved away from the numerical musical notation he had invented in the 1920s in an attempt to simplify the performers' role as mediators between the score as an archive of sonic information and the instrument as the technology to retrieve that information. This is ironic since this notation system was also central to the Sonido 13 exhibit at Expo 58 (one can see the notation prominently featured in the lower part of the cover of the booklet advertising the show; see figure 6.4). I argue that Carrillo's scores for these pianos are sources for trying to unravel what it is that these instruments store as archives of possibilities. The Carrillo Pianos as archives not only contain the acoustic information and properties embedded in their materiality but also store ideas, desires, and aspirations directly connected with these acoustic properties as well as others they may only obliquely refer to in their extended archive, the material culture produced in relation to and instigated by them.

Carrillo composed the Concerto for Third-Tone Piano and Orchestra, *Balbuceos* (Babblings, 1958) for sixteenth-tone piano and orchestra, *Capricho* (1959) for quarter-tone piano, and *Estudios* (Studies, 1959) for fifth-tone piano under the effervescent spell spawned by the Expo 58 exhibit of his instruments. As mentioned earlier, Carrillo opted not to use the numerical notation he had invented as part of his Sonido 13 revolution, choosing instead to write the parts for the microtonal pianos using standard notation. In these particular cases, the notation does not represent pitches but rather keys to be played. As such, Carrillo's notation in these works takes practical advantage of the source of the somatic-sound dissonance explained above to simplify the performer's engagement with the music. It also presents itself as a window into exploring the relationship between physical gestures and musical production. In other words, it puts in evidence the connections among aesthetics, musical production, and the kinetic action necessary to produce sound in these late works by Carrillo.

This is evident in the opening sequence of *Balbuceos* (figure 6.7). Here, the melodies in the orchestral part (Piano II in this reduction) unfold over the pitches of the whole-tone set (wT0, C–D–E–F-sharp–G-sharp–A-sharp). When the piano enters (Piano I in this reduction), it seems to share the intervallic content of the orchestral part (mm. 11–15). However, since the notation is intended for the pianist to know what keys to play and not what sounds the instrument will produce, it does not show that what

FIGURE 6.7. Julián Carrillo, *Balbuceos* (1958), for sixteenth-tone piano and orchestra, mm. 1–15. Piano reduction. Archivo Julián Carrillo.

looks like a series of whole-tone sequences in the sixteenth-tone piano part sounds in fact like a series of eighth-tone progressions.

If the notation is very obscure about the relationship between score and sonic experience, what it does very well is reveal the type of physical gestures required from the pianist to play the passage. The score shows the piano part to be very idiomatic: The music fits perfectly well in the hands of the pianist. This is clear throughout the piece, but it is particularly informative in the cadenza, when the pianist alternates between what are notated as chromatic and whole-tone scale passages. Regardless of the sonic estrangement they produce when played, these sequences look on paper and feel in the hands like technical exercises meant to train pianists in speed, precision, and agility. These are the kinds of exercises most pianists spend hours with when developing their technique and are conducive to gestures and motions they are very comfortable with since they have been incorporated from the very beginning of their musical training. Similar types of hand and musical gestures characterize Carrillo's works for other Carrillo Pianos. If the use of scales lets Carrillo show the possibilities of the sixteenth-tone piano, it is the use of chordal sequences that in notation seem to move chromatically down the keyboard that allows him to show the possibilities of the fifth-tone piano in his *Estudios* for that instrument (figure 6.8). This music's status as a catalog of possibilities is even tacitly acknowledged by Carrillo in the title of this work: He seems to be literally studying and exploring the material, physical, technical, and mechanical possibilities of both instrument and performer.

By highlighting the technical and material potential of his microtonal pianos while keeping in mind the need to make the performer's interaction with their interface accessible, Carrillo's works for these instruments reveal how materiality, technique, gesture, and kinetic action operate in tandem to show the music and the instruments as organized catalogs of possibilities. They remind us that, as Roger Moseley suggests, "the keyboard is material and ideal, manipulable and manipulative, obfuscatory and revelatory."[45] Indeed, the sonic affordances of the Carrillo Pianos are unveiled by the idiomatic movement of the pianist's hands over the keyboard, but as my first encounter with these instruments showed, the piano also has an uncanny agency over the pianist's body. It is significant that just as the sequences that characterize Carrillo's works for the Carrillo Pianos entail mechanical patterns and repetitions that are usually cataloged and stored in piano-technique methods, they translate into sonic explorations of these instruments that could also be described as inventories; they are

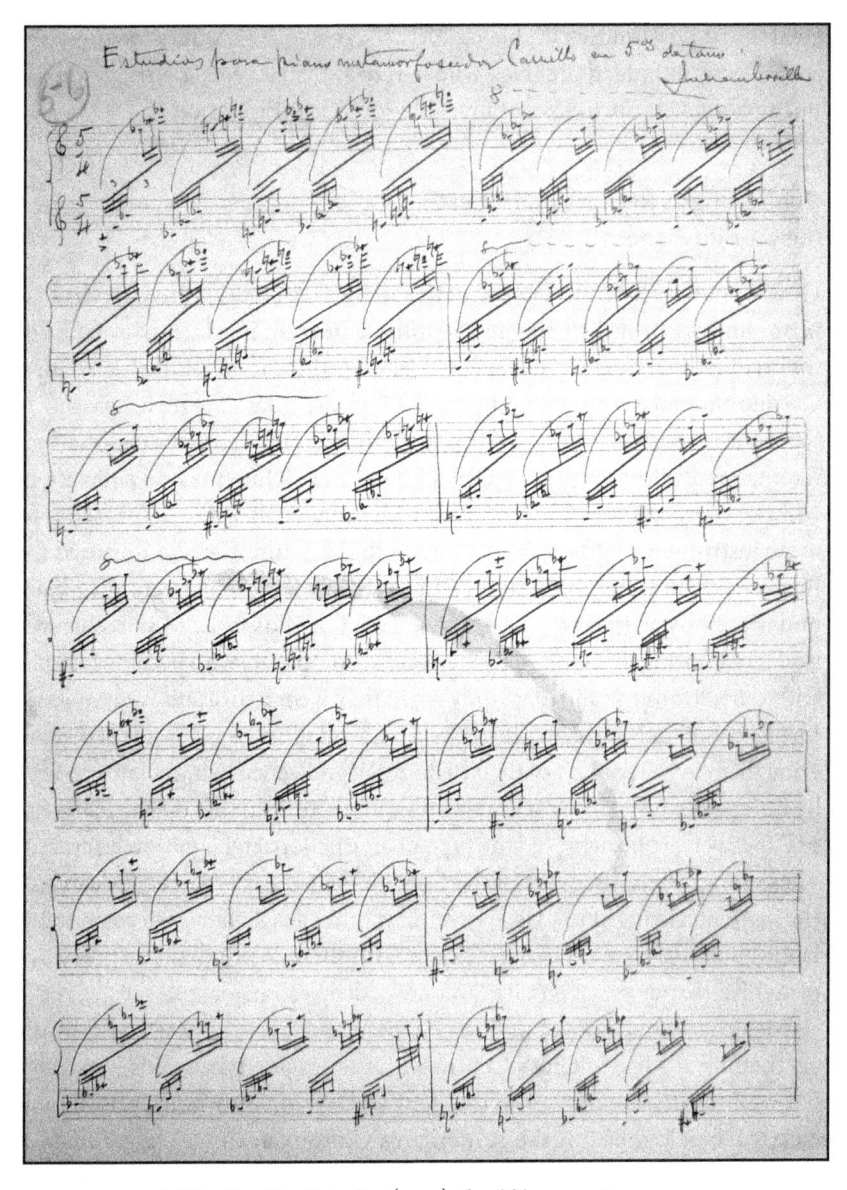

FIGURE 6.8. Julián Carrillo, *Estudios* (1959), for fifth-tone piano, mm. 1–14. Archivo Julián Carrillo.

catalogs of possibilities in this case. The scores of the works for these instruments are indeed archives whose retrieval strategies are triggered by, but also awaken, an archive of invisible somatic information.

Composing as Retrieval: Reactivating the Carrillo Pianos

How do people listen to or engage the archival content of the Carrillo Pianos? How does this listening reinvent these instruments by retrieving from them something new and unexpected (the "spiritual food for the future" that Carrillo alluded to when writing about them a few days before his passing)? Since the turn of the twenty-first century, there have been a few national and international efforts to put the Carrillo Pianos back into musical circulation. Most of these initiatives have started with individuals whose approaches to these instruments have nothing to do with the future Carrillo dreamed for the instruments and his Sonido 13. In a few cases, these projects have been undertaken by artists who were oblivious to Carrillo's legacy. I start this section by exploring the works of two Mexican composers who became familiar with Carrillo's music and ideas only when they went to live and study abroad. I believe that their sudden interest in the Carrillo Pianos is a result of discovering them in the context of their historical significance being acknowledged by European musicians. In a way, it is a similar setting to that in which, by way of Expo 58, Carrillo achieved the type of international recognition he craved. These composers' discovery of the Carrillo Pianos as a type of Mexican patrimony when living abroad is a powerful way to engage their own cosmopolitan aspirations. I see their engagement with these instruments as processes of archival retrieval that entail moments of deep estrangement.

Mexican-Dutch composer Juan Felipe Waller (b. 1971) started his musical training at Mexico City's Centro de Investigación y Estudios Musicales (CIEM). In 1994 he moved to the Netherlands to study composition and electroacoustic music at the Rotterdams Conservatorium. On graduation in 1999, he studied computer and electronic music composition for a year at the Institut de Recherche et Coordination Acoustique/Musique (IRCAM) in Paris. Waller's interest in Carrillo and the Carrillo Pianos came about when in 2011 Sander Germanus, the artistic director/president of the Huygens-Fokker Foundation's Centre for Microtonal Music in Amsterdam and a former classmate, commissioned him to write a piece for his foundation's sixteenth-tone Carrillo Piano. According to Waller, Germanus sold him on the idea of composing a piece for the foundation's Carrillo Piano and its 31-TET Fokker organ by appealing to identity concerns and stating that he was the perfect composer to write for a Mexican piano and a Dutch organ since that combination reflected his own multicultural experience.[46] The result was *Lhorong, 31°N 96°E* (2011–15), whose first movement was premiered at the BAM Zaal of the Muziekgebouw aan 't IJ in Amsterdam.

Waller, whose early training was as a pianist, states that his first encounter with the sixteenth-tone Carrillo Piano was one of estrangement. He says that he was particularly struck by the piano's extremely narrow range and the fact that it lacks the standard piano's higher and lower registers. Extending his arms as wide as possible, as if getting ready to embrace someone, Waller explains that "you have the extension of the [arms and] hands, but it is only one octave.... It would be interesting to have seven pianos with each of the octaves because sometimes you crave the higher or lower registers. Although nowadays you can replicate that with piano samplers. It is very easy to reproduce that type of tuning nowadays ... although it is not exactly the same, right?"[47] Waller started his process of composition by first becoming familiar with the instrument: He recorded a series of improvisation sessions in which he tried to explore the musical possibilities of the instrument while trying to avoid the rhetorical recurrences of Carrillo's music. The kinetic action of his improvisation studies of the Carrillo Piano led him to find clusters as almost the natural feature to explore given the disposition of pitches within the pianist's hand positions. Nevertheless, he also determined that in order to provide a certain variety of timbre that the instrument lacks, he would prepare the piano in a manner similar to John Cage's. The preparation consists of "muting the strings with the use of masking tapes" to change the sound color of the piano "to a filtered [timbre] with a quasi-electronic quality."[48]

Like in Carrillo's *Balbuceos*, the score of *Lhorong, 31°N 96°E* is a hybrid that has the Fokker organ written in actual pitches and the sixteenth-tone piano in pitches that represent the keys to play (see figure 6.9). Waller explains that he did this purely for practical reasons but that he also devised a notation system for sixteenth tones with the intention of producing a score showing the actual pitches played by the instrument. His main concern with the way the score is currently notated is that it does not show the musical relation between the Carrillo Piano and the Fokker organ. This is also the main objection he has to using Carrillo's numerical notation. In his opinion, the arbitrary relation between numbers and pitches in Carrillo's notation is aurally very disorienting. As he explains, "If it is number 86 and 87, I just do not have a reference about what is the actual pitch at stake."[49]

FIGURE 6.9. Juan Felipe Waller, *Lhorong, 31°N 96°E* (2011–15), mm. 73–90. Courtesy of Juan Felipe Waller.

The title of the piece came about by taking the materiality and intervallic affordances of the instruments Waller was writing for (the 31-TET in Fokker's organ and the 96-TET in Carrillo's sixteenth-tone piano) as the latitude and longitude components in the geographic coordinate system. He discovered that those are the GPS coordinates of Lhorong, a village in the Tibet Autonomous Region that, according to Wikipedia, "is most famous for having the most well documented cases of human levitation."[50] Waller liked that, in a way, this legend intersected with his own impressions of microtonality. He feels that microtonality "does not have that gravity that you have in tonal music. It is in a way more ethereal."[51] In *Lhorong, 31°N 96°E*, the composer pairs this unearthly, weightless character of microtonality with an almost hypnotic rhythmic and structural organization based on the extreme repetition of melodic and harmonic motives and the acoustic resonances they generate (figure 6.9). The aural result also reinforces the sense of uncanniness and otherworldliness that randomly but poetically connects microtonality and the village of Lhorong.[52] Thus, although evidently based on the materiality of the instrument and connected to some of the ideas behind the creation of these pianos, the aesthetic archive in Waller's work noticeably moves away from that of Carrillo. As such, *Lhorong, 31°N 96°E* shows a composer digging into the archive of the Carrillo

Pianos to extract from them acoustic information to be used in the production of a type of aesthetic knowledge that transcends the boundaries of Carrillo's aesthetic and theoretical project.

Mexican composer Arturo Fuentes (b. 1975) studied composition at Mexico City's CIEM before moving to Paris to continue his musical studies. He received PhD and master's degrees in composition from the Université Paris 8 and, like Waller, also attended courses at IRCAM. Although Fuentes grew up in San Luis Potosí, Mexico, the region where Julián Carrillo was from, he only learned about the Carrillo Pianos on a visit to this city when he had already finished his studies in France and was living in Austria. As he learned about the instruments, he also realized that "there were no works [for the Carrillo Pianos]. No composer was interested ... or if they were, it was only very superficially. I could not find any works for the Carrillo Pianos. So, I set myself to compose for those pianos."[53]

Agua (Water, 2017), Fuentes's first work for the Carrillo Pianos, is a massive undertaking for a keyboard ensemble that brings together all fifteen Carrillo Pianos and a standard piano tuned in semitones. The thirty-minute-long piece is still a work in progress. It was meant to be premiered in 2021 at the Centro Julián Carrillo in San Luis Potosí, but budgetary problems and the COVID-19 pandemic prevented the performance from taking place.[54] His second project with these instruments is a cycle of Preludes (2021), one for each Carrillo Piano, also in progress. Fuentes's description of his approach to these instruments likens them to archives:

> I believe that what these pianos offer is a tactile, sonic, and mechanical possibility, to enter deeply into a history and Carrillo's way of thinking. You start reflecting on him as well as the piano and the theory that is behind [it] ... although the theory is probably the simplest thing of it all. Carrillo's great achievement was to have them made. So, my first approach was a little bit theoretical. I was trying to decipher the microtonal aspect of the sixteenth tones, the thirteenth tones, etc., but I soon realized that was not the way to go about it. I realized that it had to be more intuitive, like playing any other piano. That's when I came into a creative process and was able to see the nature of each of the pianos, to understand what each piano offered.[55]

Fuentes's approach to the Carrillo Pianos, like Waller's, was guided by the physicality of playing the instruments. Unlike Waller, Fuentes has some basic piano training, although he is not a pianist (his main instrument is the classical guitar). This fact allowed him to engage the pianos and the material shortcoming they present due to their inadequate maintenance in a different way. While Waller realized that the physicality of the hands over the keyboard facilitated certain sonorities, Fuentes understood that his lack of standard piano technique could be an advantage to overcome the physical problems of a set of instruments that "nobody has played" for several decades. Fuentes describes the condition of the pianos in the following way:

> Unfortunately, the pianos are not in optimal playing condition. It is not like they are unplayable. Each of them maintains its own unique personality. Each of them sounds differently. [But] they all have different mechanical problems: Some of them have keys that get stuck, the pedal does not work for another one, some other one may have problems with the sound of some strings, etc. But since I was looking for the "historical" sound [of the Carrillo Pianos], they worked well for me. They were like sound documents . . . with a sound quality that made them work artistically. . . . I realized that, with all of those [mechanical] problems, you cannot really play them using a [traditional] piano technique . . . especially if [like me] you are not a pianist. . . . But it is better if you are not a pianist because your creativity exponentially increases the less of a pianist you are.[56]

Fuentes started exploring the Carrillo Pianos through a tactile and physical approach to the keyboard. However, his exploration was not strictly constricted by the physical disciplining of a standard piano technique. This allowed him to resort to unconventional ways of producing sounds with his hands over the keyboard that were guided by the particular sonorities he had in mind. Because of the dimensions of the instrument and its limited extension range, he understood that he had to use his whole body through strenuous movements to get the sounds he wanted. For example, in order to play a glissando, he could not simply use the physical gesture of a standard piano technique since that would cover a very limited register range. Instead, he needed to explore and find out what kind of kinetic action would allow him to produce the desired sonic result.[57]

Fuentes was trained as a composer within a tradition heavily influenced by French spectralism, and his exploration of the Carrillo Pianos takes as its point of departure an interest in sound and its intrinsic qualities. However, instead of the general exploration of "the psychoacoustic properties of sound [that would lead to] a heightened appreciation for the interdependence of its constituent parameters" that characterizes spectralist composers, Fuentes's approach is very specific.[58] He focuses on the particular sonorities and sound properties inherent to the current material status of the Carrillo Pianos and the kinetic action needed to feature those sonorities. Fuentes does not take an ideal abstract sound as a point of departure; instead, he dwells on the specific sound of the shortcomings and mechanical deficiencies of the Carrillo Pianos as they relate to performatic modes of action. The physical exploration of the instruments' materiality enables Fuentes to engage in an "immanent listening" that allows the sonic affordances of the machine to manifest and guide his artistic creativity.[59] In that sense, he takes advantage of the semichaotic status of the Carrillo Pianos, as determined by the mechanical problems generated by their lack of use and adequate maintenance, as well as the problems in their design, to generate a composition. This approach is a very good example of the need to develop new "rules of engagement and codes of conduct" as retrieval strategies in order to produce new creative knowledge out of the acoustic information stored in an instrument.

Besides Waller's and Fuentes's recent efforts in composing for the Carrillo Pianos, there have been other initiatives to write music for these forgotten instruments. In 1988 Jean-Étienne Marie asked French pianist Martine Joste to participate in a concert featuring the Carrillo Pianos at the Darmstadt Ferienkurse für Neue Musik (Darmstadt New Music Summer Course). Besides Carrillo's *Balbuceos*, the concert included some of the few works composed for the pianos since Carrillo's death up to that date. It featured performances of *Hélicoïde* (Helical, 1975) for sixteenth-tone piano and tape, by Fernand Vandenbogaerde (b. 1946); *Mémoires* (Memoirs, 1986) for sixteenth-tone piano, by Pascale Criton (b. 1954); and Marie's *Le Tombeau de Carrillo*. After this encounter with the Carrillo Pianos, Joste continued not only to play recitals on the sixteenth-tone piano but also to encourage composers to write for it. This activity led to a renaissance of sorts of the Carrillo Piano in French microtonal circles. In 1997, at the request of several French composers, Sauter reissued the sixteenth-tone Carrillo Piano, which was acquired by the Conservatoire National Supérieur de Musique (figure 6.10). This piano was reintroduced to European audiences in a concert featuring music by French composers performed by Sylvaine Billier and Martine Joste at the Internationale Pianoforum Antasten in Heilbronn, Germany. According to German composer Ernst Helmuth Flammer, "the appearance of this remarkable instrument ... created a stir throughout Europe and prompted Sauter to manufacture a small series of Carrillo [P]ianos available for purchase. There being no music for this piano in Germany, the piano firm encouraged the composers Flammer, Grimmel, and Pröve to write works for the extraordi-

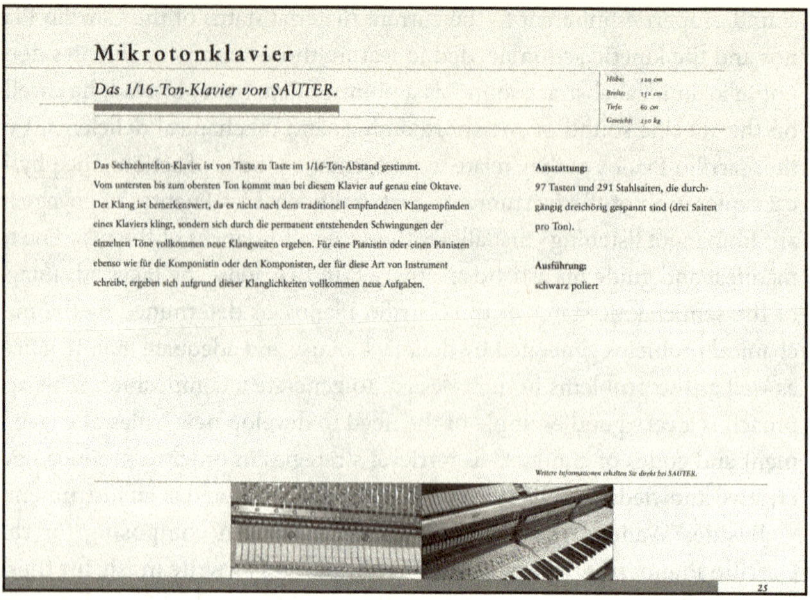

Mikrotonklavier

Das 1/16-Ton-Klavier von SAUTER.

Höhe: 129 cm
Breite: 152 cm
Tiefe: 60 cm
Gewicht: 250 kg

Das Sechzehntelton-Klavier ist von Taste zu Taste im 1/16-Ton-Abstand gestimmt. Vom untersten bis zum obersten Ton kommt man bei diesem Klavier auf genau eine Oktave. Der Klang ist bemerkenswert, da es nicht nach dem traditionell empfundenen Klangempfinden eines Klaviers klingt, sondern sich durch die permanent entstehenden Schwingungen der einzelnen Töne vollkommen neue Klangweiten ergeben. Für eine Pianistin oder einen Pianisten ebenso wie für die Komponistin oder den Komponisten, der für diese Art von Instrument schreibt, ergeben sich aufgrund dieser Klanglichkeiten vollkommen neue Aufgaben.

Ausstattung:
97 Tasten und 291 Stahlsaiten, die durchgängig dreichörig gespannt sind (drei Saiten pro Ton).

Ausführung:
schwarz poliert

Weitere Informationen erhalten Sie direkt bei SAUTER.

25

FIGURE 6.10. Page from the catalog of Carl Sauter Pianofabrik in Germany about its reissued sixteenth-tone Carrillo Piano.

nary instrument."[60] This music was recorded in *The Carrillo 16-Tone Piano* (2003), a CD cosponsored by the Sauter Pianofortemanufaktur.[61]

In 2000, with support from Canada's Fondation Émile-Nelligan, Montreal-based composer Bruce Mather (b. 1939) acquired a sixteenth-tone Carrillo Piano and loaned it to the Montreal Conservatoire "so that it would be accessible to composers and performers."[62] Besides composing for this instrument himself, Mather also commissioned works from several Canadian composers. A few of those compositions were recorded on the CD *Musiques en tiers et en seizièmes de ton* (Musics in thirds- and sixteenth-tones, 2009), one of the many collaborations between Mather and Martine Joste promoting new and old microtonal music.[63]

Both *The Carrillo 16-Tone Piano* and *Musiques en tiers et en seizièmes de ton* provide a good sense of the type of musical activity generated by the Carrillo Pianos among these Canadian, French, and German circles of microtonal musicians in the early 2000s. They feature music by composers from a wide variety of aesthetic creeds that bypass Carrillo's conceptualization of microtonality as an extension of or framed by equal temperament. Particularly interesting is that many of these works explore the contradictions within a system of microtonality based on equal temperament (see

Flammer's *A la recherche de l'autre* [In search of the other, 2002] and Marc Kilchenmann's *Vertrautheitsselig auf Eis* [Blessed familiarity on ice, 2001]), use the materiality of the piano and its unique effect on the instrument's vibrating strings to erase the boundaries between traditional musical parameters (see Flammer's *Klavierstück VII* [Piano piece VII, 2001], Alain Bancquart's *Habiter l'ambre* [Living in amber, 2001], or John Burke's *Persistance de la mémoire* [Persistence of memory, 2005]), highlight the contradictory ways in which equal-temperament microtonality relates to the type of natural acoustic phenomena privileged by spectralist composers (see Bernfried E. G. Pröve's *Echo a Gérard* [Echo to Gérard, 2001] or Michel Gonneville's *Naturel tempéré* [Natural tempered, 2003]), or problematize traditional forms of pitch organization in Western art music (see Marc Patch's *À l'affaire en seize* [About the case in sixteen, 2003], John Oliver's *Hot Tempered Clavier* [2005], or Vincent-Olivier Gagnon's *Litographies* [Lithographs, 2005]). In that sense, these composers, like Waller and Fuentes, use the acoustic information stored in the Carrillo Pianos archive to express what the instruments were not meant to express. These are all artistic interventions and aesthetic experiences of the archive that allude to the performatic reordering that Valeria Luiselli argues is "the only way to record the soundmarks, traces and echoes" that the archive renders invisible.[64]

The Carrillo Pianos as Open-Source Archives

In "El piano que nadie toca," Jaime Moreno Villarreal indulges in the materiality of a piano that nobody plays in order to walk us through the moments that define how "the piano is confined and postponed."[65] Nonetheless, the Carrillo Pianos as archives whose stored information becomes knowledge in a wide variety of retrieval exercises show us that their story, as much as it is about confinement and postponement, is also, and maybe most important, about affordances and agency, about what Cristina Rivera Garza calls the aesthetic recognition and restitution of the archival document condemned to immobility.[66] Thor Magnusson argues that as "technologies of human expression," musical instruments "equally serve as prosthetics of the human body, and as tools for understanding the world, both the objective physical world, and the subjective world of the mind."[67] In that sense, the materiality of the Carrillo Pianos stores a type of musical affordance that reflects the mind of their designer, his desires, and his aspirations to belong in a musical world whose past haunted him but whose futurity he aspired to fully occupy. In their design, the Carrillo

Pianos offer a great deal of freedom to those musicians who engage them as long as they do so within the epistemological boundaries of the physical and subjective worlds these instruments represent. Nevertheless, as it happened when I first encountered one of these instruments, the pianos may also impose a type of agency over the body of those who play them. The agency of these instruments lies in the complex interrelation among tradition, training, interface, and representation, which can trigger a series of archival memories in the performer, memories that may in turn dictate the musician's physical, sensorial, kinetic, and affective interaction with the instrument. In other words, the story of the Carrillo Pianos as archives shows us how ideology and design provide archives with an agency over those who interact with them. By guiding the pathways through which any given individual is able to navigate them, archives are enabled to successfully reproduce the ideologies that inform their design and organization. That is how an archive's agency operates, and in that sense, that agency is all about performativity. In those instances, we are dealing with archives that are manifestly closed to being reinvented from the outside and whose information, by way of recurring repetition, is meant to ineluctably lead to the same type of knowledge.

On the other hand, as Roger Moseley suggests, "human actions can still invest play [the playing of an instrument] with transformative power.... The play of the keys demonstrates how a system operates, but also probes its limits. Whether we choose to play along or to rewrite the rules of ludomusical engagement remains up to us."[68] This type of transformative agency has the potential to help us out of the closed discursive systems that instruments as archives may entail. In the case of the Carrillo Pianos, it is the archives themselves—in the limitations that their design imposes on standard musical parameters and in their decaying materiality and the chaos it provokes—that hold the key to their opening and to the transfiguration of the information they store. Whether in the creative practice that moves through and takes advantage of the obstacles implied in the shortcomings of the instrument's design and the physical conditions it imposes on the musician, or in the epistemological estrangement that ensues from the development of new rules of engagement and codes of conduct, the Carrillo Pianos, as open-source technology, demonstrate the possibility of alternative retrieval strategies. These strategies not only establish new and unforeseen mechanical, aesthetic, and affective networks but also force the Carrillo Pianos' own reconfiguration as archives. In other words, the story of the Carrillo Pianos as archives also shows us a way out of the ideological dictums that archives are often designed to reproduce. As such, the story of the Carrillo Pianos is one of hope, an optimism that arises from the mysterious potential of mishearing, of breaking the rules, of bypassing normative disciplining efforts, of metaphorically taking an ax and smashing the beloved instrument.

From "El piano que nadie toca" to "El piano que un día ha de tocarse"

Carrillo's microtonal crusade was received with a sense of revolutionary hope and incredulity, as well as mockery, by different constituencies in the Mexican music scene of the mid-1920s. The polemic generated was very public and occupied the central pages of some of Mexico City's most important newspapers. It was a discussion that put in evidence the crisis of a music scene that sought to reinvent itself in the chaos that immediately followed the armed phase of the Mexican Revolution and that also signaled the advent of an incipient proto–Aural City. Figure 6.11 shows "Los Estragos del Sonido 13" (The ravages of Sonido 13), one of the many newspaper cartoons that humorously illustrated and commented on the moment.

Here we see Julián Carrillo in the process of destroying a piano, to the dismay of three musicians who invoke tradition in the form of formality, folk music, and European elite culture. On the other hand, we also witness an ecstatic crowd cheerfully welcoming and encouraging Carrillo's obliterating efforts and the chaos they generate. The cartoon is not a direct reference to the Carrillo Pianos since it was published more than two decades before the presentation of Carrillo's first microtonal piano. Nevertheless, its comedic interpellation of Sonido 13 as a disruptive but also welcome type of estrangement that seeks to radically transform the status quo into something new and exciting could be used to describe what the composer imagined his pianos would do. And in fact, the Carrillo Pianos can be seen as that, a type of estrangement of tradition that seeks to replace it with a system that expands and transcends it. In their design and construction, the Carrillo Pianos provide a perfect example of how the composer imagined this to happen: by prescribing a type of music theory and sonic experience. However, the estranging power of the pianos was diminished by their history of invisibility and their quick transformation into museum objects and *archivos muertos*. If, in Carrillo's vision of the future, his microtonal pianos made the standard piano into a superseded instrument "que no ha de tocarse" (that would not be played), in reality, the Carrillo Pianos

FIGURE 6.11. "Los Estragos del Sonido 13" (The ravages of Sonido 13). Newspaper cartoon depicting Julián Carrillo destroying a piano.

would have to wait decades before they could be awakened from a long sleep to be further estranged beyond Carrillo's dream of futurity.

Beyond their particular historical trajectory, the story the Carrillo Pianos tell us is that archives and the information they may or may not store are important not as instruments in a normatively reproducing economy of the "being" as fixed and essential but rather as triggers of a libidinal economy of the "yet-to-come." Therefore, the Carrillo Pianos as archives are not as relevant as the embodiment and materialization of what Carrillo imagined as they are as prompts for that which he could not imagine. Their relevance as instruments and archives lies not in the desires and aspirations they kept hidden while they were not played; their importance lies in how those fantasies were creatively articulated and transformed when they were finally played. As Moreno Villarreal puts it, sometimes the piano nobody plays is "el piano que un día ha de tocarse" (the piano that one day will be played); it is an instrument and an archive full of "emotional experience and possibility," as Polina Barskova would suggest.[69] May these poetic divinations be the mandate to interrogate nihilism in hopeful, productive, and transformative ways.

In Search of the Aural City
Collective Action and the Invisible Sound Archive

> Our model of the cosmos must be as inexhaustible as the cosmos. A complexity that includes not only duration but creation, not only being but becoming.
> —Ursula K. Le Guin, *The Dispossessed* (1974)

Shevek, the main character in Ursula K. Le Guin's classic science fiction novel *The Dispossessed*, is a physicist from Anarres, the utopian communitarian anarchist moon of Urras, an Earth-like planet where the dominant materialistic system has created a huge wealth gap between a small economic elite and the vast majority of the population at the bottom of the economic pyramid. Shevek is working on a General Temporal Theory that seeks to reconcile two diametrically opposed understandings of time: sequencing (the idea that time flows as a succession of events) and simultaneity (the deterministic idea that everything happens at once; therefore, there is no past or future, only a sort of eternal present). Eventually, the utilitarian potential of Shevek's promising research gains him an award to continue and publish his work in Urras, to the dismay of his Anarresti peers, who are concerned with the possible political consequences of his trip. In trying to explain his complex unconventional ideas to an audience of skeptical Urrasti colleagues concerned with the moral implications of his work, a despairing Shevek states, "Neither pure sequency nor pure unity will explain it. We don't want purity, but complexity, the relationship of cause and effect, means and end," concluding, "Our model of the cosmos must be as inexhaustible as the cosmos. A complexity that includes not only duration but creation, not only being but becoming, not only geometry but ethics. It is not the answer we are after, but only how to ask the question."[1]

Shevek's concern with a model of the cosmos "as inexhaustible as the cosmos" itself reverberates with Jorge Luis Borges's library as a simulacrum in "La biblioteca de Babel," an "unlimited and periodic" repository that, in its hopeless unlimitedness, "will prevail: illuminated, solitary, infinite, perfectly immobile, full of precious volumes, useless, incorruptible, secret."[2] Both Shevek's inexhaustible model of the cosmos and Borges's infinite library speak of archival entities whose vastness homologizes them with the universes they attempt to encapsulate. The incommensurability of such spaces makes them impossible to navigate and renders them "perfectly immobile, . . . useless, secret." In "The Persistence of Memory," episode 11 of *Cosmos: A Personal Voyage* (1980), Carl Sagan touches on this intrinsic paradox of the modern archive. There, Sagan walks into the New York Public Library and informs us that if we were to read one book a week for our entire lives, we would probably end up reading less than 0.03 percent of this library's holdings. After showing the viewers how small the section of the library containing those books is, he states the obvious: "The trick is to know which books to read."[3] Sagan's statement highlights a fundamental concern in any monumental archival project, including those models imagined by Le Guin and Borges. One can only fathom such immensity through the establishment of tangible relations that make them both manageable and intelligible. It is only in such mobilization that the objects in the archive(s) can also be read or listened to in ways that transcend the fixity that traditional archival endeavors attempt to impose on the materials they store. Thus, as Borges suggests, the justification for such infinite, motionless universes can only exist in the instants when individuals are able to grasp their vastness through the articulation of specific constellations that, by becoming momentarily meaningful, come to metaphorically encapsulate the significance of the universe as a whole. In other words, as Le Guin's character argues, what makes such a model significant is not only that it acknowledges the incommensurability of its being but, most important, that it recognizes that it is only in the process of becoming a constellation, when external gazes make sense of its information and arrange it in meaningful ways, that it becomes a knowable reality.

This idea may be true for any archive as they are often hidden—needing to be sought out as we determine what we want to get out of them—and collective in nature. In this chapter I use this notion to explore and articulate a series of archival networks that are invisible, intangible, and elusive. One could argue that given the appropriate political circumstances, any archive could be rendered discursively invisible. Nevertheless, I take the notion of invisibility to speak about archives that are not just metaphorically or discursively invisible but in fact physically invisible due to their dispersed and virtual nature. Here, I use the label *invisible* to designate an archival network made of collections and repositories that lack a material presence and live digitally on the internet but that are also largely the product of marginal, alternative, and semi-alternative projects. Surely, the visibility or invisibility of any archive is a matter of positionality. For those who have spent part of their lives collecting and organizing these materials, these archives are anything but invisible. However, for folks outside of these archivists' networks and unaware of the archives' logics of legibility, they live invisible lives. Without articulating the right network, even those looking for their traces may never positively know of their existence. The collections and repositories I focus on in order to trace and render visible a partial and decidedly nondefinitive archive of the Aural City are generated by alternative experimental music and sound art collectives and projects who precariously use these collections in the vastness of the internet. These elusive digital archives are reminiscent of the social networks that Baird Campbell has called "archives of the self." These are digital archives that allow for the construction of identities by "interacting with other archives, creating affective bonds and mutually changing each other.... They confirm a place of resignification at the same time that they represent a 'non-place' and a place made of many places."[4] Living at the triple intersection of being discursively invisible, analogically invisible, and representing an invisible nonplace, these inherently unstable archival networks are rendered visible by their users, who invariably make them "change with every interpersonal interaction as well as with the intersections with different social networks."[5]

If locating the semiprivate personal archives of sound artists who wish to stay away from mainstream commercial networks is a difficult task, identifying, arranging, and delineating a larger, exhaustive archive of the alternative music, sound art, and experimental practices of artists and researchers equally weary of the mainstream, even if circumscribed to a single country or a particular cultural region, is even more of an almost impossibly utopian project. Evidently, as immense as it may be, the totality of this virtual archival network or constellation cannot be considered an infinite repository in the way that the universe and the library are in Le Guin's and Borges's stories. Nevertheless, their invisible character, as well as the fact that they continue to grow unstoppably, dispersedly, and exponentially as we speak, makes them very similar to the fantastic spaces imagined by these authors. In both cases, their totality exists solely as that of virtual beings because it is impossible to grasp or even imagine their inclusive wholeness. One can infer their existence by the oblique traces of them we find or conjecture about as we follow actors that may sporadically come into contact with them. But until rhizomatically articulated, identified, and arranged into archival constellations, the infinite archive and the invisible archive exist only as possibilities. In other words, their archival beings can only be grasped as they become something tangible or audible for the gazing eye or the eavesdropping ear to recognize and make sense of. As such, the invisibility of this archival network of semiprivate personal repositories is similar to what Patricia López-Gay calls Clarice Lispector's "invisible archive of the quotidian," a "hybrid archive [whose] 'dissonant' fragments obliquely signal the apparent disorder of the anarchic under which lies the creative potential of *other* possible orders."[6] In its randomness, the archival constellation I trace and articulate here also signals the potential of other possible archives that may also speak of the Aural City. As in Shevek's model, this infinite network only becomes a tangible reality in the process of looking for it and in relation to how we ask questions about its possible content. Nevertheless, it is essential that these archives exist as virtual or abstract fantasies because that is precisely what triggers our desire to look for them, articulate them, and thus render them partially and contextually visible or audible. In the end, one must remember that archives and archival networks are not ontologically visible or invisible. They may appear one way or the other depending on whether we do or do not know how to look at, look for, hear, or listen for them. Thus, these archival explorations are also excuses for us to learn how to look and listen anew.

In the past, while writing about the Nortec Collective and Sonido 13, I faced a variety of ways of experiencing, recognizing, and understanding invisible archives.[7] Researching the Nortec Collective, which did not have a central repository, was a process that started as an exercise in following actors and identifying networks in a way that resembles the method that Benjamin Piekut has suggested in his rearticulation of Bruno Latour's actor-network theory.[8] By following musicians, artists, producers, patrons, fans, critics, journalists, scholars, and venues through their on-the-ground as well as online activities, exchanges, and communications, I was able to assemble an archive. This archive only became evident and visible when I conceptualized it as such by linking together the personal repositories of these individuals and groups—encompassing both the actual physical collections of materials related to their individual and collective activities and also their personal memories and repertoires stored in their bodies and collective performances—with those of virtual entities, amalgamations, and individuals who supported the collective's endeavors or openly criticized them. In a way, the recognition of the actions, paths, and networks that ended up shaping my archive was the result of my engagement with the persons who first gave me access to their individual archives and repertoires. Those early, somehow random encounters determined what I was able to find in the field and how I was able to recognize spaces whose importance went beyond simply allowing for processes of identification to take place. These were spaces whose principal mission was to enable individuals and groups to become. In the case of my previous work about the Nortec Collective, putting together an archive was similar to the kind of archiving labor I explore in chapter 1 of this book. Therefore, my intervention was fundamental in the creation and articulation of the archival constellation I ended up writing about because it did not exist as such before I encountered its constituent nodes. However, this intervention was not my own imposition from the outside; it was dialogically prompted and guided by the individuals and moments I wanted to write about as well as those I serendipitously discovered and realized I needed to write about as my fieldwork and cyber-fieldwork unfolded. Reflecting on the particularities of this project led me to theorize about the nature of these cultural circuits in very similar ways to how Le Guin describes the goal of Shevek's model of the cosmos in *The Dispossessed*, not just as networks of identification but also as spaces for becoming. That early theorization led me to propose the notion of performance complexes as sites for the exploration of a postnational imagination as well as the study of the larger temporal, spatial, and performative implications of developing such a model.[9]

My project about Carrillo and Sonido 13 was slightly different from that of the Nortec Collective in that there was a central, very visible archive that, physically housed by the composer's family and morally guarded by some of his most outspoken followers, projected a certain degree of authority over the subject I intended to write about. However, my entry into Carrillo and Sonido 13 happened through alternative performances and artistic actions that had very little to do with the materials and information guarded at the Carrillo Archive. They were powerful performances and media frenzies that spoke about the contemporary reinventions and relevance of Carrillo's music and ideas and the collective desires triggered by them in ways that necessarily colored my later work in the Carrillo Archive and the way I read the information I found there. It took me several years to fully grasp not only that my approach to Sonido 13 as a cultural phenomenon could not end with an in-depth exploration of the Carrillo archive but that the repository itself had to be read through and against the network of musicians, audiences, advocates, critics, spaces, and mythologies that I was able to identify and articulate during my fieldwork. In sum, the archive was much larger and less hierarchical than the central repository whose control seemed to validate the authority of certain individuals and gatekeepers. This realization allowed me to theorize my archive as a performance complex in a more sophisticated way than I had done in reflecting on my work for the Nortec Collective project. Thus, I refined my understanding of a performance complex as "a transhistorical space that allows for performative processes to occur as the networks of relations between experiences, events, and actors in the past, present, and future are continuously established and re-established."[10] More than the space itself as a fixed entity in which one experiences fixed relations or an archive one enters to simply retrieve information, what is important in a performance complex is the possibility for the space and those who enter it to become something else. In sum, a performance complex could be understood as an archive in which information and subjects perform each other relationally as interactions between them ensue.

A good way to describe how these archival networks come into existence in relation to the researcher's gaze, ear, and labor is by referring to Gilles Deleuze and Félix Guattari's notion of the rhizome.[11] As rhizomatic networks, archival constellations can be understood as decentered and interconnected formations without an end and without a central point of unity. They are networks that encourage nonhierarchical and nonteleological modes of thinking and analyzing. They sprout in a wide variety of configurations as the observer looks at, identifies, and articulates them or as the listener lends an ear to them. It is in that sense that the labor of researchers is performative; it configures the rhizomatic archive as they see and hear its individual nodes and link them into meaningful social networks beyond the specificity of their individual significance. Needless to say, the researcher's own educational, ideological, or cultural biases play a role in what they are able to see and hear, as in any other archival venture. Yet understanding archives and archival networks as performance complexes allows us to activate their hidden potential by seeing and listening to *lo inaudito*, the novel social and cultural relations in the information stored within them.

In this chapter I trace the archiving/archival labor of the constituencies who put together and maintain certain sound repositories as well as the labor of the researcher who creates an archival network or constellation—a performance complex—by articulating some of these archives, collections, and repositories, which may otherwise remain unconnected. I do this in order to tackle the invisible and often intangible archive generated by the Aural City through projects and actions that take place beyond the walls of some of the institutions explored in previous chapters, or that engage these institutions only temporally in tangential or strategic ways. In focusing on collective projects developed outside the aegis of these institutions, I do not intend to imply that there is an institutional Aural City and a noninstitutional one. That reductionist dichotomy would misrepresent a more complex and fluid reality. Instead, the notion of the institutional Aural City refers to a specific type of labor done by individuals I have identified as citizens of the Aural City that is channeled through formal institutions and that could lead to the formation of material as well as virtual archives. "Institutional Aural City" is not a label to be attached to specific individuals univocally or in perpetuity. In fact, following the actors behind some of the collective aural and sound initiatives explored here makes it blatantly evident that individuals develop networks fluidly while traversing a variety of institutional and noninstitutional settings in order to strategically articulate their personal and collective artistic and intellectual agendas. Rather than embracing a misguided notion of intellectual or political purity by taking inflexible anti-institutional stances, citizens of the Aural City tend to be much savvier. They are often very tactical in their interactions and would not hesitate to engage institutional projects when doing so helps them advance their own personal and collective agendas.

Here, I follow some of these actors and map out their endeavors and networks by identifying the virtual archives and repositories they have generated and by assessing the type of labor that the structure, character, and content of these archival nodes suggest. In doing that, I am able to locate the Aural City more explicitly, making it less abstract and more palpable. To that effect, the central section of this chapter introduces a large list of actors and initiatives and explains in detail the type of labor and ideas behind the creation of their repositories. Thus, I take these individual and collective archives as evidence of the type of aural labor that defines the Aural City. By searching for the Aural City within this performative listening mode, I put together a rhizomatic archival constellation of individuals, groups, circuits, and projects that may help the reader have a more grounded understanding of what the Aural City is and does. However, as in Clarice Lispector's invisible archive, the performance complex I articulate here is just one of many possible constellations that may render visible or audible larger, analogically immaterial archives. Furthermore, while the artistic and intellectual labor of most of the people I follow here would make them citizens of the Aural City, not all of its citizens are part of this narrative. It would be practically impossible to account for everyone. Thus, my archive is the result of my very particular path through the field and of my encounters with the individuals that path enabled me to meet. In this universe, each individual is an archive, and their knowledge and repertories, as well as their individual and collective repositories, work as interfaces to engage and articulate a larger sound archive yet to be seen and heard as such. Like discovering celestial bodies by observing how their gravity bends the light around them rather than actually looking at them, putting together this specific archival constellation provides an indication that there is something else out there that ought to be pursued and rendered visible and audible.

My intention is to trace the internet footprint of a series of sound and art collectives through the location of their elusive immaterial repositories. Since musical style, genre, and format are not considerations, there is a wide stylistic diversity of sound practices illustrating this exploration, including experimental musics of both classical and popular origin as well as in-between practices, sound art and sound installation, electronic music, Noise, and improvisation. Despise their heterogeneity, what these projects have in common other than their alternative artivist profile is their ephemeral character and the digitally dispersed nature of their archival production.[12] While this dispersion and ephemerality makes it difficult to locate these archives of the Aural City, it also makes it urgent to record their existence in order to dilucidate what they have been and imagine what they may become as we engage and make sense of them as part of novel rhizomatic circuits. However, since the presentation of these archival projects and their unfolding into rhizomatic networks is the main concern of this section of the chapter, in order to keep it simple, I leave aside a conceptual and theoretical discussion of these archives until the final section of the chapter. There, I take the words and logics these practitioners and scholar-practitioners have used to refer to their own practice, to offer a theoretical assessment of the type of archival labor that characterizes these stylistically, politically, and institutionally heterogeneous projects and repositories. This assessment is done in the manner of a montaged transhistorical conversation in which I put these citizens of the Aural City into dialogue among themselves but also establish connections between them and current critical scholarship about archives as well as work about feminist anti-monumentality and disappropriation that resonates with their labor in producing and maintaining these archives.

Archives of Noise as Feminist and Queer Collective Action

In the 2010s, live coding and Noise became artistic and political vehicles for liberating feminist projects among young sound music practitioners in Mexico and Latin America. *Live coding* refers to performance practices in which people seek to interact in real time with their audiences and fellow performers through the manipulation of computer software. In a live coding performance, visuals or sounds are generated by computer programs as practitioners write them live onstage. Usually, the projection of the coders' computer screens is part of the performance, thus showing audiences how they are manipulating the code live. This type of improvisation-based technological strategy allows for the incorporation of chance and "errors" into the generation of visual or sonic materials. Live coding is a way of estranging and reengineering available software live, while "allowing us to see beyond routine practices and interpretations of code."[13] Thus, live coding is a tactic for artists to engage technology in ways that refurbish it for their own aesthetic and political goals rather than simply accepting what the technology's design allows them to do.

Noise is "a sonic creative practice featuring glitches, hiss, computer generated sounds, distortion, hum, and other unconventional sound materials [that] make this practice into an act of interference, an act that questions the conceptual boundaries between genres using electroacoustic research."[14] As an aesthetic element, noise has been invoked and used by many avant-garde and experimental music movements since the beginning of the twentieth century. Noisers are rarely interested in reforming the Western art music tradition or canon to make it more inclusive. Instead, they crave and dwell in the freedom and creativity that Noise allows them as well as its potential to collectively exert social change. As Mexican curator and researcher Andrea Ancira argues, Noise is "a cathartic form of inhabiting the error," or that which is anathema in a music tradition that has made many feel unwelcome.[15]

Irene Noy points out that for a long time, sound art as a field of practice and theorization was dominated by gendered discourses that privileged men as founders, actors, and creators. With noticeable exceptions, women were largely rendered invisible by this rhetoric, which often resorted to the infamous stereotype about the arguably "problematic relationship between female practitioners and technology."[16] Noy argues that dichotomies such as men/women, sex/gender, reason/emotion, and looking/listening have traditionally supported the social naturalization of gender difference and have worked to marginalize women from certain fields of practice (including sound art). The historical trajectory and development of sound art and Noise collectives in Mexico is no different. As in other Noise and live coding scenes, in Mexico these kinds of dichotomies were fiercely enforced among the country's early Noise collectives. This translated into a gendered division of labor in which men were in charge of live-coding sound and women in charge of live-coding visuals.[17] However, already having access to the technology motivated women not only to master its specialized sonic features and change that situation but also to actively promote and document the creation of spaces where they, as well as LGBTQ allies, could express themselves more fully.[18]

Live coding as an artistic practice was first introduced to Mexico in 2004 via the work of mU, an experimental audio/visual trio formed by Eduardo Meléndez, Ezequiel Netri, and Ernesto Romero, whose work favored software hacking techniques, circuit bending, and live coding, as well as the use of free software. However, it was not until 2010, with the creation of the Taller de Audio (Audio Workshop), a multidisciplinary initiative housed by the Centro Multimedia at Mexico City's Centro Nacional de las Artes (CENART), and the development of a series of regular workshops led by Hernani Villaseñor, Luis Navarro Del Angel, and members of mU at the Taller, between 2010 and 2014, that the live coding scene in Mexico truly took off. The activities at the Centro Multimedia also included tutorials for programming languages like SuperCollider and Pure Data, live coding environments like Fluxus, and open-source software like Ardour and Arduino, as well as the organization of conferences and concerts. La Noche de las Luciérnagas (The Night of the Fireflies), considered to be the first live coding concert open to the public in Mexico, was held on November 8, 2013, at the parking lot of the Instituto Francés de América Latina (IFAL) in Mexico City. This event led to the formation of the LiveCodeNet Ensamble, a "group of people who improvise music on their computers using code while being interconnected to a 'local network,'" whose members included Emilio Ocelótl, Libertad Figueroa, José Carlos Hasbun, Hernani Villaseñor, Eduardo H. Obieta, Luis Navarro Del Angel, and Katya Álvarez.[19] Archives of these activities can be found on Hernani Villaseñor's and the Taller de Audio's websites.[20] They document the workshops, courses, tutorials, and projects developed through CENART's Centro Multimedia, as well as the history and chronology of these initiatives. The Centro Multimedia continues to provide a space for research about open-source technologies, the development of code for experimental visual-sonic performance, and the promotion of recycling technologies. However, as the original cohort of live coders nurtured by the institution left to foster more independent, individual, and collective projects, live coding expanded aesthetically, stylistically, and politically.

A large majority of the founding members and early collaborators of the Taller who explored sound were individuals with formal academic training in composition or music performance, including Katya Álvarez, Alexandra Cárdenas, Alberto Cerro, Juan Sebastián Lach, Sergio Luque, Eduardo Meléndez, Luis Navarro Del Angel, Emilio Ocelótl, Ernesto Romero, and Hernani Villaseñor. However, most of the folks who studied and worked with them at the Taller came from nonmusical fields such as sociology, communications, graphic design, philosophy, or literary studies, including Malitzin Cortés, Libertad Figueroa, Mitzi Olvera, Jessica Rodríguez, Heliodoro Santos, Marianne Teixido, and Rodrigo Velasco. The diversification of coders beyond music academia had an impact on how their work was presented to the world. Since many of these practitioners were closer to the world of popular music, instead of simply abiding by the concert conventions and dynamics of the Western art music tradition, many of them embraced music styles and performance strategies derived from electronic dance music (EDM). Thus, the structures of their music and their performance came to be characterized by the presence of regular beats and repeated rhythmic patterns in loop for an audience eager to dance rather than to sit and observe a performance onstage. This type of EDM performance setting was already evident in La Noche de las Luciérnagas. However, it was only after they moved away from their institutional sponsorship that these events came to be categorized as *algoraves*.

The algorave is a practice that merges computer science and EDM into an experience in which people "gather together to watch and perform live coding, or the act of exploring and editing code as performance. [Thus, it brings] the mundanity of computer programming to a social context where expertise is lived and live, where immediacy is skill and labor is performed."[21] The term itself is a neologism that highlights two central issues in the practice: that algorithmic techniques are central to the type of virtual studio software used for live coding and that the events resemble EDM raves more than academic experimental music concerts. Furthermore, as scholar and practitioner Joanne Armitage notes, nowadays the term *algorave* stands for more than simply a live coding EDM party; it has become synonymous with an international movement that stages algoraves worldwide, from Europe and North America to Asia, Australia, and Latin America. The members of this movement, although highly decentralized, share a virtual forum that provides basic guidelines about the organization of algoraves.[22] Some of these guidelines include focusing on individuals and local communities rather than on sponsoring institutions, pushing for a collapse of hierarchies in the spirit of semianonymous collaborations, developing diverse lineups (in terms of ethnicity, gender, age, and class), and implementing codes of conduct that make the events into safe spaces.[23]

Composer Alexandra Cárdenas and communications major Libertad Figueroa both became live coders at the Centro Multimedia sessions. Originally from Colombia, Cárdenas combined her work at the center with her activities, classes, and collaborations with well-known figures of the Latin American contemporary art music scene. However, her disappointment with the rigid conceptualizations of music and composition in the academy led her into live coding, a practice that allowed her to recapture the joy of making music, being creative, and discovering ways to build community around issues of gender equality that were very significant for her.[24] In 2013 Cárdenas moved to Berlin, where her work refocused on live coding, and she became one of the most prominent members of the international algorave scene.[25] Figueroa converted to live coding due to the practicality of the technique when she was first exposed to it at the Centro Multimedia in 2012. She was particularly attracted by its collaborative ethos and the culture of open sharing around it. Figueroa was one of the founding members of the LiveCodeNet Ensamble, where she predesigned simple individual codes and structures and their variables in order to trigger them live during the collective performance that constitutes the ensemble as a communal instrument.[26] Despite their separate paths and the brevity of their overlapping time at the Centro Multimedia, Cárdenas and Figueroa coincided again in 2015, when they became founding members of the Orchestra for Females and Laptops (OFFAL), a nonhierarchical collective of female laptop performers from around the world who are interested in multilocation collaborative improvisation.[27] By providing the opportunity for young performers to collaborate in collective online performances with some of their role models, the OFFAL allowed them to develop a sense of sorority and solidarity that inspired and provided a space of belonging and self-worth that was particularly important for beginners entering a scene that has been very male oriented. The project was born out of collective discussions about the role of gender in defining the live coding movement as well as the type of aggression that female and LGBTQ individuals faced in that scene. In fact, regardless of its name, the collective welcomes individuals who identify as any gender other than male.

The creation of the OFFAL not only recognized a pervasive problem in the live coding scene but also provided a blueprint for how to respond to it through the development of a strategic sense of sorority. Following on the feminist experimental work of the OFFAL, Libertad Figueroa and Erika Cruz (aka Piaka Roela), who had been members of the feminist collective Sororidad, formed *Todas las Anteriores, a mixed ensemble devoted to live improvisation in the manner of dialogues between live coding with SuperCollider (Figueroa) and the sound of electric guitar and other objects hacked through circuit bending (Roela).[28] Their collaboration in this ensemble was one of the foundational nodes for the creation of Híbridas y Quimeras, a collective formed in 2017 with the intention of "promoting the work of artists who identify as women as well as transgender and nonbinary [and for] the creation of safe spaces for reciprocal learning, and the sharing of tools of creation and collaboration that reflect the Latin American reality."[29] The project brought together Figueroa and Roela with Chilean sound artist and dancer Constanza Piña (aka Corazón de Robota), Mexican singer and producer Itzel Noyz (aka Naerlot), and Guatemalan composer and cellist Mabe Fratti.[30] As Ana Mora Flores explains, the name of the collective comes from a quotation from Donna Haraway's influential "A Cyborg Manifesto" (1985) that states: "Insofar as we know ourselves in both formal discourse ... and in daily practice ..., we find ourselves to be cyborgs, hybrids, mosaics, chimeras."[31] Inspired by Haraway's ideas and the cyberfeminist movement that sprang out of them in the 1990s, the collective takes technology as a liberating tool and understands Noise as "a form of expression to awaken the senses, open perception and intuition, and generate a [type of] sensorial enjoyment that even retains aspects of rituality."[32] Resonating with the work of other artists and collectives concerned with a feminist critique of the dance floor and the experimental music scene, Híbridas y Quimeras soon not only became a space for collaboration among its founders but also attracted allies such as artists Alexandra Cárdenas, Malitzin Cortés (aka CNDSD), Emilia Bahamonde, Marianne Teixido, Leslie García (aka Microhm), Liliana Rodríguez Alvarado (aka Nabora Carrillo), and Manitas Nerviosas (aka Valis); scholars and scholar-practitioners like Lizette Alegre González, Ana Mora Flores, Ana María Romano, and Susan Campos Fonseca; other overlapping collectives and initiatives, like Chingona Sound, TopLapMX, WOMXN, and PiranhaLab; and projects such as RGGTRN and La Generación Espontánea, among many others. As such, the collective's work moved beyond Noise and refocused on the organization and production of experimental music events, the sharing of resources, and a creative critique of systems of gender and class oppression. Among the initiatives developed within Híbridas y Quimeras or in collaboration with other collective endeavors are concerts at institutional, private, alternative, and semialternative venues that include Ex Teresa Arte Actual, the UNAM's Instituto de Investigaciones Económicas, Centro Cultural Border, Mutek Mx Festival, Centro de Artes Vivas, Laboratorio 118, Multiforo Alicia, and others. The organized events included Hello World 1.0, Feminoise Latin America, Mercenarias del Fango, and several instances of the international techno-feminist encounter Cyborgrrrls.[33]

In 2019 Híbridas y Quimeras released *Compílame'sta* (Compile me this one), a digital album that includes tracks by thirty-two Latin American female experimental projects (figure 7.1), and curated *Feminoise México*, a compilation of work by twenty-nine female sound artists living in Mexico City that articulated locally the goals of the Feminoise Latinoamérica network, for whom Híbridas y Quimeras curated their first digital compilation, *Feminoise Latinoamérica Vol. 1*, also in 2019.[34] Their work producing digital compilations was a direct response to the growing presence of female artists in the scene and an attempt to record their activities in a more lasting way. This led to a larger collaborative project with Leslie García in the creation of the Oris label in 2020. Defined as "the origin and meeting point of a complex rhizome created by frequencies and sounds with much to say," Oris is an independent Mexican label that "operates from the principle of sorority" and "is made up of women, trans and non-binary identities dedicated to sound composition and production from multiple fronts."[35] The label has released two digital compilations that include works by many of the members and collaborators of Híbridas y Quimeras, as well

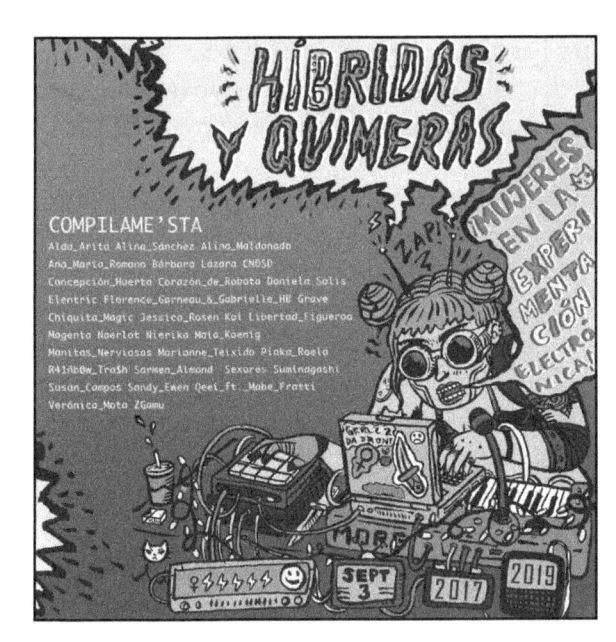

FIGURE 7.1. Flyer for Híbridas y Quimeras' digital album *Compílame'sta* (Compile me this one) (2019).

as individual digital albums by Microhm, Manitas Nerviosas, *Todas las Anteriores, and Chiquita Magic.[36]

Marianne Teixido first arrived at the Centro Multimedia in 2017, to fulfill the social service requirement for her undergraduate degree in communications. At that time, the live coding boom of the early 2010s was already fading. However, her contact with the folks still working there got Teixido interested in the production of computer-based visual and sound art using open-source software. At the time, Emilio Ocelótl, who was working on a master's degree in music technology, and Jessica Rodríguez, who was working on a master's degree in visual arts, invited Teixido to document their participation at the 2017 International Symposium on Electronic Art in Manizales, Colombia, where they were presenting their work.[37] Their visit to Colombia put them in touch with members of the Algo0ritmos Collective (the group that organized the first algorave in South America in 2013). This encounter introduced Teixido, Ocelótl, and Rodríguez to the practice of *cacharreo* (testing and trying or fiddling), the process of playing with technology regardless of whether one has fully mastered it or not, with the goal of creating new things based on trial-and-error experimentation. Discovering this practice and being introduced to live coding via a TidalCycles workshop taught by Alexandra Cárdenas, also in Manizales, made Teixido's trip to Colombia a life-changing experience

In Search of the Aural City   241

for her.[38] This motivated her to fully immerse herself in sound art and to challenge the gender stereotypes that linked women's labor in the scene to visual production. Her first sound project saw Teixido and Rodríguez join RGGTRN, a mixed music duo that Ocelótl and Luis Navarro Del Angel had formed in 2012. On their joining, the ensemble was revamped into an algorithmic dance music collective interested in exploring Latin American popular musics (such as *cumbia sonidera* and tribal) that are often frowned on as unsophisticated by algorave participants. In their experimentation RGGTRN aimed at eroding the boundaries between the music styles most valued by algoravers (classical, jazz, rock, and techno) and the type of working-class popular musics most Mexicans grow up listening to.[39]

In 2018 RGGTRN offered a series of live coding workshops in Colombia, Peru, and Ecuador. The tour was sponsored by Ibermúsicas, an international joint program supported by the culture ministries of several Latin American countries.[40] The aim of the workshops was to incentivize the local development of textual interfaces for live coding that responded to the linguistics of their native tongues.[41] Motivated by the positive reception and results of this experience, Teixido, Ocelótl, and Dorian Sotomayor formed PiranhaLab, a collective focused on building communities interested in developing and sharing open-source software for audiovisual creativity, taking into consideration Mexico's socioeconomic reality. Funded by a Mexican government PADID (Programa de Apoyo a la Docencia, Investigación y Difusión de las Artes) grant and housed by the Centro de Cultura Digital, PiranhaLab sponsored a series of workshops, presentations, and collective conversations aimed at learning strategies to develop software and environments and discussing the artistic, political, and financial possibilities these technologies may afford.[42] Unlike most of the archives of the projects and collectives discussed so far, which work mainly as repositories of creative output and registries of activities, the PiranhaLab website also features sections devoted to connections with like-minded individuals and projects, as well as reflections on technical aspects of their craft and discussions about historical, cultural, and social issues affecting the Noise, computer music, EDM, live coding, and algorave scenes.[43] In that sense, PiranhaLab's website is a space conducive to fostering collaborations as well as pedagogical and intellectual conversations rather than simply a repository.

Besides her work with RGGTRN and PiranhaLab, Teixido has collaborated with LivecoderA as well as the digital networks Redes Autónomas de Memoria (RAM; Autonomous Memory Networks) and Generx

Experimentación Latinoamérica (GEXLAT; Latin American Gender/Genre Experimentation). LivecoderA is a global community of live coding women and nonbinary individuals created by Alexandra Cárdenas. The collective seeks to use code to fight situations of gender violence and patriarchal oppression and hostility through art.[44] LivecoderA is linked to Toplap, one of the largest live coding communities worldwide, and uses Toplap's website to provide its members with a space for networking, chatting, and discussing technologies as well as activist strategies within the collective and beyond.[45] In turn, RAM, formed by Teixido and Argentinean live coder Florencia Alonso (aka Flor de Fuego), is a critical lab whose goal is to provide women and nonheteronormative individuals with knowledge about cybertechnology and AI that could be used in projects of self-liberation and defense against gender violence (figure 7.2). It also focuses on the use of algorithms and the hacking of machine-learning technology in the process of generating databases and archives that could help with deconstructing the way in which these technologies reflect and reproduce normative gender and class relations.[46]

Marianne Teixido explains that the goal of these archives and repositories "is not only to render us visible but also to recognize ourselves, build bridges for collaborations, demand a larger percentage of female presence in festivals, etc."[47] In that sense, the work of these Noise and live coding collectives and the political and aesthetic noise they introduce into Mexican culture and society is akin to the work of a number of Mexican artivist collective efforts that, in remodeling "the relationship between art and that which is common, collective, public, and assembled," as Mara Polgovsky Ezcurra argues, have "blurred distinctions between art and activism ... to create nonofficial spaces of remembrance and memorialization."[48] In a way, these virtual archives metaphorically operate as the type of *antimonumentos* (antimonuments) many of these collectives use in their protest against normalized gender and racial violence.[49] They are communitarian responses to the state's inability to address their needs; as such, they take the form and structure of the traditional celebratory archive/monument and fill it with noise and a sense of punk anarchy. Thus, as Marianne Hirsch argues, these projects are "alternative practices for mobilizing the memory of past inequities to spur progressive change."[50] Furthermore, the creation of these virtual archival projects speaks to a type of redistribution of the sensible that, according to Jacques Rancière, creates "a form of common expression or a form of expression of the community."[51] These archives work as spaces whose existence, even if ephemeral, provides a loudspeaker

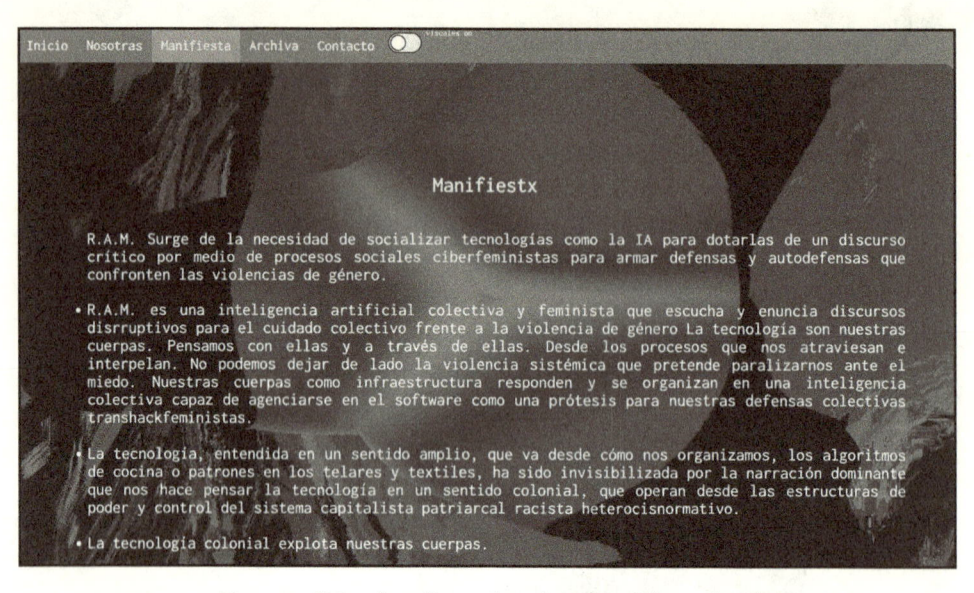

> **Manifiestx**
>
> R.A.M. Surge de la necesidad de socializar tecnologías como la IA para dotarlas de un discurso crítico por medio de procesos sociales ciberfeministas para armar defensas y autodefensas que confronten las violencias de género.
>
> - R.A.M. es una inteligencia artificial colectiva y feminista que escucha y enuncia discursos disruptivos para el cuidado colectivo frente a la violencia de género La tecnología son nuestras cuerpas. Pensamos con ellas y a través de ellas. Desde los procesos que nos atraviesan e interpelan. No podemos dejar de lado la violencia sistémica que pretende paralizarnos ante el miedo. Nuestras cuerpas como infraestructura responden y se organizan en una inteligencia colectiva capaz de agenciarse en el software como una prótesis para nuestras defensas colectivas transhackfeministas.
>
> - La tecnología, entendida en un sentido amplio, que va desde cómo nos organizamos, los algoritmos de cocina o patrones en los telares y textiles, ha sido invisibilizada por la narración dominante que nos hace pensar la tecnología en un sentido colonial, que operan desde las estructuras de poder y control del sistema capitalista patriarcal racista heterocisnormativo.
>
> - La tecnología colonial explota nuestras cuerpas.

FIGURE 7.2. The RAM "Manifiestx" as archived on the collective's website, https://ram-lab.glitch.me/manifiestx.html. Screenshot, April 20, 2023.

for voices that have been silenced, neglected, or marginalized to come together and noisily make themselves heard.

Archives and Repositories as Technologies of Visibilization

To keep up with the explosion of female collective and individual experimental music projects and to provide a way to record their often-ephemeral existence and keep their work and memory alive, a group of Latin American experimental musicians created GEXLAT. Coordinated by Alma Laprida from Argentina, Marianne Teixido and Jessica Rodríguez from Mexico, Emilia Bahamonde from Ecuador, Mariana Carvalho and Vanessa de Michelis from Brazil, and Ana María Romano from Colombia, GEXLAT is a database of Latin American women and gender-dissident experimental sound artists. This database was started in 2015 by Alma Laprida as a Google document listing the names and addresses of women working on experimental sound art projects and initiatives.[52] A few years later, the coordinating team, led by Teixido and Rodríguez, created a website with a General Public License. This is an open and collective site to which anyone can contribute and whose information can also be modified or freely shared by anyone interested.⁵³ The multimedia database includes not only the names and personal archives and repositories of female composers and sound artists from seventeen countries in Latin America but also links to sound maps and blogs as well as other, non-female-based archival projects.

Of similar Latin American scope but not restricted to artists who identify as female or nonheteronormative, MUSEXPLAT (Música Experimental Latinoamericana) is an internet platform developed by Ecuadorian experimental musician Emilia Bahamonde in the context of her master's and doctoral work in music technology at UNAM (figure 7.3). As a member of the independent experimental pop music band Sexores, Bahamonde toured throughout Latin America, which allowed her to encounter many alternative, DIY music projects largely unknown beyond their local communities with whom she and her musicians shared many artistic interests and practical concerns. In informal conversations with them, she noticed that they also shared many musical friendships and professional acquaintances. In fact, Sexores's ability to tour internationally enabled the band to become a common link between these unconnected musical projects. Bahamonde realized then that, without knowing it, she was part of a larger invisible network of experimental musicians. That realization led her to think about the need to develop a space or sound map of Latin American experimental music projects to make her Latin American colleagues aware of each other's work.

Bahamonde conceived MUSEXPLAT in 2019 as a sustainable digital community that would help facilitate the growth and interconnection of the Latin American experimental music scene.⁵⁴ Like GEXLAT, MUSEXPLAT was originally planned as a database that would render Latin American experimental projects visible and provide users with a space for South-to-South transnational networking.⁵⁵ However, the platform quickly grew into a more complex archival project that connects artists, cultural managers, and music critics alike. Besides rendering musicians and their projects, activities, and ideas visible, MUSEXPLAT also offers a space for critical reflection on issues that affect the experimental music communities in the region, from aesthetics and poetics to practical concerns regarding DIY technology and the organization of events. The website also hosts an archive of interviews with practitioners and scholars, reviews of concerts and recordings, and more.⁵⁶

The inclusion of these types of self-reflective conversations and discussions makes MUSEXPLAT an archive of ideas and a space for intellectual exchange that follows the example set by the Centro Experimental Oído

FIGURE 7.3. Home page of MUSEXPLAT's website, https://musexplat.com. Screenshot, April 20, 2023.

Salvaje (Savage Ear Experimental Center), a comparable collective archival project led by Ecuadorian researcher and cultural manager Mayra Estévez Trujillo, one of Bahamonde's intellectual mentors.[57] Estévez Trujillo's project started in 1995 as an experimental radio initiative sponsored by Radio Luna, an Ecuadorian community radio station, and morphed into an artistic collective and an archival initiative in 2001. El Oído Salvaje's initial goal was to support critical artistic creations and their dissemination via the publication of books, catalogs, and limited editions of recordings. Central to the collective's mission was the generation of a local theory of sound studies that responded to the shortcomings of the nascent Anglo-Saxon field of sound studies by focusing on the "colonial regimes of sonority" that continue to operate in the context of Latin American societies.[58] Estévez Trujillo and Oído Salvaje's pioneering work, with their highly political and activist labor, has influenced and foreshadowed the sophisticated and committed critiques of neoliberalism and patriarchy as well as the pro-ecological artivism of many of the alternative initiatives developed in Latin America in the second decade of the twenty-first century. This includes the intellectual and artistic work of the young generation of Mexican scholars, artists, activists, and cultural managers whose cultural labor I have linked through the notion of the Aural City.

In the 1990s, radio broadcasting rose as an alternative to the hegemony of television as a communications media in Mexico. The development of Indigenous community radio initiatives as well as the alternative programming of public radio stations like Radio UNAM, Radio Educación, and the Instituto Mexicano de la Radio (akin to NPR in the United States), but also that of private stations like Rock 101 and WFM, exemplifies a trend that became more prominent with the arrival of internet radio broadcasting in the 2010s.[59] Two podcasts from the late 2010s and early 2020s, *Islas Resonantes* and *Bulla*, are particularly significant nodes in the network of the Mexican Aural City. Cinthya García Leyva, a cultural manager and researcher of interdisciplinary artistic practices with a background in music and comparative literature, created *Islas Resonantes: Pensar el Mundo a Través del Sonido* (Resonant islands: Thinking about the world through sound), a radio podcast for Mexico City's Radio UNAM. The show started in 2019 with the intention of exploring sound and aurality in relation to a wide variety of ideas and topics (e.g., power, objectuality, borders, nonsense, territory, fiction, improvisation, speculation, poetics, absence, etc.). The show's episodes feature conversations between García Leyva and experts on each particular topic. These experts are intellectuals from a wide variety of fields, including literary scholars, astrophysicists, ecologists, psychologists, linguists, sound artists, composers, writers, arts curators, classical music performers, DJs, and so on. Each conversation is accompanied by a playlist that works as a sonic "illustration" of the episode's theme but also as a springboard for the discussion of related topics and ideas. At the time of this writing, the *Islas Resonantes* archive includes over 120 fifty-minute-long programs.[60] Each episode's log provides detailed information about the guest, the conversation topic, and the playlist contents (figure 7.4).

Bulla (Loud collective celebration) is an interview-centered podcast produced and presented by Laura Balboa for Radio Nopal, a collective internet radio station that broadcasts from Mexico City. Balboa, a Mexican digital designer, media artist, linguist, open-source technology advocate, and independent researcher based in Malmö, Sweden, developed *Bulla* as an outlet for her research about experimental music, sound, and gender in Mexico. Twice a month, *Bulla* broadcasts two-hour-long conversations between Balboa and guest female sound artists that focus on their aesthetic and activist work. Each podcast features a playlist curated by the interviewee that works as an excuse to explore the affective world of Balboa's guests in relation to their aesthetic labor. Since the show started in 2020, Balboa has produced more than eighty podcasts, each one to two hours

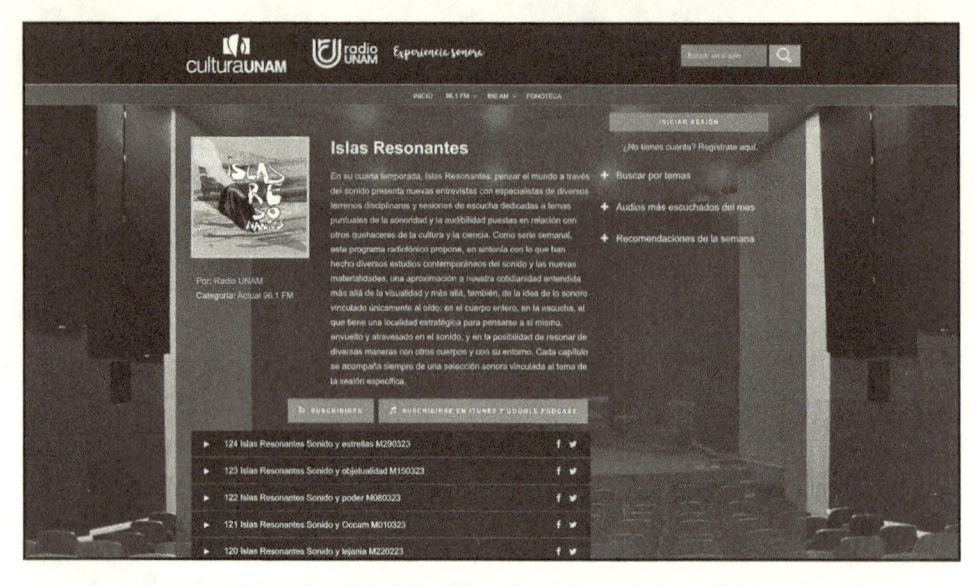

FIGURE 7.4. Main page of the *Islas Resonantes* archive at the Radio UNAM website, https://www.radiopodcast.unam.mx/podcast/verserie/319#. Screenshot, April 20, 2023.

long, featuring experimental Latin American composers and sound artists unrestricted by any particular aesthetic or stylistic tendency (figure 7.5).[61]

On November 5, 2020, Mexican musicologist Ana Mora Flores and Colombian sound artist Jenny Ramírez launched *Minga* (referring to a friendly gathering to work on a common goal) out of the Argentinean internet station Radio Caso.[62] Similar in format to *Bulla*, *Minga* was a radio podcast that started broadcasting during the COVID-19 lockdown; it lasted until mid-2022. Its goal was to promote the work of female and gender-expansive interdisciplinary sound artists, composers, and practitioners. The podcast produced forty-five sixty- to ninety-minute shows.[63] With few exceptions, each of them features a conversation with a particular artist and provides an opportunity to listen to their work. Planned as a way to "think, act, and strengthen ties through sound," this radio venture branches out of Mora Flores's academic activities documenting the work of Mexican Noisers and live coders.[64] Thus, the show was a strategy to preserve and promote the activities of a group of artists whose work is often overlooked on the basis of gender.

The archives of *Islas Resonantes*, *Bulla*, and *Minga* not only are repositories of artistic work but also store the ideas that inspire these creative

FIGURE 7.5. Main page of the *Bulla* archive at the Radio Nopal website, https://www.radionopal.com/?/programas/bulla#/bulla. Screenshot, April 20, 2023.

outputs and the types of intellectual conversations they generate. Rather than restricting their work based on national affiliations, they all emphasize the transnational connections that interdisciplinary sound artists develop in order to promote and circulate their work. By offering access to intellectual discussions and reflections about sound and aurality, as well as the ideas surrounding the work of specific artists and thinkers, the archives of these radio podcasts transcend the mission that characterizes most of the repositories explored until now (which serve as virtual storehouses of creative output), making them closer in character to the archival type of intellectual ventures engaged in the following section.

Archives and Intellectual Reflection on Sound, Listening, and Archival Labor

As a professor in UNAM's Facultad de Música and in the social anthropology graduate program of the Escuela Nacional de Antropología e Historia (ENAH), Jorge David García (aka Sísifo) has been one of the most prominent scholars and pedagogues in Mexico City's sound art and sound studies scene since the early 2010s. Trained as a composer and musicologist, García is a scholar/practitioner whose work focuses on a critique of the cultural logic of neoliberal capitalism through alternative creative uses of sound and noise. He is also interested in the impact of new technologies on music creativity, perception, and listening practices, as well as the archival possibilities they may offer. As a professor at UNAM's music and technology graduate program, García has been particularly influential in guiding undergraduate and graduate students through the design and development of many alternative multimedia research and artistic projects—for instance, Bahamonde's MUSEXPLAT was originally developed under his academic advising. Furthermore, his labor as a member and collaborator in a number of historically significant scholarly initiatives—including the Seminario de Arte y Sonido (Art and Sound Seminar) and the Red de Estudios sobre el Sonido y la Escucha (Sound and Listening Studies Network)—makes him a key figure in the unfolding of the field of sound studies in Mexico and a fundamental node in rendering visible crucial aspects of the Mexican Aural City's rhizomatic archive. García's work, linked to both institutional spaces such as UNAM and the Fonoteca Nacional (with which he has collaborated on a number of initiatives) and also independent projects, is particularly illuminating as it shows the porosity of the border between these two types of spaces, especially for those who savvily "resignify the [expected] listening, thus opening paths for new ways of approaching and occupying these spaces."[65] For him, it is important to bypass dichotomies that simplistically assert that "the dynamic of the institution is one of discipline while the dynamic of independent spaces is one of freedom."[66] Instead, García has no problem working through institutional snags and structures and using their resources when institutional projects overlap with his and his community's agenda. Sometimes these collaborations are used to jumpstart independent initiatives, but often they hijack the institution's mission to temporarily make it resonate with and respond to the labor of more progressive and critical artivist ventures. In that sense, García's strategy takes advantage of the transitory character of institutional leadership in Mexico. The fact that these leadership positions often change in response to volatile political circumstances opens the possibility for him and other members of the Aural City to network their way into having an impact on how the mission of these institutions is implemented on an everyday basis. These concerns inform García's involvement with two influential alternative archival platforms, Armstrong Liberado and Rancho Electrónico, as well as his creative labor with Epifonías.

Founded in early 2014, Armstrong Liberado is a collective focused on research about free software and culture and their creative and pedagogical uses. Its members, most of whom interact with the collective's virtual platforms using pseudonyms, come from backgrounds in both hard and social sciences as well as the humanities. García, who presents his work in the collective as Sísifo Pedroza, is one of the founding and most active members of the group. The members' contributions range from reflecting on the possibility of hacking the internet through political artistic practices, reflecting on the ontology of the musical work in the digital age, or simply sharing their artistic production—sound art, poetry, and music.[67] However, some of the most important texts García has published in the collective's blog refer to his conceptualization of a "radically subjective" transversal listening practice as the foundation for a project of "global cognitive justice" based on the establishment of resonant empathy with marginalized Others and their knowledge.[68] This alternative social justice project lies at the core of the intellectual and artistic labor of the collective members. Thus, the Armstrong Liberado website works as an archive of its members' theoretical reflections; their individual and collective artistic production; free audio files for creative use; links to free software; tutorials; information about workshops, concerts, and performances; readings about hacking and free culture; and links to download the collective's four albums.[69]

Rancho Electrónico is an independent collective project started in 2013 as a way to explore the pedagogical potential of free software and to agglutinate artists and activists involved in community radio and digital archive initiatives who were interested in the organization of technology co-ops.[70] Although most members active in this collective effort participate under pseudonyms, one of its most visible founding members is Estrella Soria, a feminist sound producer and human rights activist with a background in communications. It was Soria who circulated the open call that jumpstarted the project and delineated the group's initial horizontal rules of engagement.[71]

The members of Rancho Electrónico, self-defined as a "hackerspace," are particularly interested in developing AI strategies to counter surveillance, electronic espionage, and attempts at internet blockage that have increased with the political rise of ultraconservative constituencies in Mexico, Latin America, and beyond in the 2010s and 2020s. Rather than simply celebrating open-source and free culture, Rancho Electrónico seeks to critically assess its affordances while pushing for a progressive political agenda that works against neoliberal capitalism and in favor of co-op initiatives, local action, and community labor. The collective's website is a repository of

FIGURE 7.6. Home page of the Rancho Electrónico website, https://ranchoelectronico.org. Screenshot, April 25, 2023.

the conversations, workshops, and initiatives generated within the group as well as a space for the sharing of free software (figure 7.6).[72]

Jorge David García joined Rancho Electrónico early on and brought with him Armstrong Liberado as a sibling project of sorts. The cases of Armstrong Liberado and Rancho Electrónico show us that for many of the folks involved in the Aural City, the archive is no longer a source of authority but rather a space for the sharing of materials and experiences with the intention to challenge the neoliberal ethos. This concern with the democratization of archival labor is an issue that also permeates the creative work of García and his collaborators.

Epifonías is a creative duo formed by Jorge David García and Carlos Hernández (aka Arsan) that focuses on audiovisual technology, glitch, and noise and is also interested in expressing the concerns of nonnormative gender communities. The project maintains three internet archives that store audio recordings of some of their sound experimentations as well as videos of concerts, album presentations, and sessions of Transónica, a collaborative initiative between Epifonías and MUSEXPLAT.[73] Much of Epifonías' work deals with issues of memory and the transmission of information. García is not only interested in the archive as a repository that works as a memory of events and activities in the past but also concerned with developing strategies for the free circulation and use of that information by the changing communities who have produced that knowledge and those who share their aesthetic and political agendas. Likewise, Hernández is particularly interested in working around the structural shortcomings of the archive as an institution meant to simply document and preserve. For Hernández, crucial to the archival labor is the subversion of the traditional archiving logic in order to "listen to the archive in different contexts [and] map out new experiences onto that recording labor."[74]

Trained in conservation and restoration science under the guidance of Perla Olivia Rodríguez Reséndiz (a researcher of sound preservation at UNAM and formerly at the Fonoteca Nacional), Hernández obtained a master's degree in ethnomusicology in UNAM's Facultad de Música.[75] His research has focused on the archival strategies of experimental LGBTQ sound collectives in Latin America. At UNAM, Hernández met García, which led to a series of collaborative projects. One of them, Mnemozine, is particularly illustrative of their academic, artistic, and political goals. Planned as a memory of Transferencias Aurales, a Latin American encounter of sound and technology hosted by García through the music and technology graduate program at UNAM in November 2021, Mnemozine could be better described as a collective time capsule. Besides providing a repository for the digital preservation of conversations, workshops, sessions, and concerts presented during the event, Hernández asked participants the following question, "What do we want to share with current and future generations for the building of the Latin American music environment we dream of?"[76] This strategy "generated a sort of collection of remains that people were interested in having preserved and circulated, as well as an inventory. We used an [Internet] Archive platform to safeguard, organize, and catalog [the materials]. It was very interesting because what they sent were web pages, pictures taken with their cell phones, voice recordings they made when learning about the call—they sent their reflections."[77] Thus, the platform ended up storing materials presented at the encounter as well as materials triggered by Hernández's call for documents, and links to preexisting scores, compositions, web pages, blogs, and sound archives.[78] Thus, rather than simply documenting the event, Mnemozine works as a space that engages and encapsulates the spirit behind the intellectual discussions and artistic experimentations featured and generated by Transferencias Aurales. As such, this particular archive is the result of a process of democratization that not only reconceptualizes archival labor and archival design but also collectively refigures the very meaning of the archive. This is an archiving practice that, by opening the process of archival configuration, seeks to decenter the authority of the archive in productive and auspicious ways. In its profound critique of authority and reelaboration of authorship, this type of archival labor is in tune with what Cristina Rivera Garza has conceptualized as disappropriation, a form of community-based writing practice "that questions the legitimacy or political usefulness of a notion of authorship without community connections."[79] This comparison shows the links between writing and archiving/archival labor—both as types of performative discursive writing. It also shows this labor's potential in delineating and protecting a common good beyond the narcissistic and intimidating dynamics that characterize conventional conceptualizations of authorship and authority.

Sound Archives as Traces of the Aural City

This section focuses on the archives of three sound projects that, rather than working as repositories, point toward intellectual and political concerns that have generated a number of literary editorial and publishing initiatives. In that sense, these archives signal processes of entextualization that, as in the case of the books studied in chapter 1, encapsulate very specific acts of listening. In these cases, the listening epistemologies they invoke and the theoretical reflections they garner are also unequivocal traces of the labor of the Aural City in contemporary Mexican culture.

PoéticaSonoraMX is an artistic and academic research group founded in 2016 by Susana González Aktories, a professor of comparative literature at UNAM's Facultad de Filosofía y Letras, and a group of graduate and undergraduate students interested in exploring the intersection of vocality, poetry, and sound, including Cinthya García Leyva and Ana Cecilia Medina. The group was born as an independent project out of an academic initiative sponsored by the Laboratorio de Literaturas Extendidas y Otras Materialidades (LALOM) but was later supported by several instances at UNAM (among others, Radio UNAM, the Museo Universitario de Arte Contemporáneo and UNAM-Morelia, which houses the group's website) as well as influential museums, such as Ex Teresa Arte Actual and Laboratorio Arte Alameda, and the Fonoteca Nacional. Thus, PoéticaSonoraMX articulates the interest in sound of several intellectual and artistic constituencies in Mexico City, both independent and institutional. González Aktories herself has coordinated several interdisciplinary academic projects at the intersection of music and literature, including UNAM's Seminario de Semiología Musical (Music Semiology Seminar) and Grupo de Investigación de Literatura y Música (Literature and Music Research Group).

González Aktories's early work in semiotics was the result of an interest in studying the meaning of literature and music as independent languages that intersected in specific music-literary genres. However, her increasing attention to sound art, radio art, sound poetry, and spoken word, as well as a variety of electroacoustic practices where the boundaries between the literary and the musical are erased, led her to create PoéticaSonoraMX as a research group focused on the intermedial articulation of sound and listening through new technologies. The phrase *poéticas sonoras* (sound poetics) in the group's name invokes Norman Bryson's notion of "visual poetics" and the promise "that between poetry and the visual there is a kinship or affiliation which allows us to cross from one domain to the other with some kind of ease or sense of natural right of way."[80] González Aktories's conceptualization explores a similar epistemic bridge between poetry and sound. However, she explains that rather than simply translating Bryson's idea into a sonic context, she is interested in defining it more loosely and in relation to the mapping out of institutions and individuals whose work engages these types of interpenetrations of word, sound, and listening.[81] In that sense, González Aktories's metanarrative focus on artists, cultural brokers, and researchers, as well as the ways in which PoéticaSonoraMX seeks to generate networks with individual and collective creative, scholarly, and artivist projects, recognizes the rise of a Mexican Aural City, and acknowledges its importance in the country's contemporary intellectual and cultural life.

Due to the marginality and ephemeral character of many of the practices that the members of PoéticaSonoraMX are interested in researching, the creation of a repository and questions about its location, relevance, and authority have been central concerns for the group since its inception. Thus, the location and definition of the types of materials to be housed in the archive; the identification, curation, and compilation of specific works and documents to be stored; the digitalization of these materials; and the creation of metadata files and additional information for each of them have been priorities of the group's members. González Aktories and Aurelio Meza Valdez argue:

The value of the current survey done by the PoéticaSonora MX [sic] team lies not only in the identification of these practices but in directly working with the audio records—as well as considering, most recently, video materials—, recognizing their invaluable contribution to the country's immaterial patrimony.... Up until now, more than the discoveries we have attained due to the processes of field research, compilation, contrast, selection, and analysis of these materials, the most important challenge to our project has been the very conception of the *Repositorio Digital en Audio* [RDA; Digital Audio Repository], especially regarding questions about the rights and responsibilities involving the digital preservation and circulation of these archives.[82]

The PoéticaSonoraMX website is a repository that features reviews and criticism about activities in the Mexican sound scene; academic and curatorial texts, commentaries, and field notes written by group members; interviews and conversations with practitioners and researchers; listening sessions; a calendar of upcoming activities and a registry of past events; links to national and international sound initiatives, magazines, and research groups; and the RDA, which houses digital recordings of sound poetry, sound art, radio art, soundscapes, and experimental music (figure 7.7).[83] However, González Aktories states that due to the many legal issues and copyright restrictions that the sponsorship of a national institution like UNAM imposes on a project whose registry was originally the result of independent networking, it was decided that rather than an exhaustive archive, RDA would be "something that is representative of those [sound] practices, and within that representativity, shows links that we had not noticed before."[84]

Due to the work of its members as well as the intellectual project behind it, PoéticaSonoraMX is itself connected to several other Aural City initiatives past and present. A particularly influential space was Modos de Oír: Prácticas de Arte y Sonido en México (Modes of hearing: Art and sound practices in Mexico), an open archive designed as a catalog of sound art in Mexico. The project started as an exhibit curated by González Aktories, García Leyva, Tito Rivas, Carlos Prieto Acevedo, composer and sound artist Manuel Rocha Iturbide, art critic Bárbara Perea, cultural producer Tania Aedo, and musicologist and transmedia practitioner Rossana Lara Velázquez. It was presented at Ex Teresa Arte Actual and Laboratorio Arte Alameda from November 27, 2018, to March 31, 2019. Once the exhibit

FIGURE 7.7. Home page of the PoéticaSonoraMX website, https://poeticasonora .unam.mx. Screenshot, May 14, 2023.

closed, the curatorial team put together a digital repository that included 120 pieces of sound art, compositions, and music scores that had been part of the exhibit and announced a call for curators, researchers, artists, and individuals interested in the topic to further contribute to the repository. Thus, as an open digital archive housed by Mexico's Instituto Nacional de Bellas Artes (INBA), through Ex Teresa Arte Actual and Laboratorio Arte Alameda, Modos de Oír continues to accept virtual materials through a simple peer-review process. The repository also includes a research report produced by PoéticaSonoraMX about the archives of the Laboratorio Arte Alameda's Centro de Documentación Príamo Lozada.[85] As part of the project, political scientist and art critic Andrea Ancira and art historian Neil Mauricio Andrade joined the curatorial team to coordinate the publication of *Modos de oír: Una heterofonía sobre arte y sonido en México* (2019), a book that, in lieu of an exhibition catalog, features a critical theoretical reflection and conversation among the curators that reveals some of the logics behind the project as well as some of its shortcomings.[86]

An important harbinger of the artistic and intellectual conversations that originated Modos de Oír and the kind of work done by PoéticaSonoraMX was the Seminario de Arte y Sonido. This was a multidisciplinary academic and artistic initiative convened by Andrea Ancira, Rossana Lara Velázquez, and philosopher Inti Meza Villarino in 2014. The seminar was born as an

In Search of the Aural City 257

extension of an editorial collaboration among Ancira, Lara Velázquez, and Meza Villarino for the catalog of an exhibit entitled *Sonograma: Arte y tecnología del Hi-Fi al MP3*, curated by Esteban King and Daniel Garza Usabiaga for UNAM's Museo Universitario del Chopo (MUC) in 2013. The Seminario de Arte y Sonido was an independent initiative, also housed (but not financed) by the MUC. The convenors of the seminar organized a yearlong series of biweekly encounters meant to generate critical reflections about the type of sound-based work being done in Mexico City at the time. "¿Algo Resuena? Investigación y Prácticas en Torno al Evento Sonoro" (Something resounds? Research and practices around the sound event) was a public colloquium presented at the MUC at the peak of the project, on September 27, 2014. The event featured roundtable conversations as well as sound acts and concerts open to the general public.

The Seminario de Arte y Sonido was created precisely at the moment when the notion of sound studies as an academic field of critical inquiry was slowly being introduced to Mexican academic circles.[87] In a way, the recognition of this larger field of studies allowed for a series of intellectual endeavors, initiatives, and artistic ventures that had been brewing independently since at least the turn of the century to come together under the umbrella of a larger shared and academically sanctioned intellectual project. Lara Velázquez affirms that the seminar was meant to attract a variety of academic and nonacademic individuals as well as artists and practitioners whose work was largely circumscribed by the boundaries of their own disciplinary realms or scenes. Thus, the seminar was "interested in finding out common features among them and offering a space that would allow these actors to come together in order to promote [theoretical] reflections and exchanges in aesthetic terms."[88] The common project guiding the group's discussions was the writing of a book revolving around four fundamental topics: improvisation, electroacoustics, sound art, and N/noise. Ancira, Lara Velázquez, and Meza Villarino carefully designed an open call for contributions in order to promote the presence of a wide diversity of experiences in an attempt to democratize the editorial process. Lara Velázquez explains that they "did not want to make the typical anthology in which [the editors] decide whom to invite based on the idea of authority"; instead, they wanted "to question the very idea of authority [by] having a group of people writing a text in a collective manner."[89] Ancira states that this ideal was particularly successful in two of the book chapters, which, based on interviews and oral histories and organized as montages of multiple voices, became "mini-archives of what was happening [in those scenes at that particular moment]."[90]

The resulting book, entitled "La orquesta desafina: Prácticas experimentales alrededor del sonido en la Ciudad de México" (The Orchestra Goes Out of Tune. Experimental Practices around Sound in Mexico City), was itself an experimental exercise in collective research and writing that would eventually become an oral archive of the work generated within the seminar. The project brought together a group of fourteen collaborators that included some names that are ubiquitous in the Mexican Aural City: among them, Fernando Vigueras, Aimée Theriot Ramos, Emilio Ocelótl, Tania Islas, and Jorge David García. Although the book was thoroughly produced and exists as a PDF document, it was never published due to a change of leadership at MUC. Once the original institutional commitment to publish the book vanished, it became very difficult to find an alternate publisher precisely because of the text's experimental and nonacademic character: It was too experimental for academic publishers and too academic for alternative venues. Nevertheless, in its unpublished digital format, "La orquesta desafina" became an important hidden (but traceable) testimony of the inaudible/invisible archive. Paradoxically, for Ancira, this unpublished book became an early example of the type of *archivo vivo* (live archive) that she would work on extensively following her experience at the Seminario de Arte y Sonido, a kind of work in which the multivocal editorial project itself is an archive of a very particular and fleeting artistic, intellectual, and aesthetic moment (*Modos de oír: Una heterofonía sobre arte y sonido en México* is another example of this kind of editorial labor). Similarly, for Lara Velázquez, the methodology of "La orquesta desafina" was also a harbinger of the type of labor and oral testimony that shaped her documentation of Mexico City's experimental sound art scene in her doctoral dissertation.[91] Thus, although the website of the Seminario de Arte y Sonido is a repository that documents the conversations and reflections of the seminar participants, for those in the know or willing to dig beneath the digital surface of the archive, these documents also point toward other invisible or seemingly tangential archival ventures and collaborative efforts.[92]

The Red de Estudios sobre el Sonido y la Escucha (Network of Studies about Sound and Listening) is a research group founded in 2017 by social anthropologist Ana Lidia Domínguez Ruiz. Besides her, the group's founding members included musicologists Natalia Bieletto-Bueno, Jorge David García, and Gabriel Pareyón; electronic arts curator Elías Levín;

historian and semiologist Julián Woodside; acoustician Alejandro Ramos Amézquita; and sound artist and cultural manager Tito Rivas. Although an independent project, the group was informally hosted by the Fonoteca Nacional due to Rivas's long-standing relationship with that institution.

Domínguez Ruiz explains that the main reason behind the creation of the group was her desire to have a network of intellectuals whose approach to sound was similar to hers, that is, who understood that "sound is a means for the manifestation of the materiality of [human] relations and the complexity of social life."[93] Thus, the purpose of the group was to focus on the social study of sound and listening in order to understand how individuals construct worlds of meaning through sonic stimuli. Furthermore, central to the group's initial work was a critique of the increasing practice of the soundscape as a way of "documenting [sounds] but without knowing what for... without reflecting or problematizing the implications of trying to understand reality through sound."[94] Intellectual reflection was unambiguously a central concern for this research group.

The most active years for the group were 2017 and 2018. Its activities revolved around the coordination of a regular internal seminar and the organization of a public colloquium. The latter, Modos de Escucha: Abordajes Transdisciplinarios sobre el Estudio del Sonido (Modes of listening: Transdisciplinary approaches to the study of sound), hosted by UNAM's Music Department on October 10–13, 2018, brought together researchers, composers, sound and visual artists, and practitioners from many different scenes, state-sponsored institutions, and alternative projects, including UNAM, the Fonoteca Nacional, the ENAH, Armstrong Liberado, Rancho Electrónico, PoéticaSonoraMX, and *Islas Resonantes*, as well as sound designers from the film industry. The work done by the group in those years coalesced in an editorial project entitled "Modos de escucha" (2019), published as a special dossier in the Argentinean journal *El Oído Pensante*.[95] After the presentation of this dossier, the group went on hiatus, which was extended by the COVID-19 pandemic. However, the Fonoteca Nacional's appointment of Rivas as its general director and of Domínguez Ruiz as director of its department of sound promotion and dissemination in November 2022 and February 2023, respectively, provided an incentive for the reactivation of the group's activities.

As in the case of the Seminario de Arte y Sonido, the website of Red de Estudios sobre el Sonido y la Escucha points to other archival endeavors. As a repository, the website stores information about the group's internal seminars and the colloquium, as well as related sound-based activities, but no more than that.[96] However, the main role of the group's website, rather than serving as a storage space, is to redirect the user to the group's editorial projects, which act as the types of *archivos vivos* invoked by Andrea Ancira. Domínguez Ruiz is very explicit about the written word as an archive when she explains her decision not to record sound as part of her quotidian research and fieldwork activities: "I realized I became a better listener when I decided to stop recording. When I recorded, I was very concerned with doing it correctly, which distracted me from paying attention to reality. I was also just accumulating tapes and did not know what to do with them. . . . [In the end], my best ally in the project of documenting the sonorities of the world, the culture, the processes, etc., was the written word. So, I understand writing as a way of recording sound."[97] Domínguez Ruiz's approach is clearly informed by her skepticism regarding the uncritical practice of soundscape recording that concerned the group's early discussions. As such, one may consider the group's true archive, the sum of its intellectual labor, to be the dossier published in *El Oído Pensante*.

In a way, the archives and the archival labor of PoéticaSonoraMX, the Seminario de Arte y Sonido, and the Red de Estudios sobre el Sonido y la Escucha refer us to the idea of writing as a technology of transduction, as explored in chapter 1. In all these cases, the written word works as an inscription or blueprint of specific listening epistemologies or ways of understanding the world through acts of listening that are performative in many different respects. Thus, the act of writing is a process that allows for the archiving of a listening experience that transduces into written words the performative relation by which the ear relates to sound in order to politically and aesthetically mediate a variety of human relations and create a sense of reality. Ironically, it is at this intellectual contingency that the labor of the Aural City both resembles and splits away from the project of the Lettered City. On the one hand, there is a confirmation of the written word as a technology of representation. On the other, there is a clear understanding that such an act of transduction is a performative action thoroughly informed by the mediating presence of the listening body and its strategies (physiological, psychological, subjective, and cultural) to engage and make sense of the world out there. As such, the written word is the transduced articulation of an invisible archive that, as Borges puts it in "La biblioteca de Babel," justifies the very existence of that quasi-infinite universe. At the same time, these particular transduced articulations are both the *archivos vivos* that Ancira invokes in her work and distinct examples of the complexity of an archival cosmos that, although inexhaustible, paraphrasing Le Guin, becomes something as it is interpellated by the Aural City's performative listening.

The Archive and the Aural City: A Montaged Conversation

The screen is completely dark. A low electric drone fades in, slowly increasing its volume as a series of screeching sounds spark in and out intermittently, developing into a type of anxious electroacoustic counterpoint. The eerie sounds produce a somehow subdued, uncanny atmosphere. Suddenly, the screen glows with a caption in yellow letters: "Archivo: conjunto organizado de documentos que son clave para comprender la historia social, política, colectiva y personal..." (Archive: organized set of documents that are key for the understanding of social, political, collective, and personal history...). The lit screen gives way to a disturbing noise that resembles an airplane propeller in full action and to a polyphonic vocal chorale that weaves an understated dialogue whose basic premises keep repeating like a distant echo: "Escucho a los noventas" (I listen to the 1990s), "miedo" (fear), "lo escucho" (I hear it), "odio" (hate), "esa voz" (that voice), "la mía" (mine), "la mentira" (the lie), "¿Cómo sabías que me estabas hablando?" (How did you know you were talking to me?). At the same time, a slideshow featuring pictures of graffiti and political slogans found on the walls and streets of Buenos Aires unfolds like a slow visual procession. The vocal parade coming out of the loudspeakers continues, "Lo que se calla hace silencio" (What remains unspoken makes silence); it is accompanied by a new written caption that interrupts the succession of images: "Que el silencio no sea sólo callar, sino una decision colectiva" (Let silence be not simply being quiet but a collective decision). In synchronicity, the audiovisual piece comes to an abrupt halt. Seconds later, the brief pause is interrupted by a female voice that, over the resumed propeller-like rumble in the background and a series of high squeaks and shrieks that fill out the sonic spectrum's upper register, states firmly and clearly, "La vida es hacer silencios, es también escuchar para no estar solos" (Life is to make silences; it is also about listening in order not to be alone). The speech is followed by a repeated phrase, "Hay una herida" (There is a wound), which brings the sound procession to a second halt. This time, the silence is broken by a soft mantra, "La reproducción no existe" (Reproduction does not exist) that fades in and becomes louder with every repetition, which unfolds faster and faster to move into the final section of the piece. The sound intervention ends with an array of questions and phrases—"¿Qué fuiste antes de ser archivo? El archivo cura. El archivo está por todos lados" (What were you before you were an archive? The archive cures. The archive is everywhere)—over an increasingly loud screeching drone and a female voice imitating a wolf howl that leads into two repeated whispers: "escuchar el milagro invisible" (to listen to the invisible miracle) and "¿Dónde empieza y dónde termina un archivo?" (Where does the archive begin, and where does it end?).[98]

The preceding describes "El archivo inaudible" (The Inaudible Archive), a multidisciplinary artistic reflection about the meaning and continuous resignification of the archive prepared by Jorge David García and Carlos Hernández and presented at Argentina's Radio Caso on August 27, 2022, as part of an event entitled Concierto Dialogosónico (Sound-dialogue concert). The piece is the result of a multilayered listening process that took the digitized archive of the Argentinean Radio Comunitaria La Tribu (La Tribu Community Radio) as a point of departure. Hernández explains that the radio station had "organized and digitized its archive . . . and we wanted to resignify it or give its life a sense of continuity."[99] Their idea was to map new experiences out of an existing archive while assessing its transhistorical meaning. Rather than approaching it from a theoretical or empirical perspective, García and Hernández opted for an artistic intervention that would provide the basis for an aesthetic/critical evaluation of the archive. The process started with a series of listening sessions in which a group of people were asked to write their impressions on individual pieces of paper on hearing a selection of the Radio Comunitaria La Tribu's sound archive—an archive of digitized cassettes that focuses on issues of human rights, gender and sexual diversity, and historical social conflicts that, among other things, contains a wealth of documents pertaining to the Argentinian political, social, and economic crises of the 1990s and 2000s.[100] The goal of this project was to develop a palimpsest in which, as Hernández states, "sound is extracted in order to pass through the listening body, which inscribes it again on a different type of recording device—paper and writing in this case. These written notes were reinscribed into sound files through improvised performative actions in which people generated sound with their voices or bodies."[101] In the end, these recordings were used as the basic sound files to produce "El archivo inaudible," itself a reiteration of the archive that was reinterpreted by the Radio Comunitaria la Tribu's audience "not in terms of the information from the past that originally gave birth to this archive, but rather, on the basis of that which interpellates them in the present . . . according to what interests them today."[102]

In Search of the Aural City 263

García and Hernández's archival labor in "El archivo inaudible" and the agency it affords its users cannot be better described than by using Valeria Luiselli's words: "An archive gives you a kind of valley in which your thoughts can bounce back to you, transformed. You whisper intuitions and thoughts into the emptiness, hoping to hear something back."[103] As such an intervention, besides the archive's transhistorical reconfiguration, "El archivo inaudible" also engages another invisible aspect of the archive, its immaterial surplus; the archive's affective data concealed and rendered mute that cannot be accessed through conventional retrieval strategies but rather, as Polina Barskova suggestively affirms, can only be rendered audible through the "emotional experience and possibility" of poetry.[104] In this regard, Valzhyna Mort, who has collaborated closely with Barskova on the translation of her poetry, states that to grasp and communicate these nuances, one must be "a poet who listens."[105] Indeed, in their engagement of the invisible affective surplus of the archive, García and Hernández's work shows the subtlety and depth of the poet's ear.

The elusive and intangible archival projects in this chapter have been created and designed following specific political and intellectual understandings of what archives do, could do, and should do. The members of the collectives who have designed them share a democratizing attitude regarding the circulation of knowledge that is manifested in their reasons for archiving and their relationship with the archive. Carlos Hernández suggests that issues regarding visibility and invisibility arise from "the possibility of self-documenting, of self-commemorating, of generating situations for remembrance and the transmission of information from one body to another one, from one generation to the next."[106] This move toward self-documentation and self-archiving as a way of naming and rendering visible and audible gender-expansive bodies, activities, networks, and social relations that have been systematically marginalized in conventional narratives about music or artistic experimentalisms focuses on issues of commemoration and preservation. This attitude characterizes the activities of both feminist sound and Noise practitioners as well as researchers like Ana Mora Flores, who believe that "in the absence of sources, [our job is] to create those sources."[107] This type of archiving labor speaks to a type of mobilization of memory that "intervenes against daily acts of forgetting" and essentially builds up communality.[108] Thus, by delineating the visible and the invisible and what their communities share, the labor of these members of the Aural City directly responds to what Rancière describes as the distribution of the sensible in which aesthetics revolves around the politics of "what is seen and what can be said about it, around who has the ability to see and the talent to speak, around the properties of spaces and the possibilities of time."[109]

Ironically, Rossana Lara Velázquez argues that regardless of their highly politicized aesthetic ideas, "among these feminist collectives, metanarratives and critical perspectives have not coalesced because their needs are about recognition and self-validation, and a critical reflection [on their practice or their archival logics] may create tensions that would be detrimental to their political agenda."[110] Ann Cvetkovich has addressed the paradox in this uncritical work of inclusion and celebration, stating that "the critique of the archive and the creation of counterarchives exist in a necessary, and ideally productive, tension with each other. We need both—a passion for alternative collections and ongoing attention to absences that can't be filled."[111] Certainly, the lack of critical reflection among these practitioners and researchers does not mean that one should consider these archives to be simple repositories of impersonal raw data since, as Cait McKinney states, "the ability to access information is always about much more than the simple fulfillment of a query. In its movement and use, information makes promises that are much greater than 'finding things out.'"[112] Following on this statement, central to the archival labor of the Aural City is the ability to transmit information "from one body to another one, from one generation to the next," as Hernández proposes, as well as to engage with the poetic logics that not only allow for that information to become relevant in the present but also unleash the invisible and silent affective surplus of the archive. The process behind García and Hernández's "El archivo inaudible" puts in evidence that these types of transhistorical exercises of critical retrieval show that an archive's logic of reproduction can be overcome through nontraditional recovery strategies that afford archive users with critical agency. It is in this sense that the mantra repeated in "El archivo inaudible" acquires meaning; if "la reproducción no existe," it is precisely because the sound archive users' performative listening has the potential to poetically and politically reconfigure the archive and challenge the power relations its design and structure tend to reproduce intentionally or unintentionally. The epistemic consequences of this type of aural labor speak to the potential of understanding the archive as an open-source entity; it allows for the archive's creative reconfiguration in the process of retrieval according to the aesthetic and political context in which that retrieval happens.

While the accumulation of collective experiences and their preservation is still central to the archival projects of many of these groups and collective initiatives, for many others, gathering information and sounds is no longer the main goal of their archiving/archival labor. The latter attitude reverberates with Ana María Ochoa Gautier's description of how technological changes have allowed for a proliferation of dispersed private sound repositories beyond the walls of institutional sound archives, which challenges the traditional understanding of the sound archive as the "institutional preservation of sound." For Ochoa Gautier, this situation calls for a retheorization of the sound archive's mission along the lines of Rancière's distribution of the sensible; thus, she states that we must "rethink the sound archive not only through its instrumentalization as a resource for storage and cultural recognition . . . , but rather on the basis of the relations between people and sounds."[113] Therefore, rather than collecting, what many citizens of the Mexican Aural City are interested in is (re)collecting; their goal is to recall the information in the archive and do something new with it. In this context, as Rossana Lara Velázquez suggests, "you no longer accumulate to possess but rather to remix and generate new things. In this DIY culture, everybody can be a producer, and the border between producer, consumer, and archivist becomes irrelevant."[114] This is especially true in the case of the invisible alternative archives that proliferate on the internet and that are growing exponentially at this exact moment. It is impossible to assert any type of authority over such a practically infinite archive. In a context in which anyone has the potential to rhizomatically articulate an archive, the traditional authority of the archivist as a gatekeeper is called into question. Indeed, these types of open-source archives are spaces in which a variety of listening experiences and histories of the ear can coexist, synchronically and diachronically. Ultimately, this situation allows for the defiance of teleological, hierarchical, and unidirectional forms of authority. Lara Velázquez argues that it is in "the expansion of listening [experiences] that these spaces erode the borders of the groups of belonging that use them."[115] Following on that assertion, one can say that it is in the act of listening together, reflecting, making sense of the aural information, and sometimes transducing it to other storage formats (such as the written word, as proposed by Andrea Ancira and Ana Lidia Domínguez Ruiz) that these archival ventures provide a space for a more democratic circulation of information and for a more inclusive articulation of the notion of authority.

I have not attempted a thorough exploration of the individual projects, repositories, and archives featured in the constellation I articulate in this chapter. This is not the proper space for such a comprehensive study nor the goal of this virtual road map. Instead, my intention is to simply follow one of many possible rhizomatic connections between these collectives through the identification of their virtual archives while providing a passing glimpse into the Aural City via the traces their labor has left behind. Figures 7.8 and 7.9 show radial and rhizomatic visualizations of this virtual archival network.[116]

As much as the rhizomes explored in this chapter allow us to engage certain aspects of the Mexican Aural City's archival network, I believe that the true potential of articulating this archival constellation lies not in what it shows but rather in the possibilities for further research it renders audible and leaves open. In other words, as a performance complex, this archival constellation's true potential lies in the open-source character of the road map it provides and the agency it may afford for future individuals to make something out of it or use it to infer *lo inaudito*. That is, the importance of this particular performance complex lies in its potential to insinuate the presence of other still invisible or inaudible archival projects and explore new rhizomatic routes and connections.

The archival constellation I trace here focuses on the recent past. It privileges the work of collectives and projects active in the 2010s. At the moment of writing, this is a very recent iteration of the Mexican Aural City. Furthermore, this road map is necessarily partial; it is the serendipitous result of following certain actors and not others, of deciding to articulate certain nodal connections and not others. In other words, to borrow the popular Mexican saying, *Ni son todos los que están, ni están todos los que son* (Not all of those who are there belong, and not all of those who belong are there). Indeed, each of the individuals, projects, and repositories discussed here could have led to other destinations or could be traced back in time to uncover further connections into the past that would provide historical depth—both locally and transnationally—to the archival complex exposed here and to the Aural City behind it. For example, I could have chosen to follow the links between Alexandra Cárdenas and Manrico Montero (aka DJ Linga); the late 1990s collaboration of Leslie García with the Nortec Collective; Rossana Lara Velázquez's scholarship about the work of Gilberto Esparza, Mario de Vega, and Leslie García, or her connections with ecocritical projects such as Tania Rubio's Acoustic Ecology Lab; professional collaborations like those between Perla Olivia Rodríguez Reséndiz and Margarita Valdovinos, between Carlos Prieto Acevedo and Inti Meza Villarino, or between Mabe Fratti and Gibrana Cervantes (whose 2023 album *¿Cómo pasamos la eternidad?* [How did we spend

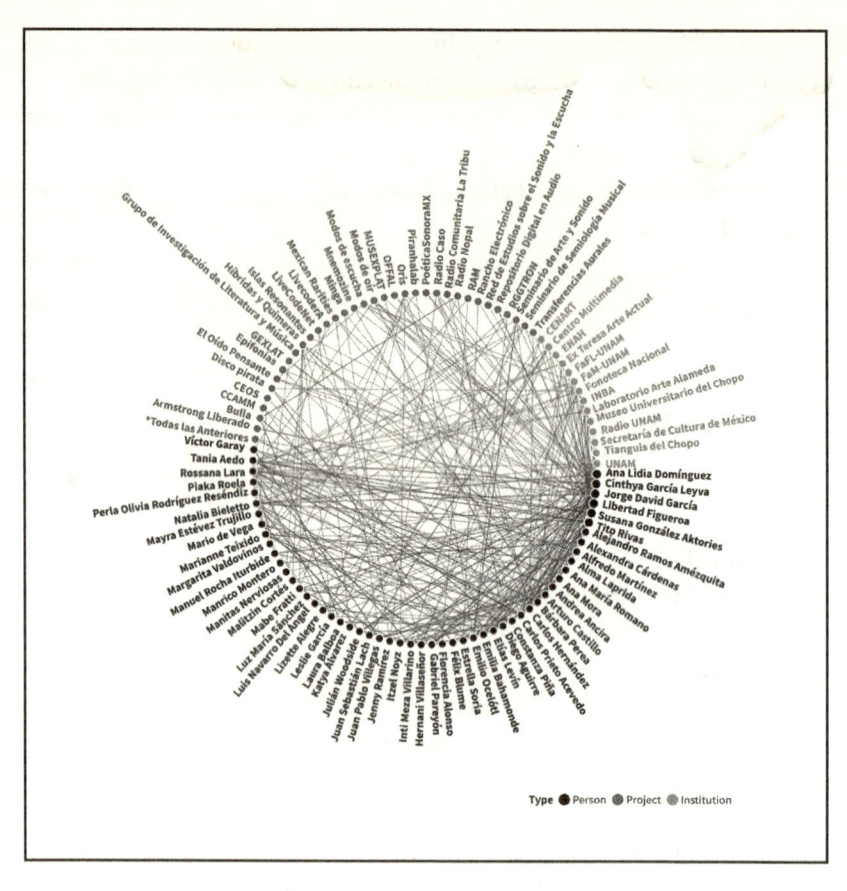

FIGURES 7.8 AND 7.9. Radial and rhizomatic visualizations of a partial network of the Mexican Aural City as rendered visible in the archive(s) articulated in this chapter.

eternity?] was released by Mexican Rarities); or overlapping radio projects like *Islas Resonantes* and *Gabinete de Curiosidades* (Cabinet of curiosities) by Frida Zaldívar Jiménez (aka Frida Revontulet). Following any of these networks at certain nodal moments would have led me to the configuration of very different archival constellations. For example, an emphasis on certain transnational connections might have led me to engage archival projects such as Microcircuitos, a Soundcloud repository that hosts the work of many Latin American experimental composers—including many of the Mexican alternative artists who are members of the collectives explored in this chapter.[117] This would have opened the possibility of taking any particular node as a point of entry into an exploration of a different articulation of a more expanded Latin American Aural City. I chose not to do that in order to keep the scope of this exploration manageable. Nevertheless, these unexplored networks are some of the many kernels left behind for the possible future recycling of the archival constellation I offer here. After all, as Le Guin proposes in *The Dispossessed*, these archival explorations are not about finding any specific answers but about asking questions. Therefore, I am content if, following Borges's dictum in "La biblioteca de Babel," this particular articulation of the universe vindicates its magnitude by providing a sense of hope in the face of the universe's overwhelmingly blinding and deafening unlimitedness.

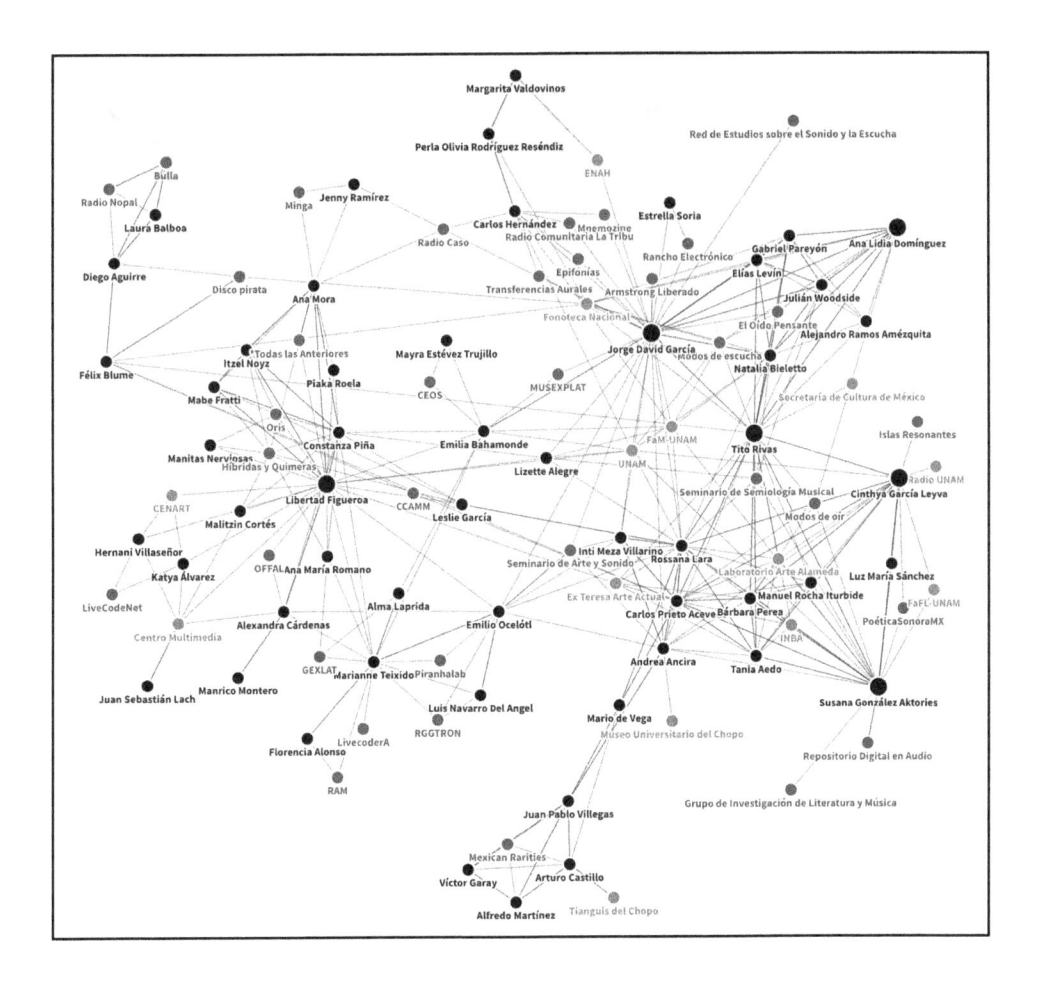

Epilogue: The Relevance of Archives in Times of Post-Truth

An Essay Against Nihilism in the Neoliberal Age

. . . ser el eco de un sonido imaginario . . .

(. . . to be the echo of an imaginary sound . . .)

—Myriam Moscona, *León de Lidia* (2022)

This essay serves as a concluding discussion aiming to unify and connect the theoretical and conceptual arguments presented across the chapters in this book. While I do not aim to delve into new case studies or research findings, I do take ongoing concerns and debates about post-truth as a contextual backdrop for this reflection. My musings on this issue are included here to guide a discussion about the importance of contemplating the relevance of archives at a critical moment when their traditional role as guardians of truth is being challenged from various ideological standpoints. The notion of post-truth and its ubiquity in contemporary media and social networks have been central to this ongoing culture war. Hence, a discussion of the issues raised in this book against this particular cultural background is beneficial beyond the obvious topicality of the subject: It not only problematizes our understanding of the archive but also sheds light on a number of timely political strategies that the interrogation of the archive in this context renders visible and audible, and it appropriately contextualizes and clarifies the intellectual questioning of the archive(s) I have outlined throughout this book. My intention is to underscore the importance of understanding the various ways in which challenges to the credibility of archives emerge in the era of post-truth, their implications for our perception of truth and falsehood, and the way these actions and ideas support the type of pervasive and often politically paralyzing cynicism that characterizes the contemporary neoliberal ethos.

Post-Truth, Anti-Intellectualism, and the Archive

The term *post-truth* has been around for several decades—at least since the early 1990s, when it was first used in journalistic and academic writing regarding the aftermath of the Iran-Contra scandal and the Persian Gulf War and its media coverage. However, the expression became widespread in the public sphere in the mid-2010s, especially in 2016, when Donald Trump's presidential campaign in the United States and the apologists of Brexit in the United Kingdom recurred to the blatant manipulation of facts as well as outright lying as a systematic strategy to achieve their political goals.

The Oxford Dictionaries define *post-truth* as an adjective that relates "to circumstances in which people respond more to feelings and beliefs than to facts."[1] Lee McIntyre argues that in the specific political climate of the Trump campaign, Brexit, and the totalitarian or proto-totalitarian regimes that have proliferated in the past decade around the world, "the post-truth era is a challenge not just to the idea of knowing reality but to the existence of reality itself." The predicament in this scenario, McIntyre continues, is "the overarching idea that—depending on what one wants to be true—some facts matter more than others [and that people] only want to accept those facts that justify their ideology."[2] Even discursive lies and alternative realities need to be based on some sort of symbolic agreement, at least between those who spout them and those who willingly consume them. Sometimes those narratives are built around conveniently selected and accepted facts or pieces of information; other times, the need to provide a sense of firmness to these deceitful narratives makes it necessary to fabricate evidence to support them. Thus, the notion of alternative facts entered the public arena when Kellyanne Conway, Donald Trump's senior counsel, tried to defend Sean Spicer's lies regarding the size of the crowd at the president's inauguration.[3] In effect, the rootedness in reality of these so-called facts is not important in order for them to be accepted as truthful. These data often refer to semidigested theories, schemes, or beliefs that have previously been circulating widely among communities who embrace them, as they seem to validate their desires, fears, and phobias, but that have no root in reality whatsoever. On other occasions, this information is simply unabashedly invented to provide a foundation and advance a particular agenda. These dynamics show us that what is truly important in the dissemination and acceptance of these so-called facts and the narratives they generate or support is the way they affectively and emotionally connect with their audiences' preconceived fantasies.[4] This articulation of emotion and affect is central to the success of these discursive lies since it allows both their producers and their consumers to bypass any type of evidential standards in reasoning, appealing instead directly to feelings and beliefs. The dogmatism and cynicism behind the generation of these post-factual narratives and post-truths have also led to an anti-intellectualist backlash and a denial of science that has taken aim at academia and other institutions that have traditionally validated the production and circulation of knowledge. Although the anti-intellectual climate and the disdainful portrayal of academia as an elitist, liberal space predates the rise of post-truth and alternative facts, their advent did exacerbate those trends.

The situation is no different in Latin America, where one can witness anti-intellectualism, science denial, and the strategic use of misinformation and pseudofacts across the political spectrum: from right-wing demagogues like the Argentinean Javier Milei and Salvadoran Nayib Bukele to left-wing dictators like the Venezuelan Nicolás Maduro and Nicaraguan Daniel Ortega. In Brazil, right-wing politician Jair Bolsonaro ascended to the country's presidency in 2018 via a campaign that denied the crimes against humanity of the Brazilian military dictatorship (1964–85) in order to portray the military as the only institution capable of saving the country from what he portrayed as threats to the moral and political order from the left-wing Partido dos Trabalhadores (PT; Workers' Party).[5] That Bolsonaro's reelection bid went awry among "insinuations of cannibalism, pedophilia, and devil worship" propagated by the campaign team of the left-wing candidate, Luiz Inácio Lula da Silva, in 2022 is but one example of the widespread use of post-truth tactics in Latin America across ideological divides.[6] In Mexico the left-wing President Andrés Manuel López Obrador's infamous phrase, "Yo tengo otros datos" (I have other data), which he cynically uttered when facing facts that contradicted his claims about the decline of violence in the country during his term in office, became the repeated waggish mantra used to characterize the misinformation and false claims propagated in *las mañaneras*, his daily morning press conferences.[7] The similarity between López Obrador's "other data" and Kellyanne Conway's "alternative facts" as ways of justifying false claims is evident. Both responses insinuate that an alternative reality is being obscured by the biased ideological rhetoric of their critics. These kinds of strategies aim at the fabrication of representations of reality based on propaganda and misinformation rather than accurate information and truthful data.

Both Bolsonaro and López Obrador have also been involved in anti-intellectual controversies. In Brazil, Bolsonaro's infamous response, on the election campaign trail, when he was asked about the fire that destroyed Brazil's Museu Nacional in 2018 ("It's already burned down. What do you want me to do about it?"), epitomizes the kind of disdain toward intellectual endeavors that came to characterize his administration.[8] In Mexico, this anti-intellectual agenda is perfectly illustrated by López Obrador's frontal attack on the Centro de Investigación y Docencia Económicas (Center for Economic Research and Teaching), one of Mexico's preeminent higher education centers—which the president accused of being a nest of neoliberal ideology—and the continuous budget cuts to cultural institutions, including a 75 percent cut to the Fonoteca Nacional's annual budget, which forced the institution to lay off over ninety workers and jeopardized its ability to fulfill its institutional mission.[9]

It is telling that both Bolsonaro's and López Obrador's contempt for facts and evidence became apparent in relation to the future of two national archives, one in each of their respective countries. Since the archive has traditionally been understood as the institution charged with storing verifiable information, the attitude of these politicians symbolizes better than anything else that in a world where dogma becomes truth, articles of faith become facts, and conspiracy theories become argumentations, archives, the documents they store, and the information they authorize become politically irrelevant. One is led to wonder whether the decimation and weakening of specific archives in this toxic ideological context metaphorically marks the end of the archive as an institution.

The End of the Archive(s)?

If the survival of archives depends on the recognition of the value of the documents they store and the information they validate, it would seem fair to say, with Daniel German, that "once that value is gone, so goes the archive."[10] The rise of alternative facts and the disregard for the value of archives that characterizes the post-truth moment—along with challenges to the archive's traditional narratives and authority prompted by postmodernist, poststructuralist, and decolonial scholars alike—would seem to signal the irrelevance of the archive. However, such an assumption implies that at their core, the anti-intellectual and post-truth assaults on the archive(s) as institution and the postmodern, poststructural, and decolonial challenges to its archival authority (such as those explored throughout this book) are somehow analogous. Lee McIntyre proposes that post-truth pundits have taken postmodern and poststructuralist theories to validate their claims that truth does not exist and that we live in a world of interpretations in which all narratives are equally valid.[11] McIntyre's statement may be true. However, this does not mean that post-truth, poststructuralism, postmodernist theory, or decoloniality claim the same thing. In fact, establishing such a parallel would be a colossal misrepresentation of what these antagonistic projects attempt to do, how they go about it, and what role the archive plays in their agendas.

As Stanley Fish argues, the postmodernist and poststructuralist "insistence on the primacy of narratives and interpretations does not involve a deriding of facts but an alternative account of their emergence. [To question] the category of *objective* fact ... is not to question the category of fact, it is to question the picture of fact as something sitting there all by itself and waiting to be discerned by clear-eyed observers."[12] This is precisely the type of critical approach that characterizes the decolonial and deconstructivist work of the Aural City explored in this book. The Aural City's goal is to estrange the archive(s) and its/their contents in order for those documents to say something different from the naturalized narratives they have conventionally validated. For them, challenging archival authority is about questioning a particular type of archival labor and its interpretation of the archival documents, not a denial of the documents stored in the archive. They seek to read the documents that support the archive's authorized information from new and illuminating angles in order to question the idea that archival documents and the facts they represent are knowledge per se and that any particular archival narrative is a natural outcome of the archive(s)' documents. Their goal is to look at archival information critically to figure out its performative power and find clues as to what has been left out (of the archive and/or its narratives) and why.

The Aural City's claim is not that facts or truth do not exist or that the archive as a concept is obsolete. Instead, they recognize that archival information can never be complete, and thus attaining an absolute truth is impossible. However, although they understand that archival information only tells part of the story, they are also aware that by taking advantage of the fact-checking that the archive affords and by furthering the scientific tradition of questioning and opening conventional knowledge up for debate, "a version of the truth—usually incomplete and often partial—will emerge," as Daniel German argues.[13] In that sense, the labor of the Aural City is scientific in the most rigorous understanding of science, as a constant open effort to get closer to the truth by continuously challenging our partial descriptions of it. That is just how scientific revolutions and paradigm shifts happen. That is how quantum mechanics relates to relativity and how the latter engaged Newtonian physics—neither denied their predecessors but instead took them as points of entry into a more detailed understanding of Nature, their object of study. And that is precisely how, as seen in chapters 3 and 4, Carlos Prieto Acevedo's labor relates to the nationalist canon of Mexican music or how Margarita Valdovinos's work relates to Konrad T. Preuss's archive. As such, the labor of the Aural City follows on Fish's assertion that "there is surely something out there [the *Ding an sich* (the thing as it really is)] not caused by our descriptions of it, but our descriptions of it provide us all the access to it that we can have."[14] One can only try to continually estrange those descriptions in an attempt to get closer and closer to the unattainable truth. Thus, knowledge is the object of desire of science as well as the humanities. This libidinal aspiration to knowledge endures in stark contrast to the type of dogma and nihilism that informs post-truth rhetoric.

The Aural City's work exhibits a fascination—even an obsession—with the archive and the archive(s) that negates the apocalyptic and dystopic post-truth elegies about their disappearance. As seen in chapter 7, the fact that efforts to document neglected artistic projects or the presence of marginalized individuals are central to the Aural City's labor speaks of the centrality of the archive(s) as part of its decolonial project. Rather than advocating for the archive's end, the Aural City argues for the introduction of productive schizzes into it, schizzes that would help one reassess what it is that one can hear and how one hears it in the archive. Its project is not about delegitimizing the documents in the archive and archive(s) but about questioning them in such a way that they tell us about how and why the archive and the archive(s) were put together and what was left out in the process. The centrality of the archive and the archive(s) is not clear only in the Aural City's development of alternative archival strategies, storage spaces, and retrieval methods; it is also evident in their savvy approach to the documents kept in traditional archives, one that implies that the seed for this defamiliarization of the archive(s) lies within the archive(s) themselves, in their documents and the information they hide or the hidden corners they point toward. The Aural City's tactic is like that of the astronomer who pays attention to the gravitational pull that announces the black hole they are interested in but would ultimately never be able to see. Therefore, the very seed of the archival decolonial project is the archive itself, not its negation; its origin is in the archive's documents and their information. The decolonial estrangement of the sound archive cannot be about the denial of its constitutive elements because it is a project firmly grounded on their productive interrogation.

Listening as an Act of Transfer

The idea of the archive continues to have a very important place in the Aural City's critical pursuit of less partial and incomplete versions of the truth—or, in other words, more inclusive versions of its narrative representation. However, one of the major epistemic differences between the Aural City and the Lettered City lies precisely within their understanding of the relationship between the labor required to form an archive and the notion of gatekeeping. While the Lettered City understood its role as that of a gatekeeper regarding the conformation of sanctioned forms of knowledge and their circulation, the Aural City has a more ambiguous relation with those power dynamics. While instances of the Aural City acting within specific institutions may still reproduce ideas about what is worth archiving and how to retrieve it in order to reproduce sanctioned narratives that align with certain institutional missions, as seen in the case of the Fonoteca Nacional in chapter 2, most of the Aural City projects developed beyond the walls of these institutions argue instead for a decoupling of archival labor and the notion of authority. This is evident in the archives explored in chapter 5 as well as most of the projects that conform the invisible archive surveyed in chapter 7. The Aural City goes about this not simply by decentering the locus of meaning from object/document to consumption/listening but also by arguing that such a process should always be the result of a collective negotiation that takes into account the actors' specific social and historical circumstances. Agency and its promise are always historically, socially, and culturally contextual. Thus, Jorge David García calls for an "epistemology of listening" to generate a type of knowledge that resonates with collective desires and aspirations, while Ana Lidia Domínguez Ruiz argues for a type of listening in tune with the Other, as a requirement for understanding and dialogue to take place.[15]

Rossana Lara Velázquez explains very clearly that within this context, the power dynamics of archival authority change dramatically, and thus "you no longer accumulate to possess but rather to remix and generate new things. . . . Authority is no longer a meaningful category for this generation [of scholar-activists]."[16] This refers precisely to the type of open-source archive that I have argued for in relation to the Carrillo Pianos in chapter 6, to the postnational dynamics at stake in the inception and circulation of *noriginales* in the cases of *Disco pirata* and Mexican Rarities in chapter 5, as well as the resocialization of sounds in Prieto Acevedo's *Critical Constellations of the Audio-Machine in Mexico* curatorial project and Valdovinos's engagement with the Preuss Collection, explored in chapters 3 and 4. These concerns with listening as an epistemic practice also relate to the dialogic exploration of *Instrumental precortesiano* and *Hacia una nueva música/Toward a New Music* in search of traces of performative listening in chapter 1.

The term *performative listening* denotes an interpretative tool that seeks to explore what happens when listening happens in relation to the sound archive. However, although the coinage of the term itself may be a theoretical novelty, performative listening as an action that makes epistemic sense of and interpellates a reality is nothing new. Ana María Ochoa Gautier's work about listening practices and their entextualization in historical written sources in and about Colombia deals precisely with "the role of listening in the constitution of acoustic ontologies and knowledges" that performative listening alludes to.[17] Likewise, Alexandra Vazquez's critique of ethnography as "the discovery of undiscovered material for the purpose of taxonomy [that] sets up the detail as an observable part of a natural order" and David Garcia's exploration of the dislocation between the act of listening and the racialized bodies that produce the sounds being listened to as a tool in the constitution of the idea of Western modernity and the Western self are essentially explorations of the performative power of listening.[18] My intention in directly appealing to performativity when proposing the notion of performative listening is to unabashedly articulate the tradition of performance studies not only to ask about what someone's listening does and has done in the past but also to emphasize the processes of circulation of knowledge that this scholarly tradition focuses on. The notion of performative listening highlights "the transmission of social knowledge, memory, and identity pre- and post-writing" that Diana Taylor calls "acts of transfer."[19] Thus, performative listening as an interpretative concept pays attention to listening as an embodied action, and as such, it attends to the multiple processes of entextualization, inscription, reinscription, and transduction by which the archive becomes repertoire, and the repertoire may eventually become part of the archive. An example of listening as an act of transfer can be found in "El archivo inaudible," Jorge David García and Carlos Hernández's multidisciplinary intervention, which documents the retrieval and affective mediation, transduction, and transformation of archival information into repertoire, and how, in turn, that repertoire is retransduced into a renewed and resignified archive, as analyzed in chapter 7. It can also be witnessed in how Juan Felipe Waller and Arturo Fuentes reactivate the Carrillo Pianos beyond the coordinates of their original design, as explored in chapter 6. These types of transformative embodied performances lie at the core of the transfers and transformations of knowledge at stake in each of the archival experiences explored in this book.

Therefore, in relation to the sound archive, the notion of performative listening addresses the sharing of experiences through embodied invocations of the stored materialities that make the listening experience possible. One could say, as Cristina Rivera Garza poetically argues in the case of what she calls "escrituras colindantes" (bordering writings), that performative listening is an interpretative concept that focuses on "the layers of experience that others have left behind or bring with themselves in a world in which we intervene together... in such a way that it can activate in the present a past that is always about to happen."[20] As such, performative listening is a type of dialectical sounding, a way of transhistorically sounding out or making sense of sonic and aural moments as dialectical constellations that render audible *lo inaudito* in the archive "via a specific articulation in the present and provide a possible place for that past in a new narrative of the present toward the future."[21]

Activated within a type of labor that seeks to challenge traditional notions of archival authority, performative listening is about estranging, queering, decolonizing, and denaturalizing. In that sense, it stands as a perfect interpretative concept to describe the Aural City's type of labor as evidenced in the case studies explored in this book's chapters. Listening as an estrangement tool privileges the way one hears sounds over the sound objects themselves. As has been expressed, the sound object is always mediated and thus is never the *Ding an sich*. It cannot be otherwise. Nevertheless, the ontological ambiguity informing this pairing is precisely the dialogic principle of listening that post-truth narratives have hijacked. By highlighting the process of mediation and the affectivity that informs it, post-truth narratives have appealed directly to the emotions that bind human experience to the representations individuals want to accept as accurate mediations of reality—and sometimes even as reality itself. In fact, this irrational appeal to emotion is precisely what leads people to willingly and uncritically accept post-truth narratives and fake news as authentic. People want to believe these representations are true because they respond to and validate their most deeply embedded primal beliefs. As discussed in chapter 5, sound designers in the Mexican film industry are well aware that emotion and affectivity are central aspects in making audiences experience certain sonic representations as authentic. Vivian Sobchack maintains that often, sound film environments privilege "an intensified sense of acoustical presence and sonic immersion" that favors sensual investment over realism.[22] However, as the work of the Mexican sound designers who have taken Félix Blume's *Disco pirata* as a source of acoustic validation shows, the sensual investment of the sonic experience could be further reinforced by also articulating the aura of authenticity of located sounds—in this case, Mexico City's—to create a larger affective experience that appeals to emotion and triggers memories. The sense of veracity in the sonic simulacra of these sound designers does not emanate from the sound files stored in *Disco pirata*; instead, it is the visceral and dialogic result of how those sounds in their audiovisual narrative arrangement evoke experiences and engage the audience's embodied knowledge, often referring to nostalgia and memory. In that sense, as Alison Walker argues, it is in the somatic, affective, and memory-resonant relationship with a listener that the sound designer's archive "becomes much more than a repository of sonic data, but becomes a living archive and a sensory palette."[23] Sound archive advocates and users may learn from this experience a way to reevaluate the relationship among archival data, nostalgia, and memory in order to subvert and counter post-truth arguments solely sustained by a visceral evocation of affect. The sound designer's strategy recognizes the value in the archive that sustains it while providing a way for the listener to estrange and reactivate their affective relationship with its content.

Against Nihilism and the Indulgence of Memory

Merriam-Webster's Dictionary defines the noun *nihilism* as "a viewpoint that traditional values and beliefs are unfounded and that existence is senseless and useless" and as "a doctrine that denies any objective ground of truth and especially of moral truths."[24] In thinking about nihilism in relation to human intellectual inquiry, Hannah Arendt argues that such an attitude is born out of "the desire to find results that would make further thinking unnecessary."[25] Although, evidently, nihilism precedes neoliberalism, one cannot help but notice the many figurative and actual links between a creed that begins with a disenchanted world drained of value or meaning and the radicalized implementation of an economic and political doctrine that has led to a post-truth world drained of value and meaning. As Wendy Brown proposes, "Neoliberalism intensified the nihilism, fatalism, and rancorous resentment already present in late modern culture."[26] In many cases, neoliberal nihilism is the result of a type of political disillusionment and cynicism that has led to an abrogation of human agency and the right to struggle for a better world, leading instead to a gloomy embrace of the status quo as the natural state of human affairs. In a world where utopias are considered naive and even dangerous, to stop thinking while uncritically settling for the status quo might seem like the reasonable thing to do.

A nihilistic attitude toward archives would argue that they do not tell us anything. That is what ultraconservative right-wing activism has taken as the basis for the unfolding of post-truth, fake news, and misinformation, and for attacks on the credibility of facts in the archive. However, as discussed earlier, that has not been my argument, nor the Aural City's reason to question archival authority. Indeed, the archive has been designed so that its information tells us a story. Nevertheless, my argument here has been that the archive can tell us more than just the story it was designed to tell us. The schizzes, defamiliarization, and estrangement at the core of the Aural City's engagement with the archive—as seen and heard in the rearticulation of the canon proposed by Carlos Prieto Acevedo's schizzes in his curatorial labor, in Margarita Valdovinos's epistemic challenges to the imperial archive, in the poetic exploration of instruments as open-source archives, in the rhizomatic articulation of an invisible and inaudible archive of alternative artivist practices, in the postnational memory of Mexican Rarities, and in the way Mexican film sound designers have affectively and effectively engaged *Disco pirata*—are just the types of strategies that seek to productively regenerate memory. If fascism is able to make a comeback, it is because its appeal to oblivion, its invocation of the no-memory, and its unethical but effective manipulation of the archive of affects work in tandem to generate a convenient alternative narrative that fuels that return. The case studies presented in this book do not simply provide strategies for a reimagination of the archive that challenges the intolerant post-truth condition at the intersection of radical neoliberalism and fascism; they also provide insights into the hacking of the affective tactics that fascism and far-right neoliberalism have used to attract and keep their constituencies emotionally invested in their radical political action as well as their post-truth narratives.

Thus, rather than arguing that archives should look one way or the other in order to be more productive, I propose that regardless of the archives' structure and design, our understanding of and relationship to them should change and become continually skeptical. This is a must if we want to avoid their reification and the potential circularity and immobility of the types of knowledge they help produce. The moral of the stories I have told here, if there is one, would be to be open to always search for the *inaudito* in the archive.

Throughout this book I have recurred to science fiction as a means to prompt and illuminate the discussion of certain matters and topics that are central to the arguments in each of the chapters. Here, I wish to invoke science fiction one last time, as a point of entry into a final reflection on the relationship among memory, selfhood, affect, representation, and truth that informs how I have chosen to tie together the case studies explored in this book. Chris Kelvin, the main character in Andrei Tarkovsky's *Solaris* (1972), is a psychologist who is sent to a scientific research space station studying an oceanic planet called Solaris in order to assess a series of hallucinations experienced by the station's crew. Soon after his arrival, Chris himself begins to endure incidents and situations similar to those affecting the members of the station. In this case, Chris experiences the apparition of Hari, his deceased wife, who committed suicide ten years earlier. The surviving scientists in the station explain to him that the source of this apparition is Solaris itself, which has created a replica of Chris's wife from his memories of her. Chris's initial reaction is to reject and get rid of the replica. However, since Hari continues reappearing, he eventually accepts her and starts reveling in the second chance at living the intimate moments that Hari's death prevented him from sharing with her but also in the opportunity to talk about his regrets regarding the mistakes he made during his relationship with his deceased wife. For Chris, these moments of intimacy with the replica are always filled with a sense of ambiguity since he is well aware that this Hari is not real. The film ends with a sequence that places Chris with his father at the family country house he visited right before traveling to Solaris. However, as the camera slowly zooms out, the audience learns that Chris and his father are actually on an island in Solaris. This image, which follows a discussion in which Chris states that he is ready to go back home, is a good indication that in Tarkovsky's *Solaris*, as in many of his other films, the boundaries among memory, reality, representation, and desire can better be described as porous.

Solaris is an exploration of memory, delusion, representation, and truth that can be encapsulated in the words of one of the members of the space station: "I must tell you that we do not want to conquer any Cosmos. We want to expand the Earth to the limits of the Cosmos. We do not know what to do with other worlds. We do not need any other worlds. We need a mirror." In line with Arendt's admonition about the desire to find answers that "make further thinking unnecessary," *Solaris* explores the dangerous affordances found in the narcissistic opportunity of swapping the search for knowledge for a nihilistic simulacrum that allows us to live in or relive the symptomatic comfort of our memories. Chris is revealed to be trapped in a maze of memory, a delusional stage that he enjoys in order to continue engaging the memories that have become beloved to him or, in a nod to the Lacanian symptom, finding pleasure in dwelling on the shortcomings that marked his relationship with Hari.[27] Thus, *Solaris* warns us about bestowing our memories with this type of agency. Indulging our memories gives them the affective power to potentially control us, derail our search for truth, replace facts with emotions, and place us in a state of nihilistic immobility.

Needless to say, there is a clear difference between this delusional indulgence of memory and the type of productive engagement with it asserted by the Aural City's archival labor and its democratic ideal. While the former works as a form of delusional escapism, the latter seeks to move us out of our comfort zone and force us to reimagine collective and even individual memory on the basis of the facts and narratives that power struggles tend to prevent from entering the archive, that marginalize these facts and narratives, or that simply render them invisible and inaudible in the process of developing hegemonic narratives.

Intriguingly, both the progressive and conservative challenges to archival authority and the role of memory in the archive of recent decades coincide with the so-called crisis of the archival model in the field of memory studies. A basic premise in this field of studies is that, as Astrid Erll explains, "the past is not a given, but must instead continually be re-constructed and re-presented."[28] The field embraced this postulate as a result of the influence of poststructuralism, which led to a focus on "modes of remembering" as a more productive paradigm than the traditional history-memory dichotomy. Thus, a focus on modes of remembering meant that the traditional archival model for understanding memory, based on the idea that memories are formed, stored, and retrieved, was replaced by more dynamic and multidisciplinary visions of remembering and forgetting within specific culturally and historically informed power dynamics and contingencies.[29]

In contrast to the kind of staticity and fixity that conventional wisdom bestows on memories, the fact is that they are formations in continuous mobility. Thus, the emotional and affective charge we place on them as the sources of narratives of selfhood and collective identity should also be understood in continuous flux. If memory studies has left the archive model behind, maybe archival studies could borrow ideas from memory studies in order to revamp its understanding of the archive and its relevance in the contested times of post-truth. As we see in the collective labor that produced the exhibit *Modos de Oír* and the seminar Modos de Escucha, as well as harbingers such as the Seminario de Arte y Sonido, the work of the Mexican Aural City resonates profoundly with this shift in memory studies. Indeed, this shift of focus from sound objects to listening practices is both a consequence of the influential aural turn in Anglo-American academia and also a way of punctuating those ideas within the specific histories and intellectual circumstances of the Global South, especially Latin America.[30] Thus, the intellectual and professional links between the Mexican Aural City and Latin American artists and scholars are conducive to a transnational conversation about modes of listening that addresses the affective structures at the core of the Latin American colonial and neocolonial, as well as neoliberal, condition. The idea is that these structures effectively hack their appeal to emotion and affect and redirect it into building community. The transnational labor of feminist Noise collectives is a clear example of how to productively channel the anger and fear produced by gender-based psychological and physical abuse into the construction of creative spaces in which respect, inclusion, equality, and democracy are the currencies for access and participation. The types of spaces, creative, intellectual, and archival, that the Aural City generates could be seen as prototypes for the ways in which we should understand archives in order to move beyond the logic of the archive that the field of memory studies has already left behind.

Following on that logic, sound archives could be understood as spaces for the exploration of the poetic ways in which affective modes of listening can activate *lo inaudito*, those surprising areas of the archive that have remained inaudible and marginalized. We can think of the sound archive as a nodal point in which we make sense of the world by making sense of ourselves. Rather than occupying itself with a hopeless attempt at retrieving any idealized objective data, archival labor should focus on elucidating the performative ways in which the user's listening gives meaning to the sounds kept in the archive while providing affective avenues for their insertion into narrative plots at a variety of individual and social levels. The success of *Disco pirata* both as an archive of sounds and as a performative action is connected to its articulation of deep senses of affectivity (in terms of belonging and place) that are somatically activated through the intimate act of listening. Understanding the nexus between affectivity and narrative is an essential step in hijacking the ways in which post-truth strategies engage affect. Mirroring and channeling these types of strategies into generating spaces for individual and collective well-being and flourishing, as the Mexican Aural City does, provides an effective way to overcome the nihilistic immobility and apathy that post-truth and the hopelessness of neoliberalism impose on our contemporary world. Hijacking the algorithm, listening to the echoes in the archive, avoiding the indulgence of memory, and rechanneling the affect that makes individuals lethargically accept hegemonic narratives: These are the brazen lessons of the Aural City that may help us turn resignation into agency and emotional capital into a liberating political resource.

Notes

INTRODUCTION. Questions About the Circulation of Knowledge at the Sonic Turn

1. Cook, "Archive(s)," 601.
2. Mbembe, "Power of the Archive," 19–20.
3. Cook, "Archive(s)," 606.
4. Tomlinson, "Evolutionary Studies," 651.
5. Rivera Garza, *Escribir con el presente*, 60. All translations are my own unless otherwise noted.
6. Jacques Derrida coined the term *anarchive* to refer to the paradox that an archive's destruction may lie within itself, in "the possibility of putting to death the very thing, whatever its name, which *carries the law in its tradition*: the archon of the archive." See Derrida, "Archive Fever," 51.
7. Weld, *Paper Cadavers*, 238.
8. Rama, *Lettered City*, 16.
9. R. Murray Schafer coined the term "schizophonia" to describe the separation of sound from its original source. See Schafer, *New Soundscape*, 43–47.
10. Johnes, "Archives," 131.
11. See Simon, "Introduction," 101–107.
12. Foucault, *Power/Knowledge*, 114.
13. Derrida, "Archive Fever," 14, 17.
14. Johnes, "Archives," 133.
15. See Ahmed, *Willful Subjects*; Arondekar, *For the Record*; Callahan, *Art + Archive*; Marshall and Tortorici, *Turning Archival*; Rosengarten, *Between Memory and Document*; Stoler, *Along the Archival Grain*; and Taylor, *Archive and the Repertoire*.
16. Marshall and Tortorici, "Introduction," 1.
17. Frohmann, "Documentary Ethics," 166.
18. O'Callaghan, *Sounds*, 3. Similar conceptualizations can be found in Pinch and Bijsterveld, "New Keys," 11–14; Sterne, "Sonic Imaginations," 9–10; and Bronfman and Wood, "Introduction," ix–x.
19. McEnaney, "Sonic Turn," 84.
20. Novak and Sakakeeny, introduction, 2.

21. Souza Lima, "Sound Beyond Hylomorphism," 49. See also Ochoa Gautier, *Aurality*; and Samuels et al., "Soundscapes."

22. Franco, *Decline and Fall*, 10.

23. I do not intend to reproduce traditional understandings regarding the relationship between civilization and sedentarism and the rise of cities. I address it only in relation to intellectual traditions that have assumed this chain of cause-and-effect events to be true and have used it as a foundational metaphor for their civilizing projects. Recent scholarship has questioned the historiographic naturalization of such dichotomies. See Graeber and Wengrow, *Dawn of Everything*, 276–440.

24. Rama, *Lettered City*, 13, 12–13.

25. Franco, *Decline and Fall*, 12.

26. García Canclini, *Consumers and Citizens*, 39.

27. See Perus, "¿Qué nos dice hoy?," 56; Spitta, "Prefacio," 13; and Beverly, "Writing in Reverse," 628–29.

28. Steedman, *Dust*, 167.

29. G. Baker, *Imposing Harmony*, 20.

30. G. Baker, *Imposing Harmony*, 22.

31. See Curcio-Nagy, "Giants and Gypsies"; Ramos-Kittrell, *Playing in the Cathedral*; Vera, *Sweet Penance of Music*; and Waisman, *Historia*.

32. Bieletto-Bueno, "Introducción," 12.

33. See Madrid, "Landscapes"; and Madrid, "Rastreando."

34. Sterne, *Audible Past*, 15; and Ochoa Gautier, *Aurality*, 16–17.

35. Ochoa Gautier, "Sonic Transculturation," 807.

36. Ochoa Gautier, "Sonic Transculturation," 808.

37. Ochoa Gautier, "Sonic Transculturation," 807.

38. Alegre González and García, "Presentación," 5. See also Erlmann, *Reason and Resonance*.

39. Alegre González, "Más allá de la abyección," 11; and Domínguez Ruiz, "Oído," 94.

40. Madrid, *In Search of Julián Carrillo*, 281.

41. Ramos, *Unbelonging*, 5.

42. James, *Sonic Episteme*, 4.

43. James, *Sonic Episteme*, 5, 6.

44. Kane, "Fluctuating Sound Object," 55.

45. Chion, "Reflections," 23.

46. Sterne, "Spectral Objects," 107.

47. "Text of the Convention for the Safeguarding of the Intangible Cultural Heritage," UNESCO, Intangible Cultural Heritage, 2003, https://ich.unesco.org/en/convention.

48. "Recommendation Concerning the Preservation of, and Access to, Documentary Heritage Including in Digital Form," UNESCO, Memory of the World, November 17, 2015, https://www.unesco.org/en/legal-affairs/recommendation-concerning-preservation-and-access-documentary-heritage-including-digital-form?hub=1081.

49. Montelongo, *No soy tan zen*, 45.

50. Ochoa Gautier, *Aurality*, 3.

51. In a similar way, Amanda Minks explores the role of sound, "between the ear and the letter," in the development of Lettered City projects in Mexico, Nicaragua, and Chile during the first half of the twentieth century. One could argue that the initiatives she studies can also be described as examples of proto–Aural City projects. See Minks, *Indigenous Audibilities*.

52. See Deleuze and Guattari, *Anti-Oedipus*; and Deleuze and Guattari, *Thousand Plateaus*.

53. See Appadurai, "Introduction"; Rozental, "On the Nature"; and Spitta, *Misplaced Objects*.

54. See Khoury, "Postnational Memory"; and Young, *Postnational Memory*.

55. See Rivera Garza, "Escrituras colindantes"; Rivera Garza, *Escrituras geológicas*; and Rivera Garza, *Restless Dead*.

56. See Magnusson, *Sonic Writing*; Maier and Schulze, "Tacit Grooves"; Moseley, *Keys to Play*; and Rehding, "Instruments of Music Theory."

57. See Barskova, *Air Raid*; and Barskova, *Живые картины*.

58. Williams, "Poetry Writing," 363.

CHAPTER 1. Performative Listening, Writing, Reading, and the Assemblage of Archival Constellations

An early Spanish version of this chapter was published as "Rastreando las huellas de la escucha performativa: La escritura como constelación archivística." *Anuario Musical* 76 (2021): 11–30.

1. "Foreword of the Mexican Department of Foreign Affairs," 10.

2. Chávez, introduction, 5, 10.

3. García Morillo, *Carlos Chávez*, 88.

4. These recordings are available at the following link: "*A Program of Mexican Music*," Columbia Masterworks, 1941, posted June 1, 2019, by Avide, YouTube, https://www.youtube.com/watch?v=bo_9OokJees.

5. Kennedy, "Alex Steinweiss."

6. At the end of the exhibit and its concerts, the MoMA donated those copies of pre-Hispanic instruments to the New York Public Library. Eventually, they became part of the collection of the Metropolitan Museum of Art when the library collection of instruments was moved to the museum due to lack of storage space.

7. For a detailed analysis of *Xochipili-Macuilxochitl*, see Burns, "Listening," 154–59.

8. Chávez, introduction, 7–8.

9. Chávez, *Xochipilli*. Although the conceptualization of the work as "imagined Aztec music" informs the way in which it was presented at the MoMA exhibit, it is only in this explanatory note that Chávez fully acknowledges it.

10. See Burns, "Listening"; Roberts, "Aztec Musical Styles"; and Saavedra, "Carlos Chávez."

11. Robinson, *Hungry Listening*, 6.

12. See, for example, the first and second sections in Galindo, *Historia*, 17–124; and the first part of Saldívar, *Historia*, 1–83.

Notes to Chapter 1   287

13. W. Benjamin, *Origin*, 34.

14. Krauß, "Constellations," 440.

15. Madrid, *In Search of Julián Carrillo*, 4n4.

16. Castañeda and Mendoza, *Instrumental precortesiano* (I cite the original 1933 edition; see also the 1991 facsimile edition, published by UNAM); and the following texts by Carlos Chávez: "Música y física"; "Producción y reproducción musical"; "Instrumentos eléctricos"; *Toward a New Music* (1937); and *Hacia una nueva música* (1992). Throughout this chapter I refer to Chávez's book mostly by its English title to address its chronological proximity to Castañeda and Mendoza's book (the Spanish version of the book was not published until 1992) since that temporal affinity is important in my argument and since I argue that they should be considered units of a single archival constellation. When I refer to it by its Spanish title, I do so to highlight the potential of transhistorically expanding this constellation to also include units that may be separated by larger temporal distances.

17. Carmona, "Prólogo," 18.

18. Weiss, "Listening to the World," 520.

19. Juliastuti writes that "an archive, or collection of materials is often referred to in Indonesian as a *dokumen*, or document. In everyday conversation in Indonesia, the words 'archive' and 'document' are largely used interchangeably." Juliastuti, "Indonesian Migrant Workers' Writings."

20. McKinney, *Information Activism*, 77.

21. McEnaney, "'Rigoberta's Listener,'" 394.

22. Ochoa Gautier, *Aurality*, 3.

23. Rocha, "Recovering Voices," 3–4.

24. *Diccionario de la lengua española*, s.v. "escuchar," Real Academia Española, accessed July 17, 2021, https://dle.rae.es/escuchar?m=form. Other authors have used the notion of performative listening in similar ways to what I propose here. In the context of interpersonal communication, Doyle Srader uses the notion of performative listening to describe how individuals are able to complete a communicative transaction via a type of listening that connects with their interlocutor and avoids "speaking into the air." See Srader, "Performative Listening," 98. Similarly, Chris McRae uses the term to refer to how listeners engage in ethical acts of learning from others across difference "as a way of creating connections, relationships, and knowledge." McRae, *Performative Listening*, 2.

25. *Costumbrismo* was a nineteenth-century Spanish and Latin American literary movement characterized by the realistic depiction of regional folklore and local everyday-life costumes and traditions. Important *costumbrista* writers include the Spaniards José María de Pereda (1833–1906) and Vicente Blasco Ibáñez (1867–1928), the Argentinean Domingo Faustino Sarmiento (1811–88), the Colombian Jorge Isaacs (1837–95), the Cuban Cirilo Villaverde (1812–94), and the Mexican Ángel del Campo "Micrós" (1868–1908), among others.

26. See Chávez, "Orquesta Sinfónica Mexicana"; and Sánchez Mejorada, "Antonieta Rivas Mercado."

27. About the concept of the organic intellectual, see Gramsci, *Formación de los intelectuales*, 28–32. About Chávez as an intellectual, see Alonso-Minutti, "Composer as Intellectual."

28. Chávez, "México no necesita," 119.

29. Chávez, "Música, la universidad," 127.

30. Aguirre Lora, "Escuela Nacional de Música," 100–101.

31. Beristáin-Cardoso, "Educación artística," 87–92.

32. See "Objeciones a la ley de autonomía," *El Universal Gráfico* (June 24, 1929), cited in Carmona, *Carlos Chávez*, 117fn.

33. García Morillo, *Carlos Chávez*, 60.

34. See Chávez, "Conservatorio en 1929."

35. Chávez, "Conservatorio en 1929," 150.

36. See Adler, "Umfang, Methode und Ziel," 17–18. Originally, Adler proposed historical musicology and systematic musicology (which included comparative musicology) as the two main branches of "musical science." Later, he further divided systematic musicology into two independent areas: systematic musicology and comparative musicology.

37. Parker, *Carlos Chávez*, 11.

38. Pareyón, "Castañeda (Soriano), Daniel."

39. See Madrid, *Sonidos*, 128–33.

40. See Madrid, *In Search of Julián Carrillo*, 138–65.

41. Pareyón, "Mendoza (Gutiérrez), Vicente T(eódulo)."

42. Madrid, *Sonidos*, 133–34.

43. Castañeda and Mendoza, *Instrumental precortesiano*, iii.

44. Although the official publication date of the book is 1933, Carlos Chávez wrote an introductory note dated April 1, 1934, for the CNM edition. See Castañeda and Mendoza, *Instrumental precortesiano*. It is unclear whether the authors started the research or writing of any of the following volumes and what these volumes were meant to document.

45. Castañeda and Mendoza use the Spanish term *percutores*, which I have chosen to translate as "percussive instruments."

46. Castañeda and Mendoza, *Instrumental precortesiano*, vii.

47. See Hornbostel and Sachs, "Systematik der Musikinstrumente."

48. Alviña, *Música incaica*; Robles Godoy, *Himno al sol*; and d'Harcourt and d'Harcourt, *Musique des Incas*. See also Simonett and Marcuzzi, "One Hundred Years," 10–11.

49. For problematizations of these evolutionist ideas see Sas, "Aperçu," 1–8; Sas, "Ensayo"; Vega, "Supuesta"; Vega, "Música incaica"; Vega, "Música de los incas"; and Valcárcel, "¿Fue exclusivamente de 5 sonidos?" For scholarship documenting how these ideas have informed European scholarship about South American music see R. Romero, "Panorama"; Wolkowicz, *Inca Music Reimagined*; Mendívil, *Cuentos fabulosos*; and Rios, *Panpipes and Ponchos*, 25.

50. Castañeda and Mendoza, *Instrumental precortesiano*, xv.

51. Chamorro, "Etnomusicología mexicana," 80. See also Alonso Bolaños, *"Invención" de la música indígena*.

52. Alonso Bolaños, *"Invención" de la música indígena*, 33–52.

53. For a study of the technological and business issues at stake in the transition from the acoustic to the electric era, see Barnett, *Record Cultures*. For a study of the cultural and social implications of this shift, see Denning, *Noise Uprising*.

54. These early recordings were reissued in 1981 in an LP entitled *Early Hi-Fi*.
55. Weinstock, "Foreword by the Translator," 7.
56. García Morillo, *Carlos Chávez*, 68–69.
57. Chávez, "Música y física," 210.
58. Chávez, "Producción y reproducción musical," 218–19.
59. Chávez, "Instrumentos eléctricos," 226–27.
60. Chávez, *Toward a New Music*, 76–77. The text in Spanish is slightly different from the English translation; it states, "Esa gran riqueza musical debe ponerse en movimiento, debe trascender una región, debe pertenecer a la humanidad entera." (This great musical wealth ought to be put in motion, it ought to transcend a region, it ought to belong to the entirety of humanity.) Chávez, *Hacia una nueva música*, 81.
61. Chávez, *Toward a New Music*, 120–21.
62. About Chávez's film and political projects in these years, see Parker, *Trece panoramas*, 69–86.
63. Chávez, *Toward a New Music*, 145.
64. Chávez, *Toward a New Music*, 148.
65. Chávez, *Toward a New Music*, 153.
66. This discussion about the possibility of using microtonal scales not only to control the past but also to mediate Indigenous Otherness also resonates with the ideas that Castañeda and Mendoza presented at the 1926 National Music Congress. See note 42.
67. Carmona, "Prólogo," 9–22.
68. Cvetkovich, *Archive of Feelings*, 7.
69. A case could be made for extending the transhistorical reach of this archival constellation by also including the reissue of the record collection with a cover designed by Andy Warhol in 1949 or by considering the concerns that informed El Colegio Nacional's publication of *Hacia una nueva música* in 1992. For the sake of keeping my argument focused here, I have chosen to exclude them from the conversation.
70. Andrade, "Manifiesto antropófago," 3, 7.
71. Bioy Casares, *Invención de Morel*, 115.

CHAPTER 2. Patrimony, Objectification, and Representation at Mexico's Fonoteca Nacional

An early version of this chapter was published as "Landscapes and Gimmicks from the 'Sounded City': Listening for the Nation at the Sound Archive," *Sound Studies: An Interdisciplinary Journal* 2, no. 2 (2016): 119–36.

1. The audiotheque was named after the well-known Mexican Nobel Prize winner Octavio Paz (1914–98), who used to live in the residence that now houses the Fonoteca Nacional.
2. Thomas Stanford (b. 1929) is an American ethnomusicologist who has lived in Mexico since 1956. Although he has published articles about colonial music in New Spain, his main research focuses on Indigenous and folk musics from Mexico. One of the main collections in the Fonoteca archive consists of field recordings of traditional

Mexican music collected by Stanford over a period of fifty years. Stanford donated it to the Fonoteca in 2007. See Pareyón, "Stanford, E. Thomas." Henrietta Yurchenco (1916–2007) was an American folklorist and music collector who visited Mexico regularly beginning in 1941 to record Indigenous music and rituals. In 1990 she donated a copy of her sound archive to Mexico's Instituto Nacional Indigenista. See Solís, "Acervo"; and Bitrán Goren, "¡Hurra a Henrietta Yurchenco!"

3. Salvador Novo (1904–74) was an influential Mexican writer and intellectual. In 1965 he was appointed Mexico City's official chronicler by President Gustavo Díaz Ordaz.

4. Porfirio Díaz (1830–1915) was president of Mexico from 1877 to 1880 and from 1884 to 1910 as well as for a brief period at the end of 1876. He met Thomas Alva Edison in New York when Díaz visited the city during his honeymoon. On July 8, 1909, the American inventor wrote a letter in Spanish to the Mexican president asking him to record a short message on an Edison cylinder as part of a series of recordings by international political figures, which included US president William Howard Taft (1857–1930) and William J. Bryan (1860–1925). Edison's letter is available in Fotografía y fonógrafo en el Porfiriato, posted December 7, 2012, https://fotografiayfonografo.wordpress.com/2012/12/07/el-fonografo-porfirio-diaz-y-thomas-alva-edison/dsc08753/. Díaz's recorded message can be heard in "Mensaje de voz de Porfirio Díaz 1909 a Thomas Alva Edison," posted by gannonmx, YouTube, June 12, 2007, https://www.youtube.com/watch?v=eKhi6OpEYv4.

5. Fonoteca Nacional, "Misión y visión," accessed December 18, 2022, https://fonotecanacional.gob.mx/index.php/fonoteca-nacional/mision-vision.

6. Ochoa Gautier, "Social Transculturation," 807, 803.

7. For a study of some of these sound art–based collectives and initiatives in Mexico City, see Lara Velázquez, "Poner la escucha"; some of the leading actors promoting these scenes and initiatives are featured in Prieto Acevedo, Variación de voltaje.

8. "Plan Nacional de Desarrollo."

9. "Programa Nacional de Turismo."

10. For a discussion of the Fonoteca Nacional as a project at the intersection of cultural heritage and tourism that follows on this argument, see Bieletto-Bueno, "Noise, Soundscape, and Heritage," 114–16.

11. "The Historical Collections (1899–1950) of the Vienna Phonogrammarchiv." UNESCO, Memory of the World web page, accessed December 18, 2022, https://webarchive.unesco.org/web/20220331183824/http://www.unesco.org/new/en/communication-and-information/memory-of-the-world/register/full-list-of-registered-heritage/registered-heritage-page-8/the-historical-collections-1899-1950-of-the-vienna-phonogrammarchiv#c191394.

12. Sound archives have emerged for a variety of reasons: in some cases, with the intention of studying the Other; in other cases, as attempts to capture "vanishing national essences"; and on many occasions simply as a matter of safeguarding the production of private recording, radio, or TV companies.

13. Molinari Junior, "Experiencia del Archivo Nacional."

14. Romano, "Accesibilidad," 137–41.

15. Choque Vaca, "Archivos sonoros," 170.

16. Patricia Velázquez, "La historia cantada y contada," Fonoteca del INAH, accessed October 10, 2014, http://www.inah.gob.mx/especiales/194-fonoteca-del-inah-.

17. "Nuestro Acervo," Filmoteca UNAM, Cultura UNAM, accessed December 18, 2022, https://www.filmoteca.unam.mx/nuestro-acervo/.

18. Rodríguez Reséndiz, *Archivo sonoro*, 192.

19. Rodríguez Reséndiz, interview by the author, July 24, 2013.

20. Rozental, "On the Nature," 237.

21. Rodríguez Reséndiz, *Archivo sonoro*, 203.

22. Rodríguez Reséndiz, *Archivo sonoro*, 204–6. Armando Pous Escalante is a Mexican collector of recordings and radios. He oversees the Instituto de Conservación y Recuperación Musical, a nongovernmental initiative that reissues rare and out-of-print recordings. John C. Lilly (1937–2007) was an American ethno-cinematographer and an expert on Wixárika culture from Western Mexico. See "John C. Lilly Jr.," Legacy, accessed August 7, 2016, http://www.legacy.com/obituaries/name/john-lilly-obituary?pid=1000000102482093. Alfonso Muñoz Güemes (b. 1962) is a Mexican ethnologist who specializes in Indigenous Mexican music, migration, globalization, and social change. He is a professor of cultural management and public policy at the Universidad Autónoma de San Luis Potosí in Mexico. See "Dr. Alfonso Muñóz Güemes," Gestión y Políticas Públicas, Unidad Académica Multidisciplinaria UASLP Zona Huasteca, accessed December 18, 2022, http://licgestionypoliticaspublicas.blogspot.com/p/maestro2.html.

23. The Thomas Stanford Collection, the Raúl Hellmer Collection, the Henrietta Yurchenco Collection, and the Baruj "Beno" Lieberman, Enrique Ramírez de Arellano, and Eduardo Llerenas Collection, all of them compendiums of traditional Mexican music, were nominated by the Fonoteca in collaboration with the institutions that housed them before they became part of the Fonoteca archive. The other four are sound collections nominated solely by the institutions that originally housed them: *De Puntitas*, a Radio Educación show for children that ran from 1983 to 1988; *El Foro de la Mujer*, a Radio UNAM show created in 1972 by Guatemalan feminist writer Alaíde Foppa (1914–80); the Estudios Churubusco Sound Archive, which holds original recordings of Mexican film music for movies made between 1958 and 1975; and the Comisión Nacional para el Desarrollo de los Pueblos Indígenas (National Commission for the Development of Indigenous People) collection of the fifty Encuentros de Música y Danza Indígena (Encounters of Indigenous Music and Dance). Joseph Raoul Hellmer Pinkham (1913–71) was a US-born ethnomusicologist who worked as a researcher of Mexican traditional music for the Instituto Nacional de Bellas Artes. Baruj "Beno" Lieberman (1932–85) was a Mexican businessman and amateur folklorist who self-sponsored recording field trips in many regions of Mexico and Central and South America. Enrique Ramírez de Arellano (b. 1937) is a mathematician who joined Lieberman's research trips as an electroacoustics expert in 1972. Eduardo Llerenas (1945–2022) was a biochemist, folklorist, and music entrepreneur who joined Lieberman and Ramírez de Arellano's recording field trips in 1975. He was also the founder of Discos Corason, a commercial label specializing in traditional Mexican and Caribbean musics. See Becerril, "Raúl Hellmer"; and García Ranz, "¿Quién diablos?"

24. See "La memoria sonora de México en la Fonoteca Nacional," posted February 8, 2021, by Fonoteca Nacional de México, YouTube, https://www.youtube.com/watch?v=j1MSMqeSFtg&t=2s.

25. Rodríguez Reséndiz, interview.

26. Rodríguez Reséndiz, interview.

27. Rodríguez Reséndiz, interview.

28. Fonoteca Nacional, Mapa Sonoro de México, accessed December 18, 2022, https://fonotecanacional.gob.mx/index.php/escucha/mapa-sonoro-de-mexico.

29. Mapa Sonoro de México, accessed October 14, 2015, http://fonomaps.herokuapp.com/#close. By mid-2023 the sound map featured 1,206 audio files and had been moved to the following URL: Mapa Sonoro de México, accessed June 3, 2023, https://mapasonoro.cultura.gob.mx/.

30. F. Romero, *Hyperborder*, 104.

31. Bieletto-Bueno, "Noise, Soundscape, and Heritage," 119.

32. Selections from these soundscape projects are available in "Selección: Paisajes sonoros de México," Fonoteca Nacional de México, accessed December 18, 2022, https://rva.fonotecanacional.gob.mx/fonoteca_itinerante/paisajes.html.

33. Rivas, interview by the author, July 12, 2013.

34. Rivas, interview.

35. Rivas, interview.

36. Rivas, interview.

37. Rodríguez Reséndiz, interview.

38. Rodríguez Reséndiz, interview.

39. M. Wright, *Listening After Nature*, 30–31.

40. Baudrillard, *Simulacra and Simulation*, 1–6.

41. Rivas, interview.

42. Southworth, "Sonic Environment of Cities," 52.

43. Southworth calls this enjoyment "sonic delight." Southworth, "Sonic Environment of Cities," 59–60.

44. Schafer, *Tuning of the World*, 4.

45. Schafer, *Tuning of the World*, 7.

46. Schafer, *Tuning of the World*, 9–10.

47. Ingold, "Against Soundscape," 11.

48. Helmreich, "Listening Against Soundscapes," 10.

49. Pueblos Mágicos de México, Secretaría de Turismo, December 1, 2020, https://www.gob.mx/sectur/articulos/pueblos-magicos-206528.

50. Campos Fonseca, "Noise," 178, 177.

51. I would like to thank Xilonen Luna Ruiz for pointing this out to me during the question-and-answer session of a virtual keynote lecture I presented for the Coloquio Paisajes Sonoros, Música, Ruidos y Sonidos de la Frontera at El Colegio de la Frontera Norte-Tijuana on November 24, 2022.

52. Rozental, "On the Nature," 237.

53. Rozental, "On the Nature," 238.

54. Rodríguez Reséndiz, interview.

55. Westerkamp, "Soundwalking," 49.

56. McCartney, "Soundwalking," 214.

57. Hernández Cerón, "Re-aprender a escuchar," 98.

58. Fonoteca Nacional, "Caminatas y Rodadas Sonoras," accessed January 14, 2023, https://fonotecanacional.gob.mx/index.php/105-quienes-somos/222-caminatas-y-rodadas-sonoras.

59. Erika Carmen López Pérez, who oversaw the implementation of the Caminatas Sonoras project at the Fonoteca Nacional, calls it a "program of sensibilization." See Hernández Cerón, "Re-aprender a escuchar," 110.

60. Michelle Caswell coined the term "symbolic annihilation" to describe how mainstream archival practices have "far-reaching consequences for both how communities see themselves and how history is written for decades to come." Caswell, "Seeing Yourself in History," 36.

61. Arce, "Archivos," 112–13.

62. Sánchez Cardona, "*Vis.Fuerza[in]necesaria_4*," 135.

63. Luz María Sánchez Cardona, "*V.[u]nf_1, 2014–2015*," accessed November 8, 2024, https://www.luzmariasanchez.com/work/artwork-vunf1-2014.

64. See C. Baker, *Sonic Strategies*, 59–94; and C. Baker, "Affective Acoustic Territories."

65. Borges, "La biblioteca," 99, 89, 96.

66. Borges, "La biblioteca," 94.

67. Borges, "La biblioteca," 95.

68. M. García, "Archivos sonoros."

CHAPTER 3. *Critical Constellations of the Audio-Machine in Mexico and the Performativity of Archiving/Archival Labor*

1. Baudrillard, *Cultura y simulacro*, 9–10.

2. Seem, introduction, xvi.

3. Deleuze and Guattari, *Anti-Oedipus*, 122.

4. Deleuze and Guattari, *Anti-Oedipus*, 34.

5. Deleuze and Guattari, *Anti-Oedipus*, 40.

6. Rancière, *Politics of Aesthetics*, 39.

7. Barskova, *Живые картины*, 7. All translations from the Russian language are by Ekaterina Pirozhenko unless otherwise noted.

8. Villa-Flores, "Plotting a Fire," 221.

9. Promotional leaflet for *Critical Constellations of the Audio-Machine in Mexico*, CTM Festival, January 27 to March 19, 2017.

10. "CTM 2017," CTM Festival, accessed November 9, 2024, https://www.ctm-festival.de/activities/past-ctm-festival-editions/ctm-2017.

11. Prieto Acevedo, "Critical Constellations," 3.

12. Malmström's case is particularly revealing regarding the uncritical approaches that often reproduce mainstream historical narratives. He was a Swedish musicologist who visited Mexico in the early 1970s to research his book and who, unaware of the deep

local ideologies behind them, uncritically reproduced the narratives his informants fed him as they fit his own teleological preconceptions about history and music.

13. They also reflect the contradictory "hungry" and "cannibalistic" articulations of the Indigenous Other at the intersection of *indigenismo* and futurism in this type of modernist nation-building project. Here, I use the terms *hungry* and *cannibalistic* in reference to Robinson, *Hungry Listening*; and Andrade, "Manifiesto antropófago," 3, 7. Chapter 1 in this book explores these issues in detail.

14. Prieto Acevedo, interview by the author, April 19, 2021. See also Prieto Acevedo, *Variación de voltaje*. The *Variación de voltaje* project was originally developed by Prieto Acevedo in collaboration with Inti Meza Villarino, also a Mexican philosopher and cultural broker. Although the project was meant to be a three-volume series, only the first volume was published.

15. Prieto Acevedo, *Variación de voltaje*, 13.

16. Prieto Acevedo, interview, April 19, 2021.

17. Collaborators included Amanda de la Garza, Miguel Molina Alarcón, Israel Martínez, Bárbara Perea, Álvaro Ruiz, Walter Schmidt, Mario de Vega, Alexander Brück, Tania Aedo, Iván Edeza, Francisco "Tito" Rivas, Manuel Rocha, Gonzalo Macías, Siete Catorce, Félix Blume, Rogelio Sosa, Javier Toscano, Juan Pablo Villegas, Andrés Oriard, Enrique Minjarez, and Ignacio Baca Lobera.

18. Prieto Acevedo, interview, April 19, 2021.

19. See Singuhr Hoergalerie Projekte 2014–2024, accessed June 6, 2021, http://singuhr.de/projekte-2014. A record of this exhibit can be seen in Seiffarth and Steffens, *Entre límites*.

20. Seiffarth, interview by the author, February 15, 2023.

21. Prieto Acevedo, "Critical Constellations," 3.

22. Krauß, "Constellations," 439.

23. Prieto Acevedo, interview, April 19, 2021.

24. Inspired by Walter Benjamin's conceptualization of "dialectical images," the notion of "dialectical soundings" proposes that music making "could work as a medium that makes visible the invisible via a specific articulation in the present and provides a possible place for that past in a new narrative of the present toward the future." Madrid, "Transnational Identity," 186. See also Madrid, "Sonares dialécticos," 27–29.

25. I use the terms *performative* and *performatic* following the convention suggested by performance scholar Diana Taylor. Accordingly, "the performative becomes less a quality (or adjective) of 'performance' than of discourse [while the performatic] denote[s] the adjectival form of the nondiscursive realm of performance." In sum, *performatic* refers to the theatrical qualities of performance while *performative* refers to discursive performativity or the realm of "what happens when something happens." Taylor, *Archive and the Repertoire*, 6. See also Madrid, "Why Music?"

26. In his problematic essay *La raza cósmica* (1925), José Vasconcelos argued that as descendants of what he considered the world's four "primal races" (Native American, White [European], African, and Asian), Ibero-Americans were destined to develop a superior "cosmic race," the mixture of all races of the world. For a critique of Vasconcelos's problematic racist theory, see Miller, *Rise and Fall*.

27. Although *El fuego nuevo* was never premiered as a ballet, Chávez premiered just the music during the first season of the Orquesta Sinfónica Mexicana. See Parker, "Carlos Chávez," 182–85.

28. Prieto Acevedo, interview, April 19, 2021.

29. See Douglass and Wilderson, "Violence of Presence."

30. Prieto Acevedo, interview, April 19, 2021.

31. Schneider, *Estridentismo*, 37. Estridentismo was a Mexican avant-garde multidisciplinary artistic movement inspired by Italian futurism. However, unlike its European counterpart's flirtatious relationship with fascism, with time, Estridentismo's abstract and delirious fantasies often unfolded as critiques of the bourgeois culture that flourished before the Mexican Revolution.

32. Armstrong, "This Photography," 50.

33. Julián Carrillo used the label Sonido 13 (the Thirteenth Sound) to refer to his microtonal system, one of the first microtonal systems in the Western art music tradition. See Madrid, *In Search of Julián Carrillo*, 162–65.

34. See Paz, *Children of the Mire*, 1.

35. Prieto Acevedo, "Critical Constellations," 11.

36. See Ospina Romero, "Ghosts in the Machine."

37. "T.S.H. El poema de la radiofonía" was conceived to be read and broadcast over the radio during the inaugural transmission of Mexico City's CYL "La Casa de la Radio," on May 8, 1923. This was the first broadcast of Mexico's first commercial radio station. See Hayes, *Radio Nation*, 30–31.

38. See Madrid, *In Search of Julián Carrillo*, 154–62, 231–40.

39. Barskova, *Живые картины*, 13.

40. Cuevas, "Cactus Curtain."

41. Prieto Acevedo, interview, April 19, 2021.

42. Hardt and Negri, *Multitude*, 100.

43. Hardt and Negri, *Empire*, 65–66.

44. Alonso-Minutti, *Mario Lavista*, 68–71.

45. O'Hagan, "Enrique Metinides."

46. Kristeva, *Powers of Horror*, 210.

47. Hardt and Negri, *Empire*, 396.

48. Prieto Acevedo, interview, April 19, 2021.

49. Prieto Acevedo, interview, April 19, 2021.

50. Gerber Bicecci and Villoro, *Significación del silencio*, 7.

51. Ludmer, "Literaturas postautónomas 2.0," 41. For Fernández's assessment of Gerber Bicecci, see "Verónica Gerber Bicecci en Cátedra Abierta UDP—Escrituras del Compostaje," November 26, 2020, Facultad Comunicación y Letras UDP, YouTube, https://www.youtube.com/watch?v=sT-o7gWMo5E.

52. Prieto Acevedo, "Critical Constellations," 22.

53. Prieto Acevedo, "Critical Constellations," 23.

54. Deleuze and Guattari, *Anti-Oedipus*, 8.

55. Prieto Acevedo, interview, April 19, 2021.

56. Bjerregaard, "Dissolving Objects," 74–75.

57. A. Bohlman, "Sonic Anarchy on Display," 90. This piece, which explores the possible relation among document, sculpture, and sound, was commissioned by Prieto Acevedo, who worked closely with the artist as it was created. The protest recorded by Blume was part of a strike against Mexican President Enrique Peña Nieto's 2013 energy reforms. The piece is now kept at Mexico's Universidad Autónoma Metropolitana-Xochimilco campus's sculpture space. See "Mexico Crowd Protests against Energy Reform," BBC News, December 2, 2013, https://www.bbc.com/news/world-latin-america-25180472. Blume's description of this work can be found here: "Memoria del hierro," Felix Blume's website, accessed February 2, 2023, https://felixblume.com/memoriadelhierro.

58. Seiffarth, interview.

59. Prieto Acevedo, interview, April 19, 2021.

60. A. Bohlman, "Sonic Anarchy on Display," 91.

61. Piekut and Stanyek, "Deadness," 20.

62. See Suvin, *Metamorphoses of Science Fiction*, 6.

63. Prieto Acevedo, interview by the author, May 3, 2021.

64. Prieto Acevedo, interview, May 3, 2021.

65. See "Info," Germen Estudio | Museografía, accessed June 3, 2021, http://giacomocastagnola.com.

66. "Constelaciones de la Audio-Máquina en México," Germen Estudio | Museografía, accessed June 3, 2021, http://giacomocastagnola.com/en/projects/constelaciones-de-la-audio-maquina-en-mexico.

67. A. Bohlman, "Sonic Anarchy on Display," 93; and Herman, "Among Others," 79. Herman's review puts in dialogue two sound exhibits presented in Mexico roughly around the same time, *Modos de Oír: Prácticas de Arte y Sonido en México* (Modes of hearing: Art and sound practices in Mexico) and CCAMM. The former was presented at Mexico City's Ex Teresa Arte Actual from November 29, 2018, to March 31, 2019. Although Prieto Acevedo was also involved in this curatorial project, he was not the main curator.

68. Prieto Acevedo, interview, April 19, 2021.

69. Prieto Acevedo, interview, April 19, 2021.

70. Prieto Acevedo, interview, April 19, 2021.

71. Prieto Acevedo, interview, April 19, 2021.

72. Barskova, Живые картины, 7.

73. Rancière, *Politics of Aesthetics*, 39.

74. Rancière, *Politics of Aesthetics*, 12.

75. Deleuze and Guattari, *Anti-Oedipus*, 38.

CHAPTER 4. Things, Sound Objects, and Legacy at the Berliner Phonogramm-Archiv's Konrad T. Preuss Collection

An abridged version of this chapter was published as "Listening Through the Colonial Noise: Things, Sound Objects, and Legacy at the Berliner Phonogramm-Archiv's Konrad T. Preuss Collection," *Journal of the American Musicological Society* 78, no. 1 (2025): 195–240.

1. For a discussion of the problematic history and the reconciling effort informing the reconstruction of the Berlin Palace to house the Humboldt Forum, see Parzinger et al., "Im Zweifel"; and Pieken et al., "Hinter feudalen Fassaden."

2. "Rede von Kulturstaatsministerin Roth anlässlich der Eröffnung der Ostspange des Humboldt Forums," Bundesregierung, September 16, 2022, https://www.bundesregierung.de/breg-de/aktuelles/rede-von-kulturstaatsministerin-roth-anlaesslich-der-eroeffnung-der-ostspange-des-humboldt-forums-2127080.

3. See Humboldt Forum, "Mission Statement," accessed November 30, 2022, https://www.humboldtforum.org/en/about/leitbild/.

4. "Looking at Looted Art Again? Demonstration," CCWAHF, accessed December 12, 2022, https://ccwah.info/actions/1368/. For more information about the OEIN, see Herrera, "Latin America," 20–26.

5. Valdovinos, "Introducción a la edición crítica," lxxiii–lxxiv, lxxxiii.

6. Cooper, "Colonies, Empires, Nations," 7.

7. Richards, *Imperial Archive*, 1.

8. Valdovinos, "De la acción ritual," 163.

9. I have largely chosen to use the terms each of these communities uses to identify themselves as opposed to the names given to them by outsiders. Thus, I use Náayeri, Wixárika, O'dam, and Rarámuri instead of Coras, Huicholes, Tepecanos and Tepehuanos, or Tarahumaras.

10. Rutsch, *Entre el campo y el gabinete*, 223.

11. Eduard Seler, letter to Franz Boas, September 9, 1905, quoted in Rutsch, *Entre el campo y el gabinete*, 223.

12. Preuss, letter to the administration of the KMV, August 17, 1905, Acta betreffend die Reise des Dr. Preuss nach Amerika vom 17 August 1905 bis 22 August 1913 (ABRDP), Pars I. B. 59. E. No. 1487/1905, Ethnologisches Museum Berlin.

13. See Kohl, "Ethnology."

14. The terms Kaiserreich, Deutsches Kaiserreich, and German Empire denote the same political entity, the German Empire unified under the Kingdom of Prussia's monarchical rule between 1871 and the abdication of Kaiser Wilhelm II in 1918, at the end of World War I.

15. Lehmann, "K. Th. Preuß," 145. See the discussion about Johann Gottfried Herder and Wilhelm von Humboldt in Gingrich, "German-Speaking Countries," 61–75.

16. Gingrich, "German-Speaking Countries," 87.

17. For a study of the intellectual motivations of the scholars who embarked on the gathering of the objects that came to be the cornerstones of the KMV collections, see Penny, *In Humboldt's Shadow*.

18. Preuss, "Ursprung," 381.

19. See de la Mora Pérez Arce, "Investigación sobre música wixárika," 109–10.

20. Bernardelli was the teacher of two of the most prominent painters in turn-of-the-century Mexico, Gerardo "Dr. Atl" Murillo (1875–1964) and Roberto Montenegro (1885–1968).

21. For a more detailed description of Diguet's discussion of the *xaweri*, see de la Mora Pérez Arce, *Rabel de los cahuiteros*, 59–61.

22. For a discussion of the recordings made by Carl Lumholtz among the Wixáritari, see the CD booklet written by Xilonen Luna Ruiz for *Música y cantos para la luz y la oscuridad*.

23. Lira Larios, "Cilindros de cera," 216, 219.

24. Lira Larios, "Cilindros de cera," 215.

25. Rutsch, *Entre el campo y el gabinete*, 224.

26. Valdovinos, "Voces y cantos," 78–79.

27. Konrad Theodor Preuss, letter to Eduard Seler, December 24, 1905, ABRDP, Pars I. B. 59. E. No. 188/06, Ethnologisches Museum Berlin.

28. Preuss, letter to Eduard Seler, March 14, 1906, ABRDP, Pars I. B. 59. E. No. 645/06, Ethnologisches Museum Berlin.

29. The everyday details of the expedition are recounted in Preuss, report to the Ministerium der geistlichen, Unterrichts- und Medizinalangelegenheiten, May 30, 1906, ABRDP, Pars I. B. 59. E. No. 1173/06, Ethnologisches Museum Berlin.

30. Preuss, *Nayarit-Expedition*, xvi–xvii; and Preuss, *Expedición al Nayarit*, 20.

31. The *pachitas* are syncretic carnival celebrations that incorporate the life of Christ into the Náayeri cosmovision. The celebrations begin four weeks before Lent and end the night before Ash Wednesday. Like most carnival celebrations, the *pachitas* ritual is a moment of deep transgression that, in this case, highlights the transformation of a group of community girls into goddesses. See Benciolini, "Diosa entre nosotros."

32. Preuss, *Nayarit-Expedition*, xix; and Preuss, *Expedición al Nayarit*, 23.

33. Valdovinos, *Canto de la Chicharra*, 32.

34. Valdovinos states that at the time of Preuss's visit to these communities, there was an additional ceremony, the Mitote del Vino. This was a rite of passage for children that is no longer celebrated in Jesús María or San Francisco but is still observed among Náayeri groups from La Mesa del Nayar and Santa Teresa. See Valdovinos, "Introducción a la edición crítica," lxxvii–lxxviii.

35. Preuss, *Nayarit-Expedition*, xvi; and Preuss, *Expedición al Nayarit*, 20.

36. Valdovinos, "Introducción a la edición crítica," lxxi–lxxiv.

37. Preuss, *Nayarit-Expedition*, xix; and Preuss, *Expedición al Nayarit*, 23–25.

38. For a more detailed analysis of another chant in Preuss's collection and its contemporary Náayeri counterpart as recorded in the field by Valdovinos, see Valdovinos, *Canto de la Chicharra*.

39. Preuss, letter to Eduard Seler, March 14, 1906.

40. Neurath, "Tukipa Ceremonial Centers," 103; and de la Mora Pérez Arce, *Rabel de los cahuiteros*, 28.

41. Preuss, letter to KMV colleagues, December 31, 1906, ABRDP, Pars I. B. 59. E. No. 2230/07, Ethnologisches Museum Berlin.

42. Brady, *Spiral Way*, 7.

43. Valdovinos, interview by the author, October 6, 2022. See also Valdovinos, "De la acción ritual," 162; and Valdovinos, "Introducción a la edición crítica," lxxi–lxxiv.

44. Valdovinos, "Materialidad de la palabra."

45. Valdovinos, "De la acción ritual," 162.

46. Barbara Titus, quoted in Clark, "Introduction," 14.

47. Valdovinos, "Apéndices de la edición crítica," cxviii.

48. Valdovinos, "Introducción a la edición crítica," lxxii.

49. Gitelman, *Scripts, Grooves, and Writing Machines*, 1. For some of the philological implications of the phonograph, see Rehding, "Wax Cylinder Revolutions," 136–38.

50. Valdovinos, "Introducción a la edición crítica," lxxii.

51. See Valdovinos, "Chants de mitote cora."

52. For a detailed comparative study of how Preuss was able to do this with a very specific chant, see Valdovinos, *Canto de la Chicharra*.

53. On the limitations of the phonograph and the notion of the "phonograph effect," see Katz, *Capturing Sound*, 1–6.

54. Robinson, *Hungry Listening*, 6.

55. Preuss, letter to the Psychologisches Institut, November 4, 1907, copy kept at the BPA.

56. See Stumpf, "Lieder der Bellakula-Indianer."

57. Ziegler, "Historical Sound Recordings," 140.

58. Hornbostel, quoted in Ziegler, "Historical Sound Recordings," 140.

59. For a detailed account of the cultural and historical context in which Stumpf's project and the BPA developed, see Liebersohn, *Music*, 123–74. For a detailed account of the phonograph as an imperial and colonial tool, see M. Wright, *Listening After Nature*, 13–18.

60. Lehmann, "K. Th. Preuß," 147.

61. For a detailed technical description of the process of making negative galvano copies, see Hornbostel, "Phonographische Methoden."

62. Preuss, letter to Erich von Hornbostel, February 25, 1908, copy kept at the BPA.

63. Ziegler, "From Wax Cylinders," 4.

64. See Hoffman, *Knowing by Ear*; and Lange, "Archival Silences."

65. Ziegler, *Wachszylinder*, 25.

66. Ziegler, "Deutschsprachige Sammlungen," 129–30.

67. Wiedmann, interview by the author, September 13, 2022.

68. Artur Simon, Nomination of the Berlin Phonogramm-Archiv to UNESCO's Memory of the World Register, NANOPDF, accessed November 3, 2022, https://nanopdf.com/download/memory-of-the-world_pdf.

69. The German ethnologist Viola Hörbst, who was working on a PhD dissertation about healing practices among the Náayeri, was one of the very first researchers to gain access to the digitized Preuss recordings. See Hörbst, *Heilungslandschaften*.

70. Preuss, *Nayarit-Expedition*, iv.

71. The Murui-Muinanɨ are often referred to as the Witoto. However, the term *Witoto* is an Indigenous derogatory term, meaning "enemies" or "slaves," used by the Spaniards to refer to the members of these communities. Murui-Muinanɨ is a better self-designating term.

72. Kutscher, "Berlín como centro," 28.

73. Preuss, *Nahua-Texte*, vols. 1–3.

74. Valdovinos, "Introducción a la edición crítica," xcviii–xcix.

75. See Preuss, "Viaje"; and Preuss, "Nueva interpretación." The latter, in an issue of *Anales del Museo Nacional de Arqueología* dated 1931, includes a note at the end of the

article stating that the original essay had been published in the Spanish journal *Investigación y Progreso* in November 1932. It is likely that this issue of *Anales* was published in 1933 regardless of its official date.

76. See Preuss, "Concepto"; and Preuss, "Diosa de la tierra."

77. See Preuss, "Dos cantos"; and Jáuregui and Neurath, *Fiesta, literatura y magia*.

78. Hornbostel, "Zwei Gesänge der Cora-Indianer."

79. M. García, "Cómo Europa escuchó a América," 164.

80. For an in-depth explanation of the nature of the *Naturvölker* and *Kulturvölker* discussion at the turn of the twentieth century and its role in anthropology's validation of the Kaiserreich's empire-building project, see Zimmerman, *Anthropology and Antihumanism*, 149–71.

81. Preuss, letter to the administration of the KMV, September 4, 1905, ABRDP, Pars I. B. 59. E. No. 1578/1905, Ethnologisches Museum Berlin.

82. Two infamous cases in the 1880s and 1890s led the government of Porfirio Díaz to establish more explicit and stricter legislation to stop the continuous funneling of Mexican antiquities into US and European archives and museums. The first one was the diplomatic crisis that followed the Mexican Congress's refusal to allow the export of objects obtained as part of a US-sponsored expedition and excavation led by French explorer Désiré Charnay (1828–1915) in 1880–81. The second one was the buying and sacking of Chichén Itzá by Edward Thompson (1857–1935), US consul in Mérida, Yucatán, in 1894. See Sellen, *In the Shadow*; Castro, *Fabuloso saqueo*; and Sánchez Gaona, "Legislación mexicana."

83. Preuss, letter to Eduard Seler, December 24, 1905, ABRDP, Pars I. B. 59. E. No. 188/06, Ethnologisches Museum Berlin.

84. Preuss, letter to Eduard Seler, March 14, 1906.

85. Clark, "Introduction," 7.

86. Robinson, *Hungry Listening*, 36.

87. Radano and Olaniyan, "Introduction," 7.

88. Valdovinos, interview.

89. Valdovinos, "Introducción a la edición crítica," xxxiii–xxxiv. For a detailed account of the translation process, see Valdovinos Alba, "Miradas cruzadas," 186–87.

90. Valdovinos, interview. In our conversation Valdovinos did not mention Hörbst by name; she only described having met a German ethnologist who was in possession of tape copies of recordings from the BPA's Preuss Collection. See also Valdovinos, "Cilindros de cera," 227.

91. Valdovinos, interview.

92. Valdovinos, "Introducción a la edición crítica," lxxxiii.

93. "Gesetz zum Schutz," 502.

94. *Das neue Kulturgutschutzgesetz*, 14.

95. *Deutscher Bundestag* 17/13378, April 29, 2013, 8.

96. For an account about access to the objects in Preuss's Colombian collection in the Ethnologisches Museum, as well as a similar repatriation request (although regarding archaeological pieces, not ethnographic recordings), see Reyes Gavilán, "Lógicas de archivo."

97. Göbel, interview by the author, November 3, 2022.

98. Valdovinos Alba, "Miradas cruzadas," 192.

99. Koch et al., "Vorwort," 10–11.

100. Valdovinos, interview.

101. Göbel, interview.

102. Wiedmann, interview.

103. Wiedmann, interview.

104. Valdovinos, interview.

105. Koch et al., "Vorwort," 10. On issues of repatriation, collaboration with communities, and a decolonial take on sound archives, see Fox, "Repatriation as Reanimation"; Gray, "Repatriation and Decolonization"; Kummels and Cánepa Koch, "Editor's Introduction"; and Seeger, "Archives, Repatriation, Agency."

106. Valdovinos, interview.

107. Valdovinos Alba, "Miradas cruzadas," 195–96.

108. Valdovinos Alba, "Miradas cruzadas," 197.

109. On the imperial complicity of the archive, see Moreno, "Imperial Aurality."

110. For an in-depth study of how the Mexican state has defined the notion of patrimony and invoked it in similar types of repatriation claims, see Rozental, "On the Nature," 238.

111. Brulotte, *Between Art and Artifact*, 10.

112. This strategy reverberates with Alexandra Vazquez's idea that listening should "set up the detail as an observable part of the natural order" as well as Anette Hoffman's call to listen in order to make audible "that archived sound files at times speak in ways that are contrapuntal to the object status attributed to them [and that a]t times, a sharp contrast between the audible and what was registered in the written record announces the logic of archiving." Vazquez, *Listening in Detail*, 27; and Hoffman, "Introduction," 75.

113. Sterne, *Audible Past*, 219.

114. "Looking at Looted Art Again? Demonstration," CCWAHF, accessed December 12, 2022, https://ccwah.info/actions/1368; and Robinson, *Hungry Listening*, 36. See also the flyers in figure 4.2.

115. Ndikung, *Those Who Are Dead*, 47.

116. Kane, "Fluctuating Sound Object," 59.

117. Steintrager, "Sound Objects," 8.

118. Bulut, "Problem of Archiving Soundworks."

119. A more radical interpretation may argue that the wax cylinders became sound objects only owing to Náayeri and Wixáritari labor. As such, it would not be absurd to also claim them as Náayeri and Wixáritari property. I would like to thank Esteban Buch for this provocative and productive reading.

120. Mistral, *Obra reunida*, 419.

121. Bodei, *Life of Things*, 21.

122. Spitta, *Misplaced Objects*, 3.

123. Appadurai, "Introduction," 5.

124. B. Brown, "Thing Theory," 9.

125. Valdovinos, "De la acción ritual," 163.

126. Rozental, "On the Nature," 239.

CHAPTER 5. Mexican Rarities, *Disco pirata*, and the Promise of a Sound Archive of Postnational Memory

1. *Coyotek* is a neologism derived from *coyote*, the term used for those who smuggle people across the US-Mexico border. In this case, the *coyotek* specializes in connecting individuals to the technology that enables them to cross the border virtually and enter the global economy as labor force.

2. *Fichera* refers to a woman, an escort of sorts, who dances with male bar or club patrons for money. I use the neologism *cyberfichera* to place the figure of the *fichera* in this futuristic, cyborg-driven cultural environment.

3. See Rivera, *Sleep Dealer*.

4. Perales, "Critic's Choice."

5. García Canclini, *Culturas híbridas*, 293; and Montezemolo, "Cómo dejó de ser," 143.

6. See Madrid, *Nor-Tec Rifa!*, 189–204.

7. Corona and Madrid, "Introduction: The Postnational Turn," 3.

8. Khoury, "Postnational Memory," 93.

9. Young, *Postnational Memory*, 179, 279.

10. Ulibarri, *Visible Border*, 113, 115.

11. Rivera Garza, *Escrituras geológicas*, 183–84.

12. Rivera Garza, "Escrituras colindantes."

13. Martínez, interview by the author, February 16, 2024; and Mexican Rarities, accessed March 20, 2024, https://mexicanrarities.com/es/main-home.

14. For more info about El Chopo, see Monsiváis, *Rituales del caos*, 120–24; and Domínguez Prieto, "Tianguis del Chopo."

15. Martínez, interview.

16. For instance, it was at Castillo's gatherings that Villegas met experimental sound artist Mario de Vega, who was a close friend of Castillo. Eventually, de Vega became Villegas's mentor for several professional art projects.

17. Villegas, interview by the author, February 12, 2024.

18. Castillo, interview by the author, March 1, 2024.

19. Castillo, interview.

20. According to Mexico's Ministry of the Economy, a SAS is a type of commercial association that can establish a company through electronic means, without the intervention of a public notary. The benefits of this type of organization include that no minimum capital is needed, shareholder decisions can be made via electronic means, personal assets are protected by separating them from what individuals contribute to the company, and members are only responsible for the amount of their contributions represented in shares. See "Sociedad por Acciones Simplificadas (SAS)," Secretaría de Economía, accessed November 10, 2024, https://www.gob.mx/tuempresa/articulos/crea-tu-sociedad-por-acciones.

21. Martínez, interview. Started in 2000, Discogs is a user-generated database of information about audio recordings of all genres and formats, as well as a marketplace for exchanging recordings. Unlike Discogs, Mexican Rarities is not user-generated.

22. Martínez is very clear that any links or displayed materials are immediately taken down on copyright holders' requests or when copyright issues may raise a problem. Martínez, interview.

23. Rivera Garza, *Escrituras geológicas*, 186.

24. Castillo, interview.

25. Martínez, interview.

26. Derrida, "Archive Fever," 51.

27. Martínez, interview.

28. Villegas, interview.

29. *Cromometrofonía* was a term coined by Carrillo in the 1910s to replace *music*. According to Carrillo, the term *music*, from the Greek word for "Muse," was vague and nondescriptive of what the art was. *Cromometrofonía*, on the other hand, was a neologism derived from *chromo*, *metro*, and *phonos* and described the art of measuring or organizing sound and its color (the timbre of sound). See M. Mena, "Julián Carrillo," 755. *Cometa 1973* refers to the passing of comet Kohoutek close to the sun at the end of 1973.

30. Castillo, interview.

31. Villegas, interview.

32. See Mexican Rarities, accessed March 24, 2024, https://mexicanrarities.com/es/main-home/.

33. Villegas, interview.

34. Rivera Garza, *Escrituras geológicas*, 14.

35. Schloss, *Making Beats*, 93–114.

36. Rivera Garza, "Escrituras colindantes."

37. Rolando Hernández, "¿De qué hablamos cuando hablamos de México?," Mégico Máxico Musical, Mexican Rarities, accessed March 24, 2024, https://mexicanrarities.com/es/texts/megico-maxico-musical-2/.

38. Hernández, "¿De qué hablamos?"

39. Alonso Bolaños, *"Invención,"* 62.

40. Young, *Postnational Memory*, 279.

41. T. Hamilton and Hammond, "Introduction," 2.

42. See Félix Blume and Daniel Godínez Nivón, *Coro informal* (2016), Soundcloud, accessed March 24, 2024, https://soundcloud.com/losgritosdemexico/sets/coro-informal.

43. See Félix Blume, *Los gritos de México* (2014), Félix Blume's website, accessed March 24, 2024, https://felixblume.com/losgritosdemexico/.

44. See Félix Blume and Clément Janequin, *Polyphonic Chorus 'Los gritos de México'* (2016), Félix Blume's website, accessed March 24, 2024, https://felixblume.com/coro-polifonico/.

45. See Félix Blume, *Disco pirata* (2016), released March 1, 2016, Bandcamp, https://felixblume.bandcamp.com/album/disco-pirata-ciudad-de-m-xico.

46. Blume, interview by the author, February 24, 2023.

47. Since its inception in 2005, Freesound has used various types of Creative Commons licenses that allow for different types of commercial and noncommercial uses. For Blume's Freesound archive, see felix.blume, Freesound, accessed June 26, 2023, https://freesound.org/people/felix.blume/.

48. Blume, "*Los gritos de México*."

49. Rissola, "Ruidos, sonidos y políticas urbana."

50. See Blume, *Disco pirata*.

51. Aguirre Fernández, interview by the author, February 23, 2024.

52. Panagiotopoulou, interview by the author, February 19, 2024.

53. See Félix Blume, *Disco Pirata*, uploaded March 28, 2016, by Félix Blume, Vimeo, https://vimeo.com/160690792.

54. Blume, interview.

55. Blume, interview.

56. Blume, interview.

57. Medina, "Voces del Centro Histórico."

58. See Faesler Bremer, ABCDF; J. Mena, *Sensacional de diseño mexicano*; and Madrid, *Nor-Tec Rifa!*, 50–86.

59. Olalquiaga defines *third-degree kitsch* as cultural manifestations "invested with either a new or a foreign set of meanings, generating a hybrid product." Olalquiaga, *Megalopolis*, 47.

60. For a more detailed account of the *fierro viejo* phenomenon, see Rasmussen, "Resistance Resounds," 84–85. See also Usón, "'¡. . . O algo de fierro viejo que venda!'"

61. Rojo, interview by the author, February 7, 2023.

62. González Cortés, interview by the author, February 24, 2023.

63. Isabel Muñoz Cota, "La verdad es el sonido, Isabel Muñoz Cota en Cinema 20.1 con Roberto Fiesco," interview by Roberto Fiesco, *Cinema 20.1*, premiered April 23, 2020, on TV UNAM, YouTube, https://www.youtube.com/watch?v=5v4Sz3pBYds&list=PLLYD2qbK_hDuvZlcPo1ZReXMvK1Dn5rzB&index=8.

64. Panagiotopoulou, interview.

65. When I say "planned-to-fail effort," I mean that Blume knew that he was never going to be seen as a local vendor, but that cognitive dissonance in the subway passengers, having them recognize the "inauthenticity" of the situation, was an important aspect of the performance.

66. Rivera Garza, *Escrituras geológicas*, 183; and Rivera Garza, *Escribir con el presente*, 61.

67. Rancière, *Politics of Aesthetics*, 44.

CHAPTER 6. Aurality, Materiality, and the Carrillo Pianos as Archives

1. Moreno Villarreal, "Piano," 54.

2. Moreno Villarreal, "Piano," 55.

3. Luiselli, *Lost Children Archive*, 180.

4. Barskova, quoted in "International Day of Sirens," in Barskova, *Air Raid*, 146.

5. Rehding, "Instruments of Music Theory"; and Magnusson, *Sonic Writing*, 31, 56.

6. For another study of instruments as sources of microtonal music theory, see Walden, "Tanaka Shōhei's Keyboards."

7. See Miranda, "'A tocar, señoritas,'" 91–136.

8. Maier and Schulze, "Tacit Grooves," 23.

9. Maier and Schulze, "Tacit Grooves," 23.

10. Maier and Schulze, "Tacit Grooves," 34; and Hutchby, "Technologies," 444.

11. Madrid, *In Search of Julián Carrillo*, 164.

12. Magnusson, *Sonic Writing*, 5.

13. "Contienen ese manantial sonoro que ofrece perspectivas de riquezas aún insospechadas, que servirán de alimento espiritual para el futuro." Carrillo, "Invento mexicano," 436–37, originally published in *El Universal*, September 2, 1965.

14. The article was reprinted in Julián Carrillo's magazine *El Sonido 13* a month after its publication in *El Universal*. See Carrillo, "Pianos."

15. For the article on the German piano maker, see Chantavoine, "Mouvement musical." The note does not specify who the piano maker was. It could have been Grotrian-Steinweg from Braunschweig or August Förster from Leipzig since both built quarter-tone pianos in 1924. The Grotrian-Steinweg piano had three manuals or keyboards layered on top of each other and tuned a quarter tone apart. The Förster piano had two manuals or keyboards also tuned a quarter tone apart. See Davies, "Developments and Modifications," 71; and Good, "Keyboards," 204.

16. Moritz Stoehr had designed an altered version of the traditional piano keyboard that allowed for an instrument half the size of a regular piano to play the same number of pitches as a regular instrument. In essence, it was a keyboard that "afford[ed] two pianos in the space of one." See "Inventions New and Interesting," 38. This modification may have come in handy when building a quarter-tone piano that required twice the number of strings and keys as a normal piano.

17. Carrillo, "Pianos," 7.

18. Carrillo, unpublished personal journal, February 22, 1926, Julián Carrillo Archive, San Luis Potosí, Mexico. Carrillo was most likely referring to the multiple manuals or keyboards layered on top of each other that characterized the quarter-tone pianos made for Alois Hába and Ivan Wyschnegradsky by Grotrian-Steinweg and August Förster in Germany in 1924. See Davies, "Developments and Modifications," 71; and Good, "Keyboards," 204. For Hába's testimonial regarding Förster's pianos, see Hába, *Mein Weg*, 51.

19. Carrillo, journal, February 22, 1926.

20. Morones Hernández, "Análisis," 21. This report is kept at the Centro Julián Carrillo in San Luis Potosí, Mexico.

21. Morones Hernández, "Análisis," 22–24.

22. Morones Hernández, "Análisis," 10.

23. The tuning recounted here is that of the actual piano kept at the Centro Julián Carrillo as studied and described by Mario Morones Hernández; see Morones Hernández, "Análisis," 10. Buschmann's plan of the third-tone piano contains an important inaccuracy since it states that the lower pitch should be tuned to B while the subsequent whole tones (beginning three thirds of a tone above it) should be C–D–E–F-sharp–G-sharp. Evidently, in order for the tuning in thirds of a tone to make sense, in the tonal framework every three thirds of a tone should spell a whole tone. Thus, the B should in fact be an A-sharp.

24. This is a problem with the Carrillo Pianos tuned in fifths of a tone, sevenths of a tone, ninths of a tone, elevenths of a tone, thirteenths of a tone, and fifteenths of a tone. Since there is only one of each of these instruments, each of these partitions operates within just one of the whole-tone scales.

25. Madrid, *In Search of Julián Carrillo*, 168–69.

26. The fourteenth-tone Carrillo Piano has eighty-five keys, and the fifteenth-tone Carrillo Piano has ninety-one keys. Both of their keyboards are just large enough to span an octave.

27. Morones Hernández describes a second sixteenth-tone Carrillo Piano, housed by the Museo Francisco Cossío in San Luis Potosí, Mexico, which is tuned from F4 to F5. See Morones Hernández, "Análisis," 85–88.

28. Carrillo, "Pianos mexicanos," 180. See also Davies, "Developments and Modifications," 71; and Good, "Keyboards," 204.

29. Baron Moens de Fernig, General Commissary of the Belgian government at the Expo 58, letter to Dr. Francisco del Río y Cañedo, Mexican ambassador in Brussels, June 23, 1958, Julián Carrillo Archive.

30. G. Benjamin, "Julián Carrillo," 49.

31. Carrillo, "Pianos mexicanos," 181.

32. Carrillo, *Testimonio de una vida*, 280.

33. Moreno Villarreal, "Piano," 54.

34. See Alonso-Minutti, *Mario Lavista*, 59.

35. Carrillo, "Invento mexicano," 436.

36. Carrillo, "Invento mexicano," 436.

37. Kirshenblatt-Gimblett, *Destination Culture*, 3.

38. Kirshenblatt-Gimblett, *Destination Culture*, 3.

39. See, for example, Nava Loya, *¿Qué ha pasado?*

40. Rivera Garza, *Restless Dead*, 85.

41. Rivera Garza, *Restless Dead*, 77.

42. Moseley, *Keys to Play*, 67.

43. At the time he first presented the piano in thirds of a tone, Carrillo published a book that describes these processes of intervallic expansion and diminution precisely in terms of "metamorphosizing" melodic lines, motives, or entire pieces of music. See Carrillo, *Leyes de metamorfosis musicales*. See also Madrid, *In Search of Julián Carrillo*, 153–54.

44. Carrillo, "Invento mexicano," 436–37.

45. Moseley, *Keys to Play*, 109.

46. Waller, interview by the author, July 1, 2022.

47. Waller, interview.

48. Performance note in the score. See Waller Vigil, *Lhorong*, iii.

49. Waller, interview. Waller's notation system is an expansion of the standard Western music notation for quarter tones.

50. Wikipedia, s.v. "Lhorong Town," last modified September 21, 2018, https://en.wikipedia.org/wiki/Lhorong_Town.

51. Waller, interview.

52. The premiere of the first movement of *Lhorong, 31°N 96°E* can be watched at the following link: Juan Felipe Waller, "*Lhorong, 31°N 96°E* 1st mvt.," uploaded July 27, 2015, Vimeo, https://vimeo.com/134646358.

53. Fuentes, interview by the author, June 28, 2022.

54. The piece is entitled *Agua* in the 2017 version of the score. It was announced as *Carrillo y la microbelleza* (Carrillo and microbeauty) for its aborted 2021 premiere. See Bucio, "'Silencia' presupuesto."

55. Fuentes, interview.

56. Fuentes, interview.

57. A video of Fuentes's exploration of the Carrillo Pianos can be found in Arturo Fuentes, "Carrillón y la micro belleza (Arturo Fuentes 2020)," posted November 2, 2020, YouTube, https://www.youtube.com/watch?v=qdDIKuZwMB4.

58. Drott, "Spectralism."

59. Maier and Schulze, "Tacit Grooves," 32.

60. Flammer, "Sixteenth-Tone or 'Carrillo' Piano."

61. This CD features pianists Martine Joste, Sylvaine Billier, and Dominik Blum and contains works by Ernst Helmuth Flammer (b. 1949), Marc Kilchenmann (b. 1970), Bernfried E. G. Pröve (b. 1963), Martin Imholz (b. 1961), Franck Christoph Yeznikian (b. 1969), Werner Grimmel (b. 1952), and Alain Bancquart (b. 1934). See *Carrillo 16-Tone Piano*.

62. Mather, "Music in Thirds."

63. The CD features pianists Martine Joste, Sylvaine Billier, Bruce Mather, Pierrette LePage, and Dominique Roy, as well as Ondes Martenot player Jean Laurendeau. It contains music by Ivan Wyschnegradsky, Jean-Étienne Marie, Bruce Mather, John Burke (b. 1951), John Oliver (b. 1959), Marc Patch (b. 1958), Gilles Tremblay (b. 1932), Jacques Desjardin (b. 1962), Michel Gonneville (b. 1950), Vincent-Olivier Gagnon (b. 1975), and John Winiarz (b. 1952). See *Musiques en tiers*.

64. Luiselli, *Lost Children Archive*, 180.

65. Moreno Villarreal, "Piano," 54.

66. Rivera Garza, *Restless Dead*, 77.

67. Magnusson, *Sonic Writing*, 31.

68. Moseley, *Keys to Play*, 117.

69. Moreno Villarreal, "Piano," 55; and Barskova, *Air Raid*, 146.

CHAPTER 7. In Search of the Aural City: Collective Action and the Invisible Sound Archive

1. Le Guin, *Dispossessed*, 226.

2. Borges, "Biblioteca," 99.

3. Druyan et al., "Persistence of Memory."

4. Campbell, "Archivo del yo," 209.

5. Campbell, "Archivo del yo," 209.

6. López-Gay, "Clarice Lispector's Invisible Archive," 504.

7. See Madrid, *Nor-Tec Rifa!*; and Madrid, *In Search of Julián Carrillo*.

8. See Piekut, "Actor-Networks."

9. See Madrid, "Retos."

10. Madrid, *In Search of Julián Carrillo*, 4n4.

11. Deleuze and Guattari, *Thousand Plateaus*, 8–11.

12. I use the term *artivism* to describe a type of activism through artistic praxis that, as Martha Gonzalez proposes, is "an exercise of hope [that] is relational and deeply committed to social justice generated by feelings of love for struggling communities." Gonzalez, "Chican@ *Artivistas*," 4–5.

13. Blackwell et al., *Live Coding*, 3.

14. Campos Fonseca, "Noise," 161–62.

15. Ancira, interview by the author, March 15, 2023.

16. Noy, *Emergency Noises*, 16.

17. Ocelótl et al., "Saborítmico."

18. Teixido, interview by the author, March 15, 2023.

19. Villaseñor Ramírez, "LiveCodeNet Ensamble," 328. See also LiveCodeNet Ensamble, Soundcloud, accessed March 20, 2023, https://soundcloud.com/livecodenet-ensamble.

20. Hernani Villaseñor's website, accessed March 19, 2023, https://www.hernanivillasenor.com/index.html; and Taller de Audio, Centro Multimedia, CENART, accessed March 19, 2023, http://cmm.cenart.gob.mx/tallerdeaudio.

21. Armitage, "Spaces to Fail In," 32.

22. See Algorave, accessed March 20, 2023, https://algorave.com.

23. See "Guidelines," Algorave, GitHub, accessed March 20, 2023, https://github.com/Algorave/guidelines/blob/master/README_en.md.

24. Alexandra Cárdenas, interview by Laura Balboa for the radio podcast *Bulla*, August 5, 2020, Radio Nopal, Mixcloud, https://www.mixcloud.com/radionopal/bulla-episodio-005-08-agosto-2020/.

25. See Alexandra Cárdenas, tiemposdelruido, accessed March 20, 2023, https://cargocollective.com/tiemposdelruido; and Tiempos del Ruido, Soundcloud, accessed March 20, 2023, https://soundcloud.com/tiemposdelruido.

26. Villaseñor Ramírez, "LiveCodeNet Ensamble," 331.

27. See OFFAL: Orchestra for Females and Laptops, accessed March 20, 2023, https://offal.github.io.

28. See *Todas las Anteriores, Soundcloud, accessed March 23, 2023, https://soundcloud.com/todaslasanteriores; Todxs lxs Anteriores, Facebook, accessed March 23, 2023, https://www.facebook.com/TodasLasAnterioresTLA; Lbrtd Fgr, Soundcloud, accessed March 23, 2023, https://soundcloud.com/libertad-figueroa; Piaka Roela, Soundcloud, accessed March 23, 2023, https://soundcloud.com/piaka-roela; Piaka Roela: Coleccionista de Ambigüedades, accessed March 23, 2023, https://pollycromatica.wordpress.com; and Piaka Roela, Bandcamp, accessed March 23, 2023, https://piaka-roela.bandcamp.com/community.

29. Mora Flores, "Colectivas feministas," 60.

30. See Corazón de Robota, accessed March 23, 2023, https://corazonderobota.wordpress.com; Constanza Piña, "Corazón de Robota," accessed March 23, 2023, https://constanzapinadossier.wordpress.com/corazon-de-robota; Itzel Noyz, Bandcamp, accessed March 23, 2023, https://itzelnoyz.bandcamp.com; Mabe Fratti, accessed March 23, 2023, https://linktr.ee/mabefratti; and Híbridas y Quimeras, Soundcloud, accessed March 23, 2023, https://soundcloud.com/hibridas-y-quimeras.

31. Haraway, "Cyborg Manifesto," 177, quoted in Mora Flores, "Híbridas y Quimeras," 329.

32. Mora Flores, "Híbridas y Quimeras," 329.

33. A timeline of events and activities related to Híbridas y Quimeras can be found in Mora Flores, "Híbridas y Quimeras," 332–33.

34. See *Compílame'sta*, Híbridas y Quimeras, Bandcamp, released September 3, 2019, https://hibridasyquimeras.bandcamp.com/album/compilamesta; *Compilado Feminoise México*, Sisters Triangla Records, Bandcamp, released February 19, 2019, https://sisterstriangla.bandcamp.com/album/compilado-feminoise-m-xico-3; and *Feminoise Latinoamerica Vol. 1*, Sisters Triangla Records, Bandcamp, released August 10, 2016, https://sisterstriangla.bandcamp.com/album/feminoise-latinoamerica-vol-1. See also Feminoise Latinoamérica, YouTube channel, accessed March 23, 2023, https://www.youtube.com/channel/UCX8OvXzlpySgdSi-Lij6vDg.

35. *Oris Vol. 1 _ Sonidos en resiliencia*, Oris Label, Bandcamp, released June 17, 2020, https://orislabel.bandcamp.com/album/oris-vol-1-sonidos-en-resiliencia.

36. See Oris Label, Bandcamp, accessed March 23, 2023, https://orislabel.bandcamp.com; and Oris.label, Facebook, accessed March 23, 2023, https://www.facebook.com/oris.mx.

37. Manizales has been one of the most important experimental music hubs in Colombia since Camilo Rueda, Andrés Posada, and Francisco Iovino founded the Laboratorio Colombiano de Música Electroacústica "Jacqueline Nova" at the Universidad Autónoma de Manizales in 1990. See Cuellar Camargo, "Development of Electroacoustic Music," 10.

38. Teixido, interview.

39. Ocelótl et al., "Saborítmico." See RGGTRN, accessed March 28, 2023, https://rggtrn.github.io; and RGGTRN, Soundcloud, accessed March 28, 2023, https://soundcloud.com/rggtrn. For Teixido's personal archive, see Teixido, Soundcloud, accessed March 28, 2023, https://soundcloud.com/marianne_teixido.

40. See Ibermúsicas, accessed March 28, 2023, https://www.ibermusicas.org.

41. Del Angel et al., "Bellacode."

42. See Centro de Cultura Digital, accessed March 28, 2023, https://www.centroculturadigital.mx.

43. See PiranhaLab, accessed March 28, 2023, https://piranhalab.github.io/index.html.

44. See LivecoderA, accessed April 6, 2023, https://livecodera.glitch.me; and Livecodera, Twitter, accessed April 6, 2023, https://twitter.com/livecodera.

45. See Toplap, accessed April 6, 2023, https://toplap.org/; and toplap/livecodera, accessed April 6, 2023, https://github.com/toplap/livecodera.

46. See RAM, accessed April 6, 2023, https://ram-lab.glitch.me/index.html; and "Agenda: Inteligencias colectivas hackfeministas. Desmachini-zando el sonido con AI," Transhackfeminist 2022, April 21, 2022, https://zoiahorn.anarchaserver.org/thf2022/2022/04/21/inteligencias-colectivas-hackfeministas-desmachini-zando-el-sonido-con-ai/#more-258.

47. Teixido, interview.

48. Polgovsky Ezcurra, "Introduction," 7–8.
49. For work on *antimonumentos*, see Dossier G, "Antimonuments," 213–17.
50. Hirsch, "Introduction," 2.
51. Rancière, *Emancipated Spectator*, 56.
52. Teixido, interview.
53. See GEXLAT: Generx Experimentación Latinoamérica, accessed April 6, 2023, https://gexlat.github.io.
54. Bahamonde, interview by the author, January 6, 2023.
55. A similar South-to-South networking initiative is the Mapa Sonoro Latinoamericano de Circuit Bending, started by the Argentinean computer musician Laureano Cantarruti in 2020. See *Mapa Sonoro Latinoamericano de Circuit Bending*, Bandcamp, released December 11, 2020, https://lawcant.bandcamp.com/album/mapa-sonoro-latinoamericano-de-circuit-bending-1-edici-n-2020.
56. See MUSEXPLAT: Música Experimental Latinoamericana, accessed April 15, 2023, https://musexplat.com.
57. See "Centro Experimental Oído Salvaje (1996–2016)," Antenas-Intervenciones (2008–2014), accessed April 17, 2023, https://antenas-intervenciones.blogspot.com; and Centro Experimental Oído Salvaje, "Diálogos con Oído Salvaje," Vimeo channel, accessed April 17, 2023, https://vimeo.com/channels/dialogosoidosalvaje.
58. Estévez Trujillo, "Suena el capitalismo," 15. See also Estévez Trujillo, *UIO-BOG*.
59. See Hayes, "Radio as Medium"; and Villalobos López, "Shaping of Radio."
60. See *Islas Resonantes* (podcast), Radio UNAM, accessed April 6, 2023, https://www.radiopodcast.unam.mx/podcast/verserie/319#.
61. See *Bulla* (podcast), Radio Nopal, accessed November 11, 2024, https://www.radionopal.com/?/programas/bulla#/bulla; and Laura Balboa, *Bulla* (podcast), archive, accessed April 6, 2023, https://laurabalboa.com/Archive.
62. In Colombia the word *minga* also refers to a gathering in the context of the Cauca Indigenous movement. It is a form of political activism and resistance to the neoliberal state. Thanks to Juan Fernando Velásquez for this information.
63. See Minga_CASo, Mixcloud, accessed April 6, 2023, https://www.mixcloud.com/Minga_CASo; Minga CASo, Twitter, accessed November 11, 2024, https://twitter.com/CasoMinga; and Minga-CASo, YouTube channel, accessed April 6, 2023, https://www.youtube.com/@mingaradiocaso.
64. The description of the show is from Minga_CASo, Mixcloud.
65. García, interview by the author, February 12, 2023.
66. García, interview.
67. Some of these ideas were expanded in J. García, "Efecto Internet."
68. García Castilla, "Conocimientos en resonancia," 147–48.
69. See Armstrong Liberado, accessed April 25, 2023, https://armstrongliberado.wordpress.com.
70. De la Rosa-Carrillo et al., "Pedagogy of the Hack," 33.
71. One can trace the development of the initiative in the archive of Estrella Soria's personal blog. Estrella Soria, *Cíborg*, accessed February 11, 2024, https://caracolazul.espora.org/.

72. See Rancho Electrónico, accessed April 25, 2023, https://ranchoelectronico.org; and Rancho Electrónico, Twitter, accessed April 25, 2023, https://twitter.com/hackrancho.

73. See Epifonías Delirantes, accessed April 25, 2023, https://epifonias.net; Epifonías Delirantes, YouTube channel, accessed April 25, 2023, https://www.youtube.com/@epifoniasdelirantes2080; Epifonías, accessed April 25, 2023, https://archive.org/details/@epifonias; and Epifonías, *Mensajes de Voz*, Pueblo Nuevo, 2022, https://pueblonuevo.cl/catalogo/mensajes-de-voz.

74. Hernández, interview by the author, March 24, 2023.

75. The result of Hernández's research is published in Hernández, "Archivo sonoro."

76. See Mnemozine: Memorias Colectivas del Encuentro Transferencias Aurales, LIMME-FAM, accessed May 7, 2023, https://limmefamus.wordpress.com/mnemozine. See also Transferencias Aurales, YouTube channel, accessed November 11, 2024, https://www.youtube.com/@transferenciasaurales.

77. Hernández, interview.

78. See Mnemozine, Internet Archive, accessed May 7, 2023, https://archive.org/details/@mnemozine. See also Transferencias Aurales, LIMME-FAM, accessed May 7, 2023, https://limmefamus.wordpress.com/transferencias-aurales. The project's inventory can be accessed at Inventorio Mnemozine 2021, Google Sheets, accessed May 7, 2023, https://docs.google.com/spreadsheets/d/15AvNQs9gsgexpvBOCPBMy5wlM2igt3sWh-_MUOpP3_w/edit#gid=985507460.

79. Rivera Garza, *Restless Dead*, 4.

80. Bryson, "Intertextuality and Visual Poetics," 1.

81. González Aktories, interview by the author, January 4, 2023. For more details regarding the ideas about intermediality informing the work of PoéticaSonorasMX, see González Aktories et al., *Vocabulario crítico*.

82. Meza Valdez and González Aktories, "Logros y retos," 89–90.

83. See PoéticaSonoraMX, accessed May 10, 2023, https://poeticasonora.unam.mx. For a detailed description of the technical and organizational aspects behind the putting together of the repository and its digital website, see Meza Valdez and González Aktories, "Logros y retos."

84. González Aktories, interview.

85. See Modos de Oír: Prácticas de Arte y Sonido en México, Archivo Abierto, INBA, accessed May 10, 2023, https://modosdeoir.inba.gob.mx.

86. Ancira, interview. See Aedo Arankowsky et al., *Modos de oír*.

87. The intensive weeklong sound studies seminar I taught at UNAM's Escuela Nacional de Música one year earlier, in 2013—which is mentioned in chapter 2 in relation to the class's visit to the Fonoteca Nacional—was one of the first of its type in Mexico. It introduced the work of key sound studies scholars—such as Karin Bijsterveld, Nina Eidsheim, Josh Kun, Ana María Ochoa Gautier, Benjamin Piekut, Trevor Pinch, Holger Schulze, Jason Stanyek, Jonathan Sterne, Jennifer Stoever, and so on—to Mexican academic circles in a systematic manner. Several practitioners and scholars involved in Mexico City's experimental sound scene took the class, including Rossana Lara Velázquez and Omar Soriano. At the end of the seminar, Soriano created the Sound Studies Mexico Facebook group as a space to continue the conversations that began at

the seminar. However, it quickly morphed into a space for the encounter of Spanish-speaking persons interested in sound studies. See Sound Studies Mexico, Facebook, accessed May 15, 2023, https://www.facebook.com/groups/soundstudiesmexico.

88. Lara Velázquez, interview by the author, January 2, 2023.
89. Lara Velázquez, interview.
90. Ancira, interview.
91. See Lara Velázquez, "Poner la escucha."
92. See Seminario de Arte y Sonido, Espacio de Documentación, accessed May 17, 2023, https://seminarioartesonoro.wordpress.com.
93. Domínguez Ruiz, interview by the author, March 7, 2023.
94. Domínguez Ruiz, interview.
95. Domínguez Ruiz, "Dosier Modos de escucha." The name of the journal is taken from R. Murray Schafer's book *The Thinking Ear*. The founder and editor in chief of *El Oído Pensante* is Argentinean ethnomusicologist and sound studies scholar Miguel Ángel García.
96. See RESMEX: Red de Estudios sobre el Sonido y la Escucha en México, accessed May 19, 2023, https://resemx.wordpress.com.
97. Domínguez Ruiz, interview.
98. To watch the audiovisual intervention discussed here, see "El archivo inaudible," posted March 13, 2023, by Epifonías Delirantes, YouTube, https://www.youtube.com/watch?v=YEV5JHnWZ9c.
99. Hernández, interview.
100. See Archivo Sonoro La Tribu, Internet Archive, accessed May 29, 2023, https://archive.org/details/@archivo_sonoro_la_tribu?sort=titleSorter.
101. Hernández, interview.
102. Hernández, interview.
103. Luiselli, *Lost Children Archive*, 44.
104. Barskova, quoted in "International Day of Sirens," in Barskova, *Air Raid*, 146.
105. Mort, electronic conversation with the members of the author's graduate seminar Politics, Utopia, and Noise in the Sound Archive, Cornell University, March 21, 2022.
106. Hernández, interview.
107. Mora Flores, interview by the author, March 15, 2023.
108. Hirsch, "Introduction," 7.
109. Rancière, *Politics of Aesthetics*, 13.
110. Lara Velázquez, interview.
111. Ann Cvetkovich, in Arondekar et al., "Queering Archives," 222.
112. McKinney, *Information Activism*, 10.
113. Ochoa Gautier, "Reordenamiento," 83, 84.
114. Lara Velázquez, interview.
115. Lara Velázquez, interview.
116. In these visualizations I have also included some of the individuals and archival projects discussed in previous chapters.
117. See Microcircuitos, Soundcloud, accessed June 12, 2023, https://soundcloud.com/microcircuitos.

EPILOGUE. The Relevance of Archives in Times of Post-Truth: An Essay Against Nihilism in the Neoliberal Age

1. *Oxford Learner's Dictionaries*, s.v. "post-truth," accessed July 3, 2023, https://www.oxfordlearnersdictionaries.com/us/definition/english/post-truth.
2. McIntyre, *Post-Truth*, 10.
3. Gajanan, "Kellyanne Conway."
4. See Farkas and Schou, *Post-Truth*, 71–72.
5. See Atencio, "From Truth Commission."
6. Bodle and Brito, "Analysis."
7. See Estrada, *Imperio*.
8. Fernandes, "Brazil's National Museum Fire."
9. Mateos-Vega, "Fonoteca Nacional."
10. German, "Search for Truthiness," 179.
11. McIntyre, *Post-Truth*, 123–27.
12. Fish, *First*, 159, 170–71.
13. German, "Search for Truthiness," 187.
14. Fish, *First*, 172.
15. García Castilla, "Conocimientos en resonancia," 151–52; and Domínguez Ruiz, "Oído," 103. Both García Castilla's and Domínguez Ruiz's ideas about listening resonate with Doyle Srader's and Chris McRae's conceptualizations of performative listening as a central aspect of communication. See Srader, "Performative Listening"; and McRae, *Performative Listening*.
16. Lara Velázquez, interview by the author, January 2, 2023.
17. Ochoa Gautier, *Aurality*, 34.
18. Vazquez, *Listening in Detail*, 27; and D. Garcia, *Listening for Africa*, 74–80.
19. Taylor, *Archive and the Repertoire*, 16.
20. Rivera Garza, "Escrituras colindantes."
21. Madrid, "Rigo Tovar," 115. See also Madrid, "Sonares dialécticos."
22. Sobchack, "When the Ear Dreams," 8.
23. Walker, "Archival Resonances," 207.
24. *Merriam-Webster*, s.v. "nihilism," accessed August 2, 2023, https://www.merriam-webster.com/dictionary/nihilism.
25. Arendt, *Life of the Mind*, 1:177.
26. W. Brown, *In the Ruins*, 16.
27. See Žižek, *Sublime Object of Ideology*, 81.
28. Erll, "Cultural Memory Studies," 7.
29. See "Imagining Memory," in Brockmeier, *Beyond the Archive*, 29–62.
30. See Lara Velázquez, "¿Con quién (no) (me) escucho?"

Bibliography

Institutional Archives

Archivo Julián Carrillo. San Luis Potosí, Mexico.
Berliner Phonogramm-Archiv. Berlin, Germany.
Ethnologisches Museum Berlin. Berlin, Germany.
Fonoteca Nacional. Mexico City, Mexico.

Newspapers and Magazines

Atlantic
Bundesgesetzblatt
Deutscher Bundestag
Diario Oficial de la Federación
El Economista
El País
El Universal
Forbes
Guardian
La Jornada
La Jornada Semanal
New York Times
Reforma
Reuters
Time Magazine
Wired

Discography

Blume, Félix. *Disco pirata*. Bandcamp, 2016. https://felixblume.bandcamp.com/album/disco-pirata-ciudad-de-m-xico.
The Carrillo 16-Tone Piano. Berlin: Edition Zeitklang, 2003.

Early Hi-Fi. Wide Range and Stereo Recordings Made by Bell Telephone Laboratories in the 1930s. Leopold Stokowski Conducting the Philadelphia Orchestra, 1931–1932. New York: Bell Laboratories, 1981.

Música y cantos para la luz y la oscuridad. Mexico City: Comisión Nacional para el Desarrollo de los Pueblos Indígenas/American Museum of Natural History, 2005.

Musiques en tiers et en seizièmes de ton. Montreal: Société Nouvelle D'Enregistrement, 2009.

Preuss, Konrad Theodor. *Walzenaufnahmen der Cora und Huichol aus Mexiko 1905–1907/ Grabaciones en cilindros de cera de los coras y los huicholes de México.* Berlin: Staatliche Museen zu Berlin/Preußischer Kulturbesitz/Ibero-Amerikanisches Institut/Instituto Nacional de Lenguas Indígenas/Secretaría de Educación Pública de México, 2013.

A Program of Mexican Music. New York: Museum of Modern Art, 1941.

Interviews by the Author

Aguirre Fernández, Diego. Virtual. February 23, 2024.
Ancira, Andrea. Virtual. March 15, 2023.
Bahamonde, Emilia. Virtual. January 6, 2023.
Blume, Félix. Virtual. February 24, 2023.
Castillo, Arturo. Virtual. March 1, 2024.
Domínguez Ruiz, Ana Lidia. Virtual, March 7, 2023.
Fuentes, Arturo. Virtual. June 28, 2022.
García, Jorge David "Sísifo" or "Sísifo Pedroza." Virtual. February 12, 2023.
Göbel, Barbara. Berlin, Germany. November 3, 2022.
González Aktories, Susana. Virtual. January 4, 2023.
González Cortés, César, and Alejandro Díaz Sánchez (Estudio Hasan). Virtual. February 24, 2023.
Hernández, Carlos "Arsan." Virtual. March 24, 2023.
Lara Velázquez, Rossana. Virtual. January 2, 2023.
Martínez, Alfredo. Virtual. February 16, 2024.
Mora Flores, Ana Alfonsina. Virtual. March 15, 2023.
Panagiotopoulou, Despina. Virtual. February 19, 2024.
Prieto Acevedo, Carlos. Virtual. April 19, 2021, and May 3, 2021.
Rivas, Francisco "Tito." Mexico City, Mexico. July 12, 2013.
Rodríguez Reséndiz, Perla Olivia. Mexico City, Mexico. July 24, 2013.
Rojo, Daniel. Virtual. February 7, 2023.
Seiffarth, Carsten. Berlin, Germany. February 15, 2023.
Teixido, Marianne. Virtual. March 15, 2023.
Valdovinos, Margarita. Virtual. October 6, 2022.
Villegas, Juan Pablo. Virtual. February 12, 2024.
Waller, Juan Felipe. Virtual. July 1, 2022.
Wiedmann, Albrecht. Berlin, Germany. September 13, 2022.

Books, Articles, and Other Sources

Adler, Guido. "Umfang, Methode und Ziel der Musikwissenschaft." *Vierteljahrsschrift für Musikwissenschaft* 1 (1885): 5–20.

Aedo Arankowsky, Tania, Cinthya García Leyva, Susana González Aktories, Rossana Lara Velázquez, Bárbara Perea Legorreta, Carlos Prieto Acevedo, Francisco "Tito" Rivas Mesa, and Manuel Rocha Iturbide. *Modos de oír: Una heterofonía sobre arte y sonido en México*. Mexico City: Tumbalacasa, 2019.

Aguirre Lora, María Esther. "La Escuela Nacional de Música de la UNAM (1929–1940): Compartir un proyecto." *Perfiles Educativos* 28, no. 111 (2006): 89–111.

Aharonián, Coriún, and Fabrice Lengronne, eds. *La música y los pueblos indígenas*. Montevideo: Centro Nacional de Documentación Musical Lauro Ayestarán, 2018.

Ahmed, Sara. *Willful Subjects*. Durham, NC: Duke University Press, 2014.

Alegre González, Lizette. "Más allá de la abyección aural: Hacia una escucha híbrida de la diferencia." In *Sonido, escucha y poder*, edited by Lizette Alegre González and Jorge David García, 11–23. Mexico City: Universidad Nacional Autónoma de México, 2021.

Alegre González, Lizette, and Jorge David García. "Presentación." In *Sonido, escucha y poder*, edited by Lizette Alegre González and Jorge David García, 5–8. Mexico City: Universidad Nacional Autónoma de México, 2021.

Alegre González, Lizette, and Jorge David García, eds. *Sonido, escucha y poder*. Mexico City: Universidad Nacional Autónoma de México, 2021.

Alonso Bolaños, Marina. *La "invención" de la música indígena de México: Antropología e historia de las políticas culturales del siglo XX*. Buenos Aires: Editorial Sb, 2008.

Alonso-Minutti, Ana R. "The Composer as Intellectual: Carlos Chávez and El Colegio Nacional." In *Carlos Chávez and His World*, edited by Leonora Saavedra, 273–94. Princeton, NJ: Princeton University Press, 2015.

Alonso-Minutti, Ana R. *Mario Lavista: Mirrors of Sounds*. New York: Oxford University Press, 2023.

Alviña, Leandro. *La música incaica*. Lima: Universidad del Cuzco, 1908.

Ancira García, Andrea, Rossana Lara Velázquez, and Inti Meza Villarino, eds. "La orquesta desafina: Prácticas experimentales alrededor del sonido en la Ciudad de México." Unpublished manuscript. Mexico City, 2014.

Andrade, Oswald de. "Manifiesto antropófago." *Revista de Antropofagia* 1 (1928): 3, 7.

Appadurai, Arjun. "Introduction: Commodities and the Politics of Value." In *The Social Life of Things: Commodities in Cultural Perspective*, edited by Arjun Appadurai, 3–63. Cambridge: Cambridge University Press, 1986.

Arce, Julio. "Los archivos de andar por casa: El giro digital y la investigación de la música popular." In *Los archivos de las (etno)musicologías: Reflexiones sobre sus usos, sentidos y condición virtual*, edited by Miguel A. García, 107–24. Berlin: Ibero-Amerikanisches Institut—Preußischer Kulturbesitz, 2022.

Arendt, Hannah. *The Life of the Mind*. Vol. 1, *Thinking*. New York: Harcourt, 1977.

Armitage, Joanne. "Spaces to Fail In: Negotiating Gender, Community and Technology in Algorave." *Dancecult: Journal of Electronic Dance Music Culture* 10, no. 1 (2018): 31–45.

Armstrong, Carol. "This Photography Which Is Not One: In the Gray Zone with Tina Modotti." *October* 101 (2002): 19–52.

Arondekar, Anjali. *For the Record: On Sexuality and the Colonial Archive in India.* Durham, NC: Duke University Press, 2003.

Arondekar, Anjali, Ann Cvetkovich, Christina B. Hanhardt, Regina Kunzel, Tavia Nyong'o, Juana María Rodríguez, and Susan Stryker. "Queering Archives: A Roundtable Discussion." *Radical History Review* 122 (2015): 211–31.

Atencio, Rebecca J. "From Truth Commission to Post-Truth Politics in Brazil." *Current History* 118, no. 805 (2019): 68–74.

Baker, Christina. "Affective Acoustic Territories: Mapping and Performing Disappearance in *Zona clausurada* (2022) by Teatro Línea de Sombra." *Hispanic Review* 92, no. 3 (2024): 427–52.

Baker, Christina. *Sonic Strategies: Performing Mexico's War on Drugs, Mourning, and Feminicide.* Nashville, TN: Vanderbilt University Press, 2024.

Baker, Geoffrey. *Imposing Harmony: Music and Society in Colonial Cuzco.* Durham, NC: Duke University Press, 2008.

Barnett, Kyle. *Record Cultures: The Transformation of the U.S. Recording Industry.* Ann Arbor: University of Michigan Press, 2020.

Barskova, Polina. *Air Raid.* Translated by Valzhyna Mort. Brooklyn, NY: Ugly Duckling, 2021.

Barskova, Polina. *Живые картины.* Saint Petersburg: Ivan Limbakh, 2019.

Baudrillard, Jean. *Cultura y simulacro.* Translated by Antoni Vicens and Pedro Rovira. Barcelona: Kairós, 1978.

Baudrillard, Jean. *Simulacra and Simulation.* Translated by Sheila Faria Glaser. 1981. Ann Arbor: University of Michigan Press, 1994.

Becerril, Emiliano. "Raúl Hellmer: Antropología del ritmo." *La Jornada Semanal*, September 5, 2010. https://www.jornada.com.mx/2010/09/05/sem-emiliano.html.

Benciolini, Maria. "Una diosa entre nosotros: Objetos y relaciones sociales en un ritual Náayeri." *Trace* 73 (2018): 37–59.

Benjamin, Gerald R. "Julián Carrillo and 'Sonido Trece.'" *Anuario* 3 (1967): 33–68.

Benjamin, Walter. *The Origin of German Tragic Drama.* 1925. Translated by John Osborne. New York: Verso, 1998.

Beristáin-Cardoso, José-Ángel. "Educación artística y autonomía universitaria en México: Orígenes de la Orquesta Sinfónica de la Universidad Nacional (1929–1936)." *Revista Iberoamericana de Educación Superior* 12, no. 33 (2021): 77–100.

Beverly, John. "Writing in Reverse: On the Project of the Latin American Subaltern Studies Group." In *The Latin American Cultural Studies Reader*, edited by Ana del Sarto, Alicia Ríos, and Abril Trigo, 623–41. Durham, NC: Duke University Press, 2004.

Bieletto-Bueno, Natalia. "Introducción: Sonido y escucha en las ciudades latinoamericanas; Derecho a la ciudad, poder y ciudadanía." In *Ciudades vibrantes: Sonido y experiencia aural urbana en América Latina*, edited by Natalia Bieletto-Bueno, 9–49. Santiago de Chile: Ediciones Universidad Mayor, 2021.

Bieletto-Bueno, Natalia. "Noise, Soundscape, and Heritage: Sound Cartographies and Urban Segregation in Twenty-First-Century Mexico City." *Journal of Urban Cultural Studies* 4, nos. 1–2 (2017): 107–26.

Bioy Casares, Adolfo. *La invención de Morel.* 1953. Buenos Aires: Emecé Editores, 1968.

Bitrán Goren, Yael. "¡Hurra a Henrietta Yurchenco! Folclorista pionera." *Cuadernos de Música UNAM* 2 (2022): 8–22.

Bjerregaard, Peter. "Dissolving Objects: Museums, Atmosphere and the Creation of Presence." *Emotion, Space and Society* 15 (2015): 74–81.

Blackwell, Alan F., Emma Cocker, Geoff Cox, Alex McLean, and Thor Magnusson. *Live Coding: A User's Manual*. Cambridge, MA: MIT Press, 2022.

Boadle, Anthony, and Ricardo Brito. "Analysis: Brazil's Bolsonaro Caught Off Guard by Campaign's Ugly Closing Chapter." *Reuters*, October 18, 2022. https://www.reuters.com/world/americas/brazils-bolsonaro-caught-off-guard-by-campaigns-ugly-closing-chapter-2022-10-18.

Bodei, Remo. *The Life of Things, the Love of Things*. Translated by Murtha Baca. New York: Fordham University Press, 2015.

Bohlman, Andrea F. "Sonic Anarchy on Display." *Sound Studies: An Interdisciplinary Journal* 3, no. 1 (2017): 90–93.

Borges, Jorge Luis. "La biblioteca de Babel." In *Ficciones*, 87–100. 1944. New York: Vintage Español, 2012.

Brady, Erika. *A Spiral Way: How the Phonograph Changed Ethnography*. Jackson: University Press of Mississippi, 1999.

Brockmeier, Jens. *Beyond the Archive: Memory, Narrative, and the Autobiographical Process*. New York: Oxford University Press, 2015.

Bronfman, Alejandra, and Andrew Grant Wood. "Introduction: Media, Sound, and Culture." In *Media, Sound, and Culture in Latin America and the Caribbean*, edited by Alejandra Bronfman and Andrew Grant Wood, ix–xvi. Pittsburgh: University of Pittsburgh Press, 2012.

Brown, Bill. "Thing Theory." *Critical Inquiry* 28, no. 1 (2001): 1–22.

Brown, Wendy. *In the Ruins of Neoliberalism: The Rise of Antidemocratic Politics in the West*. New York: Columbia University Press, 2019.

Brulotte, Ronda L. *Between Art and Artifact: Archaeological Replicas and Cultural Production in Oaxaca, Mexico*. Austin: University of Texas Press, 2012.

Bryson, Norman. "Intertextuality and Visual Poetics." *Critical Texts* 4, no. 2 (1987): 1–6.

Bucio, Erika. "'Silencia' presupuesto tributo a Julián Carrillo." *Reforma*, December 27, 2021. https://www.reforma.com/silencia-presupuesto-tributo-a-julian-carrillo/ar2321832.

Bulut, Zeynep. "The Problem of Archiving Soundworks." *Ethnomusicology Review* 11 (2006). https://ethnomusicologyreview.ucla.edu/journal/volume/11/piece/508.

Burns, Chelsea. "Listening for Modern Latin America: Identity and Representation in Concert Music, 1920–1940." PhD diss., University of Chicago, 2016.

Callahan, Sara. *Art + Archive: Understanding the Archival Turn in Contemporary Art*. Manchester: Manchester University Press, 2022.

Campbell, Baird. "El archivo del yo: Activismo trans y redes sociales en Santiago de Chile." In *Antropología y archivos en la era digital: Usos emergentes de lo audiovisual*, edited by Ingrid Kummels and Gisela Cánepa Koch, 2:205–29. Lima: Pontificia Universidad Católica del Perú, 2020. EPUB.

Campos Fonseca, Susan. "Noise, Sonic Experimentation, and Interior Coloniality in Costa Rica." In *Experimentalisms in Practice: Music Perspectives from Latin America*, edited by Ana R. Alonso-Minutti, Eduardo Herrera, and Alejandro L. Madrid, 161–86. New York: Oxford University Press, 2018.

Carmona, Gloria, ed. *Carlos Chávez: Escritos periodísticos*. Mexico City: El Colegio Nacional, 1997.

Carmona, Gloria. "Prólogo." In *Hacia una nueva música: Ensayo sobre música y electricidad*, by Carlos Chávez, 9–22. Mexico City: El Colegio de México, 1992.

Carrillo, Julián. "Invento mexicano: Pianos Carrillo." In *Errores universales en música y física musical*, 434–37. Mexico City: Seminario de Cultura Mexicana, 1967.

Carrillo, Julián. *Leyes de metamorfosis musicales*. Mexico City: Julián Carrillo, 1949.

Carrillo, Julián. "Los pianos con cuartos de tono construidos en Alemania y Estados Unidos." *El Sonido 13* 2, no. 1 (1925): 7–8.

Carrillo, Julián. "Pianos mexicanos." In *Errores universales en música y física musical*, 179–82. Mexico City: Seminario de Cultura Mexicana, 1967.

Carrillo, Julián. *Testimonio de una vida*. San Luis Potosí: Comité Organizador "San Luis 400," 1992.

Castañeda, Daniel, and Vicente T. Mendoza. *Instrumental precortesiano*. Vol. 1. Mexico City: Imprenta del Museo Nacional de Arqueología, Historia y Etnografía, 1933.

Castañeda, Daniel, and Vicente T. Mendoza. "Los pequeños percutores en las civilizaciones precortesianas." *Anales del Museo Nacional de Arqueología, Historia y Etnografía* 8, no. 3 (1933): 449–576.

Castañeda, Daniel, and Vicente T. Mendoza. "Los percutores precortesianos." *Anales del Museo Nacional de Arqueología, Historia y Etnografía* 8, no. 2 (1933): 275–86.

Castañeda, Daniel, and Vicente T. Mendoza. "Los teponaztlis en las civilizaciones precortesianas." *Anales del Museo Nacional de Arqueología, Historia y Etnografía* 8, no. 1 (1933): 5–83.

Castañeda, Daniel and Vicente T. Mendoza. "Percutores precortesianos (apéndice)." *Anales del Museo Nacional de Arqueología, Historia y Etnografía* 8, no. 4 (1933): 649–66.

Castillo Ramírez, Guillermo, Julie Anne Boudreau, and Adriana Ávila Farfán. "Tianguis del Chopo: Espacio urbano de regulación/transgresión." *Revista Mexicana de Sociología* 82, no. 3 (2020): 557–85.

Castro, Pedro. *El fabuloso saqueo del cenote sagrado de Chichén Itzá*. Mexico City: Universidad Autónoma Metropolitana, 2016.

Caswell, Michelle. "Seeing Yourself in History: Community Archives and the Fight Against Symbolic Annihilation." *Public Historian* 36, no. 4 (2014): 26–37.

Chamorro, Arturo. "La etnomusicología mexicana: Un acercamiento a sus fuentes de estudio." In *Memoria del Primer Congreso de la Sociedad Mexicana de Musicología*, 80–95. Ciudad Victoria: Gobierno Constitucional del Estado de Tamaulipas, 1985.

Chantavoine, Jean. "Le mouvement musical à l'Étranger: Allemagne." *Le Ménestrel* 86, no. 22 (1924): 250.

Chávez, Carlos. "El Conservatorio en 1929." In *Carlos Chávez: Escritos periodísticos (1916–1939)*, edited by Gloria Carmona, 147–51. Mexico City: El Colegio Nacional, 1997.

Chávez, Carlos. *Hacia una nueva música: Ensayo sobre música y electricidad*. 1937. Mexico City: El Colegio Nacional, 1992.

Chávez, Carlos. Introduction to *Mexican Music*, 5–11. New York: Museum of Modern Art, 1940.

Chávez, Carlos. "La música, la universidad y el estado." In *Carlos Chávez: Escritos periodísticos (1916–1939)*, edited by Gloria Carmona, 121–27. Mexico City: El Colegio Nacional, 1997.

Chávez, Carlos. "La Orquesta Sinfónica Mexicana." In *Carlos Chávez: Escritos periodísticos (1916–1939)*, edited by Gloria Carmona, 93–94. Mexico City: El Colegio Nacional, 1997.

Chávez, Carlos. "Los instrumentos eléctricos de reproducción musical." In *Carlos Chávez: Escritos periodísticos*, edited by Gloria Carmona, 221–28. Mexico City: El Colegio Nacional, 1997.

Chávez, Carlos. "México no necesita doctores ni bachilleres en música: Una rectificación del maestro Carlos Chávez." In *Carlos Chávez: Escritos periodísticos*, edited by Gloria Carmona, 117–19. Mexico City: El Colegio Nacional, 1997.

Chávez, Carlos. "Música y física." In *Carlos Chávez: Escritos periodísticos*, edited by Gloria Carmona, 209–13. Mexico City: El Colegio Nacional, 1997.

Chávez, Carlos. "Producción y reproducción musical." In *Carlos Chávez: Escritos periodísticos*, edited by Gloria Carmona, 215–19. Mexico City: El Colegio Nacional, 1997.

Chávez, Carlos. *Toward a New Music: Music and Electricity*. New York: W. W. Norton, 1937.

Chávez, Carlos. *Xochipilli (An Imagined Aztec Music for Piccolo, Flute, E Flat Clarinet, Trombone and Six Percussion Players)*. New York: Mills Music, 1964.

Chion, Michel. "Reflections on the Sound Object and Reduced Listening." In *Sound Objects*, edited by James A. Steintrager and Rey Chow, 23–32. Durham, NC: Duke University Press, 2019.

Choque Vaca, Rubén. "Archivos sonoros y visuales de pueblos indígenas y originarios de Bolivia." In *Memorias del primer seminario internacional los archivos sonoros y visuales en América Latina*, edited by Perla Olivia Rodríguez Reséndiz, 165–70. Mexico City: Radio Educación, 2002.

Clark, Emily Hansell. "Introduction: Audibilities of Colonialism and Extractivism." *The World of Music* 10, no. 2 (2021): 5–20.

Cook, Terry. "The Archive(s) Is a Foreign Country: Historians, Archivists, and the Changing Archival Landscape." *American Archivist* 74, no. 2 (2011): 600–632.

Cooper, Frederick. "Colonies, Empires, Nations: A Twentieth Century History." In *The Cultural Legacy of German Colonial Rule*, edited by Klaus Mühlhahn, 1–22. Berlin: Walter de Gruyter, 2017.

Corona, Ignacio, and Alejandro L. Madrid. "Introduction: The Postnational Turn in Music Scholarship and Music Marketing." In *Postnational Musical Identities: Cultural Production, Distribution, and Consumption in a Globalized Scenario*, edited by Ignacio Corona and Alejandro L. Madrid, 3–22. Lanham, MD: Lexington Books, 2008.

Cuellar Camargo, Lucio Edilberto. "The Development of Electroacoustic Music in Colombia, 1965–1999: An Introduction." *Leonardo Music Journal* 10 (2000): 7–12.

Cuevas, José Luis. "The Cactus Curtain: An Open Letter on Conformity in Mexican Art." *Evergreen Review* 2, no. 7 (1959): 111–20.

Curcio-Nagy, Linda. "Giants and Gypsies: Corpus Christi in Colonial Mexico City." In *Rituals of Rule, Rituals of Resistance: Public Celebrations and Popular Culture in Mexico*, edited by William Beezley, Cheryl English Martin, and William E. French, 1–26. Wilmington, DE: Scholarly Resources, 1994.

Cvetkovich, Ann. *An Archive of Feelings: Trauma, Sexuality, and Lesbian Public Cultures*. Durham, NC: Duke University Press, 2003.

Das neue Kulturgutschutzgesetz: Handreichung für die Praxis. Rostock: Publikationsversand der Bundesregierung, 2017.

Davies, Hugh. "Developments and Modifications." In *The New Grove: The Piano*, edited by Stanley Sadie, 70–74. New York: W. W. Norton, 1988.

de la Mora Pérez Arce, Rodrigo. *El rabel de los cahuiteros: Unidad y diversidad en la expresión musical wixárica; El caso del xawari y el kanari*. Guadalajara: La Zonámbula y Secretaría de Cultura de Jalisco, 2018.

de la Mora Pérez Arce, Rodrigo. "La investigación sobre música wixárika: Un recuento comentado a más de un siglo de sus inicios." In *Investigación musical desde Jalisco: Crítica, paralaje y memoria . . . ¿local, regional, universal?*, edited by Gabriel Pareyón, 105–40. Guadalajara: Secretaría de Cultura de Jalisco, 2020.

Del Angel, Luis N., Marianne Teixido, Emilio Ocelótl, Ivanka Cotrina, and David Ogborn. "Bellacode: Localized Textual Interfaces for Live Coding Music." Paper presented at the International Conference on Live Coding in Madrid, Spain, January 17, 2019. https://iclc.toplap.org/2019/papers/paper111.pdf.

de la Rosa-Carrillo, León, in conversation with Estrella Soria, Gato Viejo, Hacklib, and Jorge David García (AKA Sísifo Pedroza). "Pedagogy of the Hack: El Rancho Electrónico and the Culture of Surveillance." In *Makers, Crafters, Educators: Working for Cultural Change*, edited by Elizabeth Garber, Lisa Hochtritt, and Manisha Sharma, 33–38. New York: Routledge, 2019.

Deleuze, Gilles, and Félix Guattari. *Anti-Oedipus: Capitalism and Schizophrenia*. Translated by Robert Hurley, Mark Seem, and Helen R. Lane. 1972. Minneapolis: University of Minnesota Press, 2003.

Deleuze, Gilles, and Félix Guattari. *A Thousand Plateaus: Capitalism and Schizophrenia*. Translated by Brian Massumi. 1980. Minneapolis: University of Minnesota Press, 1987.

Denning, Michael. *Noise Uprising: The Audiopolitics of a World Musical Revolution*. London: Verso, 2015.

Derrida, Jacques. "Archive Fever: A Freudian Impression." *Diacritics* 25, no. 2 (1995): 9–63.

d'Harcourt, Raoul, and Marguerite d'Harcourt. *La musique des Incas et ses survivances*. 2 vols. París: Librairie orientaliste Paul Geuthner, 1925.

Domínguez Prieto, Olivia. "El tianguis del Chopo: Un espacio alternativo en la ciudad." *Cuicuilco: Revista de Ciencias Antropológicas* 8, no. 22 (2001): 59–70.

Domínguez Ruiz, Ana Lidia M., ed. "Dosier Modos de escucha." *El Oído Pensante* 7, no. 2 (2019): 92–217.

Domínguez Ruiz, Ana Lidia M. "El oído: Un sentido, multiples escuchas; Presentación del dosier Modos de escucha." *El Oído Pensante* 7, no. 2 (2019): 92–110.
Dossier G, "Antimonuments: The Brigade for Memory." In *The New Public Art: Collectivity and Activism in Mexico Since the 1980s*, edited by Mara Polgovsky Ezcurra, 213–17. Austin: University of Texas Press, 2023.
Douglass, Patrice, and Frank B. Wilderson III. "The Violence of Presence: Metaphysics in a Blackened World." *Black Scholar* 43, no. 4 (2013): 117–23.
Drott, Eric. "Spectralism." In *Routledge Encyclopedia of Modernism*, edited by Stephen Ross. New York: Routledge, 2016. https://www.rem.routledge.com/articles/spectralism.
Druyan, Ann, Carl Sagan, and Steven Soter, dirs. "The Persistence of Memory." Episode 11 of *Cosmos: A Personal Voyage*. Los Angeles: Public Broadcasting System, 1980.
Erll, Astrid. "Cultural Memory Studies: An Introduction." In *Cultural Memory Studies: An International and Interdisciplinary Handbook*, edited by Astrid Erll and Ansgar Nünning, 1–15. Berlin: Walter de Gruyter, 2008.
Erlmann, Veit. *Reason and Resonance: A History of Modern Aurality*. Princeton, NJ: Princeton University Press, 2014.
Estévez Trujillo, Mayra. "Suena el capitalismo en el corazón de la selva." *Nómadas* 45 (2016): 13–25.
Estévez Trujillo, Mayra. UIO-BOG: *Estudios sonoros desde la región andina*. Quito: Centro Experimental Oído Salvaje, 2008.
Estrada, Luis. *El imperio de los otros datos: Tres años de falsedades y engaños desde Palacio*. Mexico City: Grijalbo, 2022.
Faesler Bremer, Cristina, ed. ABCDF: *Diccionario gráfico de la ciudad de México*. Mexico City: Diamantina, 2001.
Farkas, Johan, and Jannick Schou. *Post-Truth, Fake News and Democracy: Mapping the Politics of Falsehood*. New York: Routledge, 2020.
Fernandes, Mauro. "How Brazil's National Museum Fire Became a Battleground of a New Culture War." *The National*, September 25, 2018. https://www.thenationalnews.com/world/the-americas/how-brazil-s-national-museum-fire-became-a-battleground-of-a-new-culture-war-1.773887.
Fish, Stanley. *The First: How to Think About Hate Speech, Campus Speech, Religious Speech, Fake News, Post-Truth, and Donald Trump*. New York: One Signal, 2019.
Flammer, Ernst Helmuth. "The Sixteenth-Tone or 'Carrillo' Piano." CD Booklet in *The Carrillo 16-Tone Piano*. Berlin: Edition Zeitklang, 2003.
Fletcher, Alice C. *Indian Story and Song from North America*. Boston: Small Maynard, 1900.
"Foreword of the Mexican Department of Foreign Affairs." In *20 Centuries of Mexican Art/20 siglos de arte mexicano*. New York: Museum of Modern Art, 1940.
Foucault, Michel. *Power/Knowledge: Selected Interviews and Other Writings, 1972–1977*. Edited by Colin Gordon. Translated by Colin Gordon, Leo Marshall, John Mepham, and Kate Soper. New York: Pantheon, 1980.
Fox, Aaron A. "Repatriation as Reanimation Through Reciprocity." In *Cambridge History of World Music*, edited by Philip V. Bohlman, 522–54. Cambridge: Cambridge University Press, 2013.

Franco, Jean. *The Decline and Fall of the Lettered City: Latin America in the Cold War.* Cambridge, MA: Harvard University Press, 2002.

Frohmann, Bernd. "Documentary Ethics, Ontology, and Politics." *Archival Science* 8, no. 3 (2008): 165–80.

Gajanan, Mahita. "Kellyanne Conway Defends White House's Falsehoods as 'Alternative Facts.'" *Time Magazine*, January 22, 2017. https://time.com/4642689/kellyanne-conway-sean-spicer-donald-trump-alternative-facts/.

Galindo, Miguel. *Historia de la música mejicana.* Colima: El Dragón, 1933.

Garcia, David F. *Listening for Africa: Freedom, Modernity, and the Logic of Black Music's African Origins.* Durham, NC: Duke University Press, 2017.

García, Miguel A. "Archivos sonoros o la poética de un saber inacabado." *Artefilosofia* 6, no. 11 (2011): 36–50.

García, Miguel A. "Cómo Europa escuchó a América." In *La música y los pueblos indígenas*, edited by Coriún Aharonián and Fabrice Lengronne, 153–66. Montevideo: Centro Nacional de Documentación Musical Lauro Ayestarán, 2018.

García, Miguel A. "El archivo sonoro y sus ausencias." In *Los archivos de las (etno)musicologías: Reflexiones sobre sus usos, sentidos y condición virtual*, edited by Miguel A. García, 65–77. Berlin: Ibero-Amerikanisches Institut—Preußischer Kulturbesitz, 2022.

García, Miguel A., ed. *Los archivos de las (etno)musicologías: Reflexiones sobre sus usos, sentidos y condición virtual.* Berlin: Ibero-Amerikanisches Institut—Preußischer Kulturbesitz, 2022.

García Canclini, Néstor. *Consumers and Citizens: Globalization and Multicultural Conflicts.* Minneapolis: University of Minnesota Press, 2001.

García Canclini, Néstor. *Culturas híbridas: Estrategias para entrar y salir de la modernidad.* Mexico City: Grijalbo, 1989.

García Castilla, Jorge David. "Conocimientos en resonancia: Hacia una epistemología de la escucha." *El Oído Pensante* 7, no. 2 (2019): 135–54.

García [Castilla], Jorge David. "El efecto Internet: Relaciones entre creación musical y tecnología digital en la 'música COVID.'" In *Algoritmos arruinados: Perspectivas situadas de tecnología musical*, edited by Jorge David García and Pablo Silva Treviño, 235–59. Mexico City: Universidad Nacional Autónoma de México, 2022.

García Morillo, Roberto. *Carlos Chávez: Vida y obra.* Mexico City: Fondo de Cultura Económica, 1960.

García Ranz, Francisco. "¿Quién diablos fue ese Beno Lieberman?" *La Manta y la Raya* 5 (2017): 13–19.

Gerber Bicecci, Verónica, and Luis Villoro. *La significación del silencio.* Mexico City: Ñ, 2018.

German, Daniel. "A Search for Truthiness: Archival Research in a Post-Truth World." In *Do Archives Have Value?*, edited by Michael Moss and David Thomas, 167–93. London: Facet, 2019.

"Gesetz zum Schutz deutschen Kulturgutes gegen Abwanderung. Vom 6. August 1955." *Bundesgesetzblatt*, part I (1955): 501–3.

Gingrich, Andre. "The German-Speaking Countries: Ruptures, Schools, and Nontraditions; Reassessing the History of Sociocultural Anthropology in Germany." In *One

Discipline, Four Ways: British, German, French, and American Anthropology, by Fredrik Barth, Robert Parkin, Andre Gingrich, and Sydel Silverman, 59–153. Chicago: University of Chicago Press, 2005.

Gitelman, Lisa. *Scripts, Grooves, and Writing Machines: Representing Technology in the Edison Era*. Stanford, CA: Stanford University Press, 1999.

Gonzalez, Martha E. "Chican@ *Artivistas*: East Los Angeles Trenches, Transborder Tactics." PhD diss., University of Washington, 2013.

González Aktories, Susana, Roberto Cruz Arzabal, and Marisol García Walls, eds. *Vocabulario crítico para los estudios intermediales: Hacia el estudio de las literaturas extendidas*. Mexico City: Universidad Nacional Autónoma de México, 2021.

Good, Edwin M. "Keyboards." In *The Piano: An Encyclopedia*, edited by Robert Palmieri, 203–5. New York: Routledge, 2003.

Graeber, David, and David Wengrow. *The Dawn of Everything: A New History of Humanity*. New York: Farrar, Straus and Giroux, 2021.

Gramsci, Antonio. *La formación de los intelectuales*. Translated by Ángel González Vega. Mexico City: Grijalbo, 1967.

Gray, Robin R. R. "Repatriation and Decolonization: Thoughts on Ownership, Access, and Control." In *The Oxford Handbook of Musical Repatriation*, edited by Frank Gunderson, Robert C. Lancefield, and Bret Woods, 723–38. New York: Oxford University Press, 2019.

Hába, Alois. *Mein Weg zur Viertel- und Sechsteltonmusik*. Düsseldorf: Gesellschaft zu Förderung der systematischen Musikwissenschaft, 1971.

Hamilton, Tom, and Nicholas Hammond. "Introduction: Voulez Ouyr?" *Early Modern French Studies* 41, no. 1 (2019): 2–6.

Haraway, Donna J. "A Cyborg Manifesto: Science, Technology, and Socialist-Feminism in the Late Twentieth Century." In *Simians, Cyborgs, and Women: The Reinvention of Nature*, 149–81. New York: Routledge, 1991.

Hardt, Michael, and Antonio Negri. *Empire*. Cambridge, MA: Harvard University Press, 2000.

Hardt, Michael, and Antonio Negri. *Multitude: War and Democracy in the Age of Empire*. New York: Penguin, 2004.

Hayes, Joy E. "Radio as Medium." In *International Encyclopedia of the Social and Behavioral Sciences*, edited by James D. Wright, 877–82. Amsterdam: Elsevier, 2015.

Hayes, Joy E. *Radio Nation: Communication, Popular Culture, and Nationalism in Mexico, 1920–1950*. Tucson: University of Arizona Press, 2000.

Helmreich, Stefan. "Listening Against Soundscapes." *Anthropology News* 51, no. 9 (2010): 9–10.

Herman, Nick. "Among Others: Sound Art in Mexico." X—TRA 22, no. 1 (2019): 74–97.

Hernández, Carlos A. "El archivo sonoro de artistas LGBTIQ+ que experimentan con el sonido." MM thesis, Universidad Nacional Autónoma de México, 2023.

Hernández Cerón, Roberto. "Re-aprender a escuchar: Las caminatas sonoras de la Fonoteca Nacional de México." BA thesis, Universidad Nacional Autónoma de México, 2013.

Herrera, Eduardo. "Latin America and the Decolonization of Classical Music." In *Open Access Musicology*, edited by Louis Epstein and Daniel Barolsky, 2:1–32. Ann Arbor, MI: Lever, 2022.

Herrera y Ogazón, Alba. *El arte musical en México*. Mexico City: Departamento Editorial de la Dirección General de las Bellas Artes, 1917.

Hirsch, Marianne. "Introduction: Practicing Feminism, Practicing Memory." In *Women Mobilizing Memory*, edited by Ayşe Gül Altınay, María José Contreras, Marianne Hirsch, Jean Howard, Banu Karaca, and Alisa Solomon, 1–23. New York: Columbia University Press, 2019.

Hoffman, Anette. "Introduction: Listening to Sound Archives." *Social Dynamics: A Journal of African Studies* 41, no. 1 (2015): 73–83.

Hoffman, Anette. *Knowing by Ear: Listening to Voice Recordings with African Prisoners of War in German Camps (1915–1918)*. Durham, NC: Duke University Press, 2024.

Hörbst, Viola. *Heilungslandschaften: Umgangsweisen mit Erkrankung und Heilung bei den Cora in Jesús María, Mexiko*. Münster: LIT Verlag, 2008.

Hornbostel, Erich Moritz von. "Phonographische Methoden." In *Methoden zu Untersuchung der Sinnesorgane*, edited by Adolf Basler, Wilhelm Brünings, Emil Budde, Carl von Eicken, Hermann Frenzel, Paul Hoffmann, E. M. von Hornbostel, et al., 419–38. Berlin: Urban und Schwarzenberg, 1930.

Hornbostel, Erich Moritz von. "Zwei Gesänge der Cora-Indianer." In *Die Nayarit-Expedition: Textaufnahmen und Beobachtungen unter mexikanischen Indianern*, vol. 1, *Die Religion der Cora-Indianer in Texten nebst Wörterbuch*, by Konrad Theodor Preuss, 367–74. Leipzig: Teubner, 1912.

Hornbostel, Erich Moritz von, and Curt Sachs. "Systematik der Musikinstrumente: Ein Versuch." *Zeitschrift für Ethnologie* 46, no. 4–5 (1914): 553–90.

Hutchby, Ian. "Technologies, Texts and Affordances." *Sociology* 35, no. 2 (2001): 441–56.

Ingold, Tim. "Against Soundscape." In *Autumn Leaves: Sounds and the Environment in Artistic Practice*, edited by Angus Carlyle, 10–13. Paris: Double Entendre, 2007.

"Inventions New and Interesting." *Scientific American* 130, no. 1 (1924): 37–40.

James, Robin. *The Sonic Episteme: Acoustic Resonance, Neoliberalism, and Biopolitics*. Durham, NC: Duke University Press, 2019.

Jáuregui, Jesús, and Johannes Neurath, eds. *Fiesta, literatura y magia en el Nayarit: Ensayos sobre coras, huicholes y mexicaneros de Konrad Theodor Preuss*. Mexico City: Instituto Nacional Indigenista/Centre d'Études Mexicaines et Centroaméricains, 1998.

Johnes, Martin. "Archives, Truths and the Historian at Work: A Reply to Douglas Booth's 'Refiguring the Archive.'" *Sport in History* 27, no. 1 (2007): 127–35.

Juliastuti, Nuraini. "Indonesian Migrant Workers' Writings as a Performance of Self-Care and Embodied Archives." PARSE 10 (2020). https://doi.org/10.70733/1suc7nwbd4fk.

Kane, Brian. "The Fluctuating Sound Object." In *Sound Objects*, edited by James A. Steintrager and Rey Chow, 53–70. Durham, NC: Duke University Press, 2019.

Katz, Mark. *Capturing Sound: How Technology Has Changed Music*. Berkeley: University of California Press, 2010.

Kennedy, Philip. "Alex Steinweiss and the World's First Record Cover." *Illustration Chronicles*, July 10, 2021. https://illustrationchronicles.com/alex-steinweiss-and-the-world-s-first-record-cover.

Khoury, Nadim. "Postnational Memory: Narrating the Holocaust and the Nakba." *Philosophy and Social Criticism* 46, no. 1 (2020): 91–110.

Kirshenblatt-Gimblett, Barbara. *Destination Culture: Tourism, Museums, and Heritage*. Berkeley: University of California Press, 1998.

Koch, Lars-Christian, Susanne Ziegler, Barbara Göbel, and Richard Haas. "Vorwort." CD booklet in *Konrad Theodor Preuss, Walzenaufnahmen der Cora und Huichol aus Mexiko 1905–1907/Grabaciones en cilindros de cera de los coras y los huicholes de México*, 10–11. Berlin: Staatliche Museen zu Berlin/Preußischer Kulturbesitz/Ibero-Amerikanisches Institut/Instituto Nacional de Lenguas Indígenas/Secretaría de Educación Pública de México, 2013.

Kohl, Karl-Heinz. "Ethnology and the Ambiguity of German Colonialism." In BEROSE—*Encyclopédie internationale des histories de l'anthropologie*, edited by Christine Laurière. Paris: Bérose, 2019. https://www.berose.fr/article1773.html.

Krauß, Andrea. "Constellations: A Brief Introduction." MLN 126, no. 3 (2011): 439–45.

Kristeva, Julia. *Powers of Horror: An Essay on Abjection*. Translated by Leon S. Roudiez. New York: Columbia University Press, 1982.

Kummels, Ingrid, and Gisela Cánepa Koch. "Editor's Introduction to Sound 'Repatriation' in South America: The Politics of Collaborative Archive Reactivations." *Journal of Latin American and Caribbean Anthropology* 28, no. 3 (2023): 185–92.

Kutscher, Gerdt. "Berlín como centro de estudios americanistas: Ensayo bio-bibliográfico." *Indiana* 7 (1976): 1–73.

Lange, Britta. "Archival Silences as Historical Sources: Reconsidering Sound Recordings of Prisoners of War (1915–1918) from the Berlin Lautarchiv." *SoundEffects* 7, no. 3 (2017): 47–60.

Lara Velázquez, Rossana. "¿Con quién (no) (me) escucho? Artes sonoras y territorios desde la ecología política." In *Prácticas artísticas, modernidades electrónicas y neoliberalismo digital*, edited by Jesús Fernando Monreal Ramírez. Mexico City: Universidad Autónoma Metropolitana, forthcoming.

Lara Velázquez, Rossana. "Poner la escucha en (corto)circuito: Arte electrónico y experimentación sonora en México; Dos décadas." PhD diss., Universidad Nacional Autónoma de México, 2016.

Le Guin, Ursula K. *The Dispossessed: A Novel*. New York: Harper, 1974.

Lehmann, F. Rudolf. "K. Th. Preuß." *Zeitschrift für Ethnologie* 71, nos. 1–3 (1939): 145–50.

Lévêque, Aurélien, dir. *El puesto*. Montreuil: Cellulo Prod and Buddy Movies, 2010.

Liebersohn, Harry. *Music and the New Global Culture: From the Great Exhibitions to the Jazz Age*. Chicago: University of Chicago Press, 2019.

Lira Larios, Regina. "Los cilindros de cera grabados por Carl Lumholtz en 1898." *Indiana* 34, no. 2 (2017): 211–32.

López-Gay, Patricia. "Clarice Lispector's Invisible Archive of the Quotidian." *Romance Notes* 58, no. 3 (2018): 497–506.

Ludmer, Josefina. "Literaturas postautónomas 2.0." *Propuesta Educativa* 32 (2009): 41–45.

Luiselli, Valeria. *Lost Children Archive*. New York: Vintage, 2019.

Lumholtz, Carl. *Unknown Mexico: A Record of Five Years' Exploration Among the Tribes of the Western Sierra Madre; in the Tierra Caliente of Tepic and Jalisco; and Among the Tarascos of Michoacan*. 2 vols. London: Macmillan, 1903.

Luna Ruiz, Xilonen. Untitled CD booklet in *Música y cantos para la luz y la oscuridad*. Mexico City: Comisión Nacional para el Desarrollo de los Pueblos Indígenas/American Museum of Natural History, 2005.

Madrid, Alejandro L. *In Search of Julián Carrillo and Sonido 13*. New York: Oxford University Press, 2015.

Madrid, Alejandro L. "Landscapes and Gimmicks from the 'Sounded City': Listening for the Nation at the Sound Archive." *Sound Studies: An Interdisciplinary Journal* 2, no. 2 (2016): 119–36.

Madrid, Alejandro L. "Listening Through the Colonial Noise: Things, Sound Objects, and Legacy at the Berliner Phonogramm-Archiv's Konrad T. Preuss Collection." *Journal of the American Musicological Society* 78, no. 1 (2025): 195–240.

Madrid, Alejandro L. *Los sonidos de la nación moderna: Música, cultura e ideas en el México posrevolucionario, 1920–1930*. Havana: Casa de las Américas, 2008.

Madrid, Alejandro L. *Nor-Tec Rifa! Electronic Dance Music from Tijuana to the World*. New York: Oxford University Press, 2008.

Madrid, Alejandro L. "Rastreando las huellas de la escucha performativa: La escritura como constelación archivística." *Anuario Musical* 76 (2021): 11–30.

Madrid, Alejandro L. "Retos multilineales y método prolépsico en el estudio posnacional del nacionalismo musical." In *Discursos y prácticas musicales nacionalistas (1900–1970)*, edited by Pilar Ramos López, 161–72. Logroño: Universidad de la Rioja, 2012.

Madrid, Alejandro L. "Rigo Tovar, Cumbia, and the Transnational *Grupero* Boom." In *Cumbia! Scenes of a Migrant Latin American Music Genre*, edited by Héctor Fernández L'Hoeste and Pablo Vila, 105–18. Durham, NC: Duke University Press, 2013.

Madrid, Alejandro L. "Sonares dialécticos y política en el estudio posnacional de la música." *Revista Argentina de Musicología* 11 (2010): 17–32.

Madrid, Alejandro L. "Transnational Identity, the Singing of Spirituals, and the Performance of Blackness among Mascogos." In *Transnational Encounters: Music and Performance at the U.S.-Mexico Border*, edited by Alejandro L. Madrid, 171–90. New York: Oxford University Press, 2011.

Madrid, Alejandro L. "Why Music and Performance Studies? Why Now?: An Introduction to the Special Issue." *Trans: Revista Transcultural de Música* 13 (2009). https://www.sibetrans.com/trans/articulo/1/why-music-and-performance-studies-why-now-an-introduction-to-the-special-issue.

Magnusson, Thor. *Sonic Writing: Technologies of Material, Symbolic, and Signal Inscriptions*. London: Bloomsbury, 2019.

Maier, Carla J., and Holger Schulze. "The Tacit Grooves of Sound Art: Aesthetic Artefacts as Analogue Archives." *SoundEffects* 7, no. 3 (2017): 21–35.

Malmström, Dan. *Introducción a la música mexicana del siglo XX*. Mexico City: Fondo de Cultura Económica, 1974.

Marshall, Daniel, and Zeb Tortorici. "Introduction: (Re)Turning to the Queer Archives." In *Turning Archival: The Life of the Historical in Queer Studies*, edited by Daniel Marshall and Zeb Tortorici, 1–31. Durham, NC: Duke University Press, 2022.

Marshall, Daniel, and Zeb Tortorici, eds. *Turning Archival: The Life of the Historical in Queer Studies*. Durham, NC: Duke University Press, 2022.

Mateos-Vega, Mónica. "Fonoteca Nacional, en 'estado de coma' tras recorte y despidos." *La Jornada*, December 31, 2020. https://www.jornada.com.mx/notas/2020/12/31/cultura/fonoteca-nacional-en-estado-de-coma-tras-recorte-y-despidos.

Mather, Bruce. "Music in Thirds and Sixteenths of Tones." CD booklet in *Musiques en tiers et en seizièmes de ton*. Montreal: Societé Nouvelle D'Enregistrement, 2009.

Mayer-Serra, Otto. *Panorama de la música mexicana: Desde la independencia hasta la actualidad*. Mexico City: El Colegio de México, 1941.

Mbembe, Achille. "The Power of the Archive and Its Limits." In *Refiguring the Archive*, edited by Carolyn Hamilton, Verne Harris, Jane Taylor, Michele Pickover, Graeme Reid, and Razia Saleh, 19–26. Dordrecht: Kluwer Academic, 2002.

McCartney, Andra. "Soundwalking: Creating Moving Environmental Sound Narratives." In *The Oxford Handbook of Mobile Music Studies*, edited by Sumanth Gopinath and Jason Stanyek, 2:212–37. New York: Oxford University Press, 2014.

McEnaney, Tom. "'Rigoberta's Listener': The Significance of Sound in *Testimonio*." *PMLA* 135, no. 2 (2020): 393–400.

McEnaney, Tom. "The Sonic Turn." *Diacritics* 47, no. 4 (2019): 80–109.

McIntyre, Lee. *Post-Truth*. Cambridge, MA: MIT Press, 2018.

McKinney, Cait. *Information Activism: A Queer History of Lesbian Media Technologies*. Durham, NC: Duke University Press, 2020.

McRae, Chris. *Performative Listening: Hearing Others in Qualitative Research*. New York: Peter Lang, 2015.

Medina, Ana Cecilia. "Las voces del Centro Histórico suenan en la Fonoteca Nacional." *El Economista*, June 14, 2016. https://www.eleconomista.com.mx/arteseideas/Las-voces-del-Centro-Historico-suenan-en-la-Fonoteca-Nacional-20160613-0057.html.

Memoria del Primer Congreso de la Sociedad Mexicana de Musicología. Ciudad Victoria: Gobierno Constitucional del Estado de Tamaulipas, 1985.

Mena, Juan Carlos. *Sensacional de diseño mexicano*. With Óscar Reyes. Mexico City: Trilce, 2001.

Mena, María Cristina. "Julián Carrillo: The Herald of a Musical Monroe Doctrine." *Century Magazine* 89 (1915): 753–59.

Mendívil, Julio. *Cuentos fabulosos: La invención de la música incaica y el nacimiento de la música andina como objeto de estudio etnomusicológico*. Lima: IFEA IDE-PUCP, 2018.

Meza Valdez, Aurelio, and Susana González Aktories. "Logros y retos del Repositorio Digital en Audio del Proyecto Poética Sonora MX a cuatro años de su creación." In *Creadores de memoria: Los archivos sonoros y audiovisuales en México*, edited by Perla Olivia Rodríguez Reséndiz, 85–99. Mexico City: Universidad Nacional Autónoma de México, 2021.

Miller, Marilyn Grace. *Rise and Fall of the Cosmic Race: The Cult of Mestizaje in Latin America*. Austin: University of Texas Press, 2004.

Minks, Amanda. *Indigenous Audibilities: Music, Heritage, and Collections in the Americas*. New York: Oxford University Press, 2023.

Miranda, Ricardo. "'A tocar, señoritas.'" In *Ecos, alientos y sonidos: Ensayos sobre música Mexicana*, 91–136. Mexico City: Fondo de Cultura Económica, 2001.

Miranda, Ricardo, and Aurelio Tello, eds. *La música en los siglos XIX y XX*. Mexico City: Consejo Nacional para la Cultura y las Artes, 2013.

Mistral, Gabriela. *Obra reunida*. Vol. 1, *Poesía*. Santiago de Chile: Ediciones Biblioteca Nacional, 2019.

Molinari Junior, Clovis. "La experiencia del Archivo Nacional de Brasil en relación con el tratamiento técnico y la atención a los encuestadores de documentos sonoros." In *Memorias del primer seminario internacional los archivos sonoros y visuales en América Latina*, edited by Perla Olivia Rodríguez Reséndiz, 39–47. Mexico City: Radio Educación, 2002.

Monsiváis, Carlos. *Los rituales del caos*. Mexico City: Era, 1995.

Montelongo, José. *No soy tan zen*. Mexico City: Almadía, 2022.

Montezemolo, Fiamma. "Cómo dejó de ser Tijuana laboratorio de la posmodernidad: Diálogo con Néstor García Canclini." *Alteridades* 19, no. 38 (2009): 143–54.

Mora Flores, Ana Alfonsina. "Colectivas feministas en México y nuestra América: Hacia otros mundos sono-sororos posibles." *Cuadernos de Música UNAM* 2 (2022): 54–79.

Mora Flores, Ana Alfonsina. "Híbridas y Quimeras: Ruido y sororidad colectiva en la Ciudad de México." *Escena: Revista de las Artes* 81, no. 1 (2021): 321–43.

Moreno, Jairo. "Imperial Aurality: Jazz, the Archive, and U.S. Empire." In *Audible Empire: Music, Global Politics, Critique*, edited by Ronald Radano and Tejumola Olaniyan, 135–60. Durham, NC: Duke University Press, 2016.

Moreno Rivas, Yolanda. *La composición en México en el siglo XX*. Mexico City: Consejo Nacional para la Cultura y las Artes, 1994.

Moreno Rivas, Yolanda. *Rostros del nacionalismo en la música Mexicana: Un ensayo de interpretación*. Mexico City: Fondo de Cultura Económica, 1989.

Moreno Villarreal, Jaime. "El piano que nadie toca." *Vuelta* 220 (1995): 54–55.

Morones Hernández, Mario. "Análisis de los 19 pianos metamorfoseadores de Julián Carrillo." Unpublished report commissioned by Centro Julián Carrillo, San Luis Potosí, Mexico, 2015.

Moscona, Myriam. *León de Lidia*. Mexico City: Tusquets Editores, 2022.

Moseley, Roger. *Keys to Play: Music as a Ludic Medium from Apollo to Nintendo*. Oakland: University of California Press, 2016.

Nava Loya, Armando. *¿Qué ha pasado con el "Sonido 13" en 100 años?* Mexico City: Editorial Herbasa, 1995.

Ndikung, Bonaventure Soh Bejeng. *Those Who Are Dead Are Not Ever Gone: On the Maintenance of Supremacy, the Ethnological Museum and the Intricacies of the Humboldt Forum*. Berlin: Archive Books, 2019.

Neurath, Johannes. "Tukipa Ceremonial Centers in the Community of Tuapurie (Santa Catarina Cuexcomatitlán): Cargo Systems, Landscape, and Cosmovision." *Journal of the Southwest* 42, no. 1 (2000): 81–110.

Nolan, Jonathan. "Memento Mori." *Esquire* 135, no. 3 (2001): 186–91.
Novak, David, and Matt Sakakeeny. Introduction to *Keywords in Sound*, edited by David Novak and Matt Sakakeeny, 1–11. Durham, NC: Duke University Press, 2015.
Noy, Irene. *Emergency Noises: Sound Art and Gender*. Bern: Peter Lang, 2017.
O'Callaghan, Casey. *Sounds*. New York: Oxford University Press, 2007.
Ocelótl, Emilio, Luis N. del Angel, and Marianne Teixido. "Saborítmico: A Report from the Dance Floor in Mexico." *Dancecult: Journal of Electronic Dance Music Culture* 10, no. 1 (2018). https://dj.dancecult.net/index.php/dancecult/article/view/1066/962.
Ochoa Gautier, Ana María. *Aurality: Listening and Knowledge in Nineteenth-Century Colombia*. Durham, NC: Duke University Press, 2014.
Ochoa Gautier, Ana María. "El reordenamiento de los sentidos y el archivo sonoro." *Artefilosofia* 6, no. 11 (2011): 82–95.
Ochoa Gautier, Ana María. "Sonic Transculturation, Epistemologies of Purification and the Aural Public Sphere in Latin America." *Social Identities* 12, no. 6 (2006): 803–25.
O'Hagan, Sean. "Enrique Metinides: Photographing the Dead for Mexico's 'Bloody News.'" *Guardian*, November 21, 2012. https://www.theguardian.com/artanddesign/2012/nov/21/enrique-metinides-photography-dead-mexico.
Olalquiaga, Celeste. *Megalopolis: Contemporary Cultural Sensibilities*. Minneapolis: University of Minnesota Press, 1992.
Orta Velázquez, Guillermo. *Breve historia de la música en México*. Mexico City: Joaquín Porrúa, 1970.
Ospina Romero, Sergio. "Ghosts in the Machine and Other Tales Around a 'Marvelous Invention': Player Pianos in Latin America in the Early Twentieth Century." *Journal of the American Musicological Society* 72, no. 1 (2019): 1–42.
Pareyón, Gabriel. "Castañeda (Soriano), Daniel." In *Diccionario enciclopédico de la música en México*, 1:198. Guadalajara: Universidad Panamericana, 2007.
Pareyón, Gabriel. "Mendoza (Gutiérrez), Vicente T(eódulo)." In *Diccionario enciclopédico de la música en México*, 2:656. Guadalajara: Universidad Panamericana, 2007.
Pareyón, Gabriel. "Stanford, E. Thomas." In *Diccionario enciclopédico de la música en México*, 2:988–89. Guadalajara: Universidad Panamericana, 2007.
Parker, Robert. *Trece panoramas en torno a Carlos Chávez*. Mexico City: Consejo Nacional para la Cultura y las Artes, 2009.
Parker, Robert L. "Carlos Chávez and the Ballet: A Study in Persistence." *Dance Chronicle* 8, nos. 3–4 (1985): 179–210.
Parker, Robert L. *Carlos Chávez: Mexico's Modern-Day Orpheus*. Boston: Twayne, 1983.
Parzinger, Hermann, Horst Bredekamp, and Neil MacGregor. "Im Zweifel für das Kreuz." *Humboldt Forum Magazin*, May 29, 2017. https://www.humboldtforum.org/de/magazin/artikel/im-zweifel-fuer-das-kreuz/.
Paz, Octavio. *Children of the Mire: Modern Poetry from Romanticism to the Avant-Garde*. Cambridge, MA: Harvard University Press, 1991.
Penny, H. Glenn. *In Humboldt's Shadow: A Tragic History of German Ethnology*. Princeton, NJ: Princeton University Press, 2021.
Perales, Jon. "Critic's Choice: New CD's." *New York Times*, July 25, 2005. https://www.nytimes.com/2005/07/25/arts/music/new-cds.html.

Perus, Françoise. "¿Qué nos dice hoy la ciudad letrada de Ángel Rama?" *Signos Literarios* 1 (2005): 55–66.

Pieken, Gorch, Lars-Christian Koch, and Paul Spies. "Hinter feudalen Fassaden." *Humboldt Forum Magazin*, May 25, 2020. https://www.humboldtforum.org/de/magazin/artikel/hinter-feudalen-fassaden/.

Piekut, Benjamin. "Actor-Networks in Music History: Clarifications and Critiques." *Twentieth-Century Music* 11, no. 2 (2014): 191–215.

Piekut, Benjamin, and Jason Stanyek. "Deadness: Technologies of the Intermundane." *TDR: The Drama Review* 54, no. 1 (2010): 14–38.

Pinch, Trevor, and Karin Bijsterveld. "New Keys to the World of Sound." In *The Oxford Handbook of Sound Studies*, edited by Trevor Pinch and Karin Bijsterveld, 3–35. New York: Oxford University Press, 2012.

"Plan Nacional de Desarrollo 2001–2006." *Diario Oficial de la Federación*, May 30, 2001. https://dof.gob.mx/nota_detalle.php?codigo=766335&fecha=30/05/2001#gsc.tab=0.

Polgovsky Ezcurra, Mara. "Introduction: Agoraphilia; Notes on the Possibility of the Public." In *The New Public Art: Collectivity and Activism in Mexico Since the 1980s*, edited by Mara Pogolvsky Ezcurra, 1–30. Austin: University of Texas Press, 2023.

Preuss, Konrad Theodor. "Der Einfluss der Natur auf die Religion in Mexiko und den Vereinigten Staaten." *Zeitschrift der Gesellschaft für Erdkunde zu Berlin*, nos. 5–6 (1905): 361–80, 433–60.

Preuss, Konrad Theodor. "Der Ursprung der Religion und Kunst." *Globus: Illustrierte Zeitschrift für Länder- und Völkerkunde* 87, no. 22 (1905): 380–84.

Preuss, Konrad Theodor. "Die Begräbnisarten der Amerikaner und Nordostasiaten." PhD diss., Albertus-Universität zu Königsberg, 1894.

Preuss, Konrad Theodor. *Die Nayarit-Expedition: Textaufnahmen und Beobachtungen unter mexikanischen Indianern*. Vol. 1, *Die Religion der Cora-Indianer in Texten nebst Wörterbuch*. Leipzig: Teubner, 1912.

Preuss, Konrad Theodor. "Dos cantos de los indios coras." In *Música y danzas del Gran Nayar*, edited by Jesús Jáuregui, 21–28. Mexico City: Instituto Nacional Indigenista/Centre d'Études Mexicaines et Centroaméricains, 1993.

Preuss, Konrad Theodor. "El concepto de la Estrella de la Mañana, según los textos recogidos entre los mexicaneros del estado de Durango." *El México Antiguo* 10 (1955): 375–95.

Preuss, Konrad Theodor. "Grammatik der Cora-Sprache." *International Journal of American Linguistics* 7 (1932): 1–84.

Preuss, Konrad Theodor. "La diosa de la tierra y de la luna de los antiguos mexicanos en el mito actual." *Boletín del Centro de Investigaciones Antropológicas* 10 (1960): 6–10.

Preuss, Konrad Theodor. *La expedición al Nayarit: Registro de textos y observaciones entre los indígenas de México; La religión de los coras a través de sus textos*. 3 vols. Edited by Margarita Valdovinos. 1912. Mexico City: Siglo XXI, 2020.

Preuss, Konrad Theodor. *Nahua-Texte aus San Pedro Jícora in Durango*. 3 vols. Edited by Elsa Ziehm. Berlin: Gebrüder Mann Verlag, 1968–76.

Preuss, Konrad Theodor. "Nueva interpretación de la llamada Piedra del Calendario Mexicano." *Anales del Museo Nacional de Arqueología, Historia y Etnografía* 7, no. 24 (1931): 424–34.

Preuss, Konrad Theodor. "Un viaje a la Sierra Madre Occidental de México." *Boletín de la Sociedad Mexicana de Geografía y Estadística* 3 (1909): 167–87.

Prieto Acevedo, Carlos. "Critical Constellations of the Audio-Machine in Mexico: Ruins and Reconstructions of a Sonic History." In *Critical Constellations of the Audio-Machine in Mexico*, 3–25. Berlin: Kunstraum Kreuzberg/Bethanien, 2017. Exhibition program.

Prieto Acevedo, Carlos. *Variación de voltaje: Conversaciones con artistas sonoros y músicos electrónicos mexicanos*. Mexico City: Universidad del Claustro de Sor Juana and Delatur, 2013.

"Programa Nacional de Turismo 2001–2006." *Diario Oficial de la Federación*, April 22, 2002. https://www.dof.gob.mx/nota_detalle.php?codigo=734655&fecha=22/04/2002#gsc.tab=0.

Radano, Ronald, and Tejumola Olaniyan. "Introduction: Hearing Empire—Imperial Listening." In *Audible Empire: Music, Global Politics, Critique*, edited by Ronald Radano and Tejumola Olaniyan, 1–22. Durham, NC: Duke University Press, 2016.

Rama, Ángel. *La ciudad letrada*. Hanover, NH: Ediciones del Norte, 1984.

Rama, Ángel. *The Lettered City*. Edited and translated by John Charles Chasteen. 1984. Durham, NC: Duke University Press, 1996.

Ramos, Iván A. *Unbelonging: Inauthentic Sounds in Mexican and Latinx Aesthetics*. New York: New York University Press, 2023.

Ramos-Kittrell, Jesús A. *Playing in the Cathedral: Music, Race, and Status in New Spain*. New York: Oxford University Press, 2016.

Rancière, Jacques. *The Emancipated Spectator*. Translated by Gregory Elliott. Brooklyn, NY: Verso, 2011.

Rancière, Jacques. *The Politics of Aesthetics: The Distribution of the Sensible*. Translated by Gabriel Rockhill. New York: Continuum, 2004.

Rasmussen, Anthony. "Resistance Resounds: Hearing Power in Mexico City." PhD diss., University of California, Riverside, 2017.

Rehding, Alexander. "Instruments of Music Theory." *Music Theory Online* 22, no. 4 (2016). https://mtosmt.org/issues/mto.16.22.4/mto.16.22.4.rehding.html.

Rehding, Alexander. "Wax Cylinder Revolutions." *Musical Quarterly* 88, no. 1 (2005): 123–60.

Reyes Gavilán, Aura Lisette. "Lógicas de archivo y circulaciones restringidas: Los materiales de la expedición de Konrad Theodor Preuss a Colombia." In *Antropología y archivos en la era digital: Usos emergentes de lo audiovisual*, edited by Ingrid Kummels and Gisela Canepa Koch, 1:55–75. Lima: Pontificia Universidad Católica del Perú, 2021. EPUB.

Richards, Thomas. *The Imperial Archive: Knowledge and the Fantasy of Empire*. London: Verso, 1993.

Rios, Fernando E. *Panpipes and Ponchos: Musical Folklorization and the Rise of the Andean Conjunto Tradition in La Paz, Bolivia*. New York: Oxford University Press, 2020.

Rissola, Ariadna. "Ruidos, sonidos y políticas urbanas: 'Los gritos de México' de Félix Blume." *Interartive: A Platform for Contemporary Art and Thought* 76 (2015). https://interartive.org/2015/09/ruidos-sonidos-politicas-urbanas-gritos-felix-blume.

Rivera, Alex, dir. *Sleep Dealer*. New York: Likely Story and This Is That Production, 2008.

Rivera Garza, Cristina. *Escribir con el presente: Archivos, fronteras y cuerpos*. Mexico City: El Colegio Nacional, 2023.

Rivera Garza, Cristina. "Escrituras colindantes." *No hay tal lugar: U-Tópicos contemporáneos*, July 10, 2004. https://cristinariveragarza.blogspot.com/2004/07/#108947489616105760.

Rivera Garza, Cristina. *Escrituras geológicas*. Madrid: Iberoamericana, 2022.

Rivera Garza, Cristina. *Los muertos indóciles: Necroescrituras y desapropiación*. Mexico City: Tusquets, 2013.

Rivera Garza, Cristina. *The Restless Dead: Necrowriting and Disappropriation*. Translated by Robin Myers. Nashville, TN: Vanderbilt University Press, 2020.

Roberts, Shawn. "Aztec Musical Styles in Carlos Chávez's *Xochipilli: An Imagined Aztec Music* and Lou Harrison's *The Song of Quetzalcóatl*: A Parallel and Comparative Study." DMA thesis, University of West Virginia, 2010.

Robinson, Dylan. *Hungry Listening: Resonant Theory for Indigenous Sound Studies*. Minneapolis: University of Minnesota Press, 2020.

Robles Godoy, Armando, ed. *Himno al sol: La obra folclórica y musical de Daniel Alomía Robles*. Lima: Consejo Nacional de Ciencia y Tecnología, 1990.

Rodríguez Reséndiz, Perla Olivia. *El archivo sonoro: Fundamentos para la creación de una Fonoteca Nacional*. Mexico City: Escuela Nacional de Biblioteconomía y Archivonomía, 2012.

Romano, Silvia O. "Accesibilidad y posibilidades de uso de materiales audiovisuales de televisión con fines académicos en Argentina." In *Memorias del primer seminario internacional los archivos sonoros y visuales en América Latina*, edited by Perla Olivia Rodríguez Reséndiz, 135–56. Mexico City: Radio Educación, 2002.

Romero, Fernando. *Hyperborder: The Contemporary U.S.-Mexico Border and Its Future*. New York: Princeton Architectural Press, 2008.

Romero, Raúl R. "Panorama de los estudios sobre la música tradicional en el Perú." *Boletín del Instituto Riva-Agüero* 14 (1986): 99–115.

Rosengarten, Ruth. *Between Memory and Document: The Archival Turn in Contemporary Art*. Lisbon: Museu Coleção Berardo, 2012.

Rothery, Gavin, dir. *Archive*. London: Independent Films, 2020.

Rozental, Sandra. "On the Nature of Patrimonio: 'Cultural Property' in Mexican Contexts." In *The Routledge Companion to Cultural Property*, edited by Jane Anderson and Haidy Geismar, 237–57. New York: Routledge, 2017.

Rulfo, Juan. *Pedro Páramo*. 1955. Mexico City: Editorial RM y Fundación Rulfo, 2005.

Rutsch, Mechthild. *Entre el campo y el gabinete: Nacionales y extranjeros en la profesionalización de la antropología mexicana (1877–1920)*. Mexico City: INAH- Universidad Nacional Autónoma de México, 2007.

Saavedra, Leonora. "Carlos Chávez and the Myth of the Aztec Renaissance." In *Carlos Chávez and His World*, edited by Leonora Saavedra, 134–64. Princeton, NJ: Princeton University Press, 2015.

Saldívar, Gabriel. *Historia de la música en México*. Mexico City: Secretaría de Educación Pública, 1934.

Samuels, David W., Louise Meintjes, Ana Maria Ochoa, and Thomas Porcello. "Soundscapes: Toward a Sounded Anthropology." *Annual Review of Anthropology* 39 (2010): 329–45.

Sánchez Cardona, Luz María. "*Vis.Fuerza[in]necesaria_4*. Práctica artística y comunidades inmersas en procesos de búsqueda de víctimas de desaparición forzada en México." In *Formas de resistencia: Siete experiencias de escucha y denuncia en las prácticas artísticas*, edited by Luz María Sánchez Cardona and Ana Paula Sánchez-Cardona, 135–92. Mexico City: Universidad Autónoma Metropolitana, 2020. EPUB.

Sánchez Gaona, Laura. "Legislación mexicana de patrimonio cultural." *Cuadernos Electrónicos* 8 (2012): 57–74.

Sánchez Mejorada, Alicia. "Antonieta Rivas Mercado en su acción cultural: Creación teatral, música y literaria." *Discurso Visual* 5 (2002). http://discursovisual.net/1aepoca/dvweb05/arto4/arto4.html.

Sas, Andrés. "Aperçu sur la musique inca." *Acta Musicologica* 6, no. 1 (1934): 1–8.

Sas, Andrés. "Ensayo sobre la música inca." *Boletín Latino-Americano de Música* 1 (1935): 71–77.

Schafer, R. Murray. *The New Soundscape: A Handbook for the Modern Music Teacher*. Scarborough: Berandol Music, 1969.

Schafer, R. Murray. *The Thinking Ear: Complete Writings on Music Education*. Toronto: Arcana Editions, 1986.

Schafer, R. Murray. *The Tuning of the World: Our Sonic Environment and the Soundscape*. Rochester, VT: Destiny Books, 1977.

Schloss, Joseph G. *Making Beats: The Art of Sample-Based Hip-Hop*. Middletown, CT: Wesleyan University Press, 2004.

Schneider, Luis M. *El estridentismo o una literatura de la estrategia*. Mexico City: Ediciones de Bellas Artes, 1970.

Seeger, Anthony. "Archives, Repatriation, Agency, and Changing Circumstances: Reflections on Shared Soundscapes, Collaborative Activations, and Repatriations in Latin America." *Journal of Latin American and Caribbean Anthropology* 28, no. 3 (2023): 230–38.

Seem, Mark. Introduction to *Anti-Oedipus: Capitalism and Schizophrenia*, by Gilles Deleuze and Félix Guattari, translated by Robert Hurley, Mark Seem, and Helen R. Lane, xv–xxiv. 1972. Minneapolis: University of Minnesota Press, 2003.

Seiffarth, Carsten, and Markus Steffens, eds. *Entre límites: Espacios sonoros en México y Alemania/Zwischen Grenzen: Klangräume in Mexiko und Deutschland*. Berlin: Lotto Stiftung Berlin, 2017.

Sellen, Adam T. *In the Shadow of Charnay: The Federal Inspector for Archaeology in Mexico, Lorenzo Pérez Castro*. Mérida: Universidad Nacional Autónoma de México, 2021.

Simon, Cheryl. "Introduction: Following the Archival Turn." *Visual Resources* 18, no. 2 (2002): 101–7.

Valdovinos, Margarita. "Voces y cantos de la Sierra Madre: Las grabaciones coras y huicholas de Konrad Theodor Preuss." In CD booklet in *Konrad Theodor Preuss, Walzenaufnahmen der Cora und Huichol aus Mexiko 1905–1907/Grabaciones en cilindros de cera de los coras y los huicholes de México*, 77–84. Berlin: Staatliche Museen zu Berlin/Preußischer Kulturbesitz/Ibero-Amerikanisches Institut/Instituto Nacional de Lenguas Indígenas/Secretaría de Educación Pública de México, 2013.

Vasconcelos, José. *La raza cósmica: Misión de la raza iberoamericana*. Mexico City: Espasa Calpe, 1948.

Vazquez, Alexandra T. *Listening in Detail: Performances of Cuban Music*. Durham, NC: Duke University Press, 2013.

Vega, Carlos. "La música de los incas." Unpublished manuscript. Buenos Aires, 1935.

Vega, Carlos. "La música incaica y el doctor Sivirichi." *Nosotros* 23, no. 239 (1929): 72–85.

Vega, Carlos. "La supuesta escala 'mestiza' de Perú y Bolivia." *La Prensa*, November 26, 1933.

Vera, Alejandro. *The Sweet Penance of Music: Musical Life in Colonial Santiago de Chile*. New York: Oxford University Press, 2020.

Villa-Flores, Javier. "Plotting a Fire: The Burning of Mexico's *Cineteca Nacional* and the Idea of a Self-Destructing Archive." In *From the Ashes of History: Loss and Recovery of Archives and Libraries in Modern Latin America*, edited by Carlos Aguirre and Javier Villa-Flores, 197–226. Raleigh, NC: Editorial A Contracorriente, 2015.

Villalobos López, José Antonio. "Shaping of Radio in Mexico in the Last Decade: 2010–2020." *Journal of Economics, Management, and Trade* 28, no. 6 (2022): 14–25.

Villaseñor Ramírez, Hernani. "LiveCodeNet Ensamble: A Network for Improvising Music with Code." In *Musical Instruments in the 21st Century: Identities, Configurations, Practices*, edited by Till Bovermann, Alberto de Campo, Hauke Egermann, Sarah-Indriyati Hardjowirogo, and Stefan Weinzierl, 327–33. Singapore: Springer, 2017.

Waisman, Leonardo J. *Una historia de la música colonial hispanoamericana*. Buenos Aires: Gourmet Musical, 2019.

Walden, Daniel K. S. "Tanaka Shōhei's Keyboards as Instruments of the Global History of Theory." With Tanaka Tasuku. *Acta Musicologica* 95, no. 2 (2023): 151–76.

Walker, Alison. "Archival Resonances: Embodied Libraries and the Corporeal Lives of Sonic Effects." *Sound Studies: An Interdisciplinary Journal* 7, no. 2 (2021): 206–24.

Waller Vigil, Juan Felipe. *Lhorong, 31°N 96°E*. Berlin: Waller Music, 2015.

Weinstock, Herbert. "Foreword by the Translator." In *Toward a New Music: Music and Electricity*, by Carlos Chávez, translated by Herbert Weinstock, 7–10. New York: W. W. Norton, 1937.

Weiss, Sarah. "Listening to the World but Hearing Ourselves: Hybridity and Perceptions of Authenticity in World Music." *Ethnomusicology* 58, no. 3 (2014): 506–25.

Weld, Kirsten. *Paper Cadavers: The Archives of Dictatorship in Guatemala*. Durham, NC: Duke University Press, 2014.

Westerkamp, Hildegard. "Soundwalking." In *Autumn Leaves: Sound and the Environment in Artistic Practice*, edited by Angus Carlyle, 49–54. Paris: Double Entendre, 2007.

Williams, Sean. "Poetry Writing as Transgressive Ethnography." *Ethnomusicology* 66, no. 3 (2022): 361–77.

Wolkowicz, Vera. *Inca Music Reimagined: Indigenist Discourses in Latin American Art Music, 1910–1930*. New York: Oxford University Press, 2022.
Wright, Mark Peter. *Listening After Nature: Field Recording, Ecology, Critical Practice*. New York: Bloomsbury, 2022.
Young, Nigel. *Postnational Memory, Peace and War: Making Pasts Beyond Borders*. New York: Routledge, 2020.
Ziegler, Susanne. "Deutschsprachige Sammlungen im ehemaligen Berliner Phonogrammarchiv (heute Musikethnologische Abteilung des Museums für Völkerkunde Berlin)." *Jahrbuch für Volksliedforschung* 40 (1995): 129–34.
Ziegler, Susanne. *Die Wachszylinder des Berliner Phonogramm-Archivs*. Berlin: Staatliche Museen zu Berlin, 2006.
Ziegler, Susanne. "From Wax Cylinders to Digital Storage: The Berlin Phonogramm Archiv Today." *Resound: A Quarterly of the Archives of Traditional Music* 13, nos. 1–2 (1994): 1–5.
Ziegler, Susanne. "Historical Sound Recordings in the Berlin Phonogramm-Archiv and the Lautarchiv." *Translingual Discourse in Ethnomusicology* 6 (2020): 136–55.
Zimmerman, Andrew. *Anthropology and Antihumanism in Imperial Germany*. Chicago: University of Chicago Press, 2001.
Žižek, Slavoj. *The Sublime Object of Ideology*. New York: Verso, 2008.

Index

Page numbers followed by *f* refer to figures.

abjection, 105, 116
Academy for Mexican Music Research, 41–43
acoustic ecology, 59, 71
Adler, Guido, 41, 289n36
Aedo, Tania, 256, 295n17
aesthetics, 136, 212, 245, 264; musical, 34, 105
affect, 105, 159, 272, 281, 283–84; post-truth and, 279, 284; sonic, 74
affectivity, 192–93, 278–79, 284; archive of, 192, 205, 208; of the archive, 7; aural, 71
agency, 2, 12, 14, 153, 282; affective, 190; archives and, 2–3, 7–8, 27, 122, 143, 224, 264–65, 267; Aural City and, 16–17, 284; Carrillo Pianos and, 210, 214, 223–24; epistemological, 173; human, 280; of instruments, 208, 210, 214, 224; of listening practices, 19; Mexican Rarities and, 173, 189; of misplaced objects, 159; of the observer, 33; of sound objects, 150
Aguirre Fernández, Diego, 182, 183*f*, 185, 188
AI, 85, 98, 243, 251
Alegre González, Lizette, 17, 240
algoraves, 237–38, 242
Alonso Bolaños, Marina, 47, 178
Alonso-Minutti, Ana, 104, 288n27
Altamirano, Santiago, 128–29
alterity, 15, 98
Álvarez, Katya, 237
American Museum of Natural History (AMNH), 123, 126

Ancira, Andrea, 235, 257–59, 261, 266
anthropology, 18, 68, 125, 136, 139, 141, 301n80
antimonumentos, 243, 311n49
Appadurai, Arjun, 24, 159
Arce, Julio, 81–82
archival constellations, 1, 22–23, 28, 32, 53–55, 105, 231, 288n16, 290n69; Aural City and, 13, 84, 230, 234, 267–69; infinite archive and, 230; invisible archive and, 230; as networks, 33, 232, 268; performance complexes and, 33–34, 56, 83, 233; performative listening and, 37; transhistorical, 34–35, 55
archival documents, 22, 164, 179, 189, 274
archival labor, 2–6, 27, 231, 233, 235, 252–54, 264–66, 284; of academic institutions, 50; of the Aural City, 265, 282; authority and, 254, 274, 276; democratization of, 252; García and Hernández's, 264; gatekeeping and, 276; Mexican Rarities and, 171–74, 176, 179; of PoéticaSonoraMX, 261; postnational memories and, 165; Prieto Acevedo's, 24, 114, 116
archival networks, 22, 27, 229–30, 232–33, 267
archival turn, 5–6
Archive (Rothery), 85–87, 114
"El archivo inaudible," 262–65, 278, 313n98
archivos muertos (dead archives), 205–6, 208–9, 225
Armstrong Liberado, 250–52, 260

art, 100, 125, 209, 303n16; activism and, 243; album cover, 30; archives, 111; code and, 243; collectives, 234; critics, 256–57; curators, 59; exhibits, 6, 31, 57; experimental, 59, 173; festivals, 181; history, 18; musical, 91, 304n29; performance, 106, 180; radio, 255–56; scene in Mexico, 105, 166; schools, 40; spaces, 88, 174; verbal, 132–33, 139; works of, 110, 149; world, 89. *See also* sound art

art music: contemporary, 59, 238; Western, 18, 26, 101–2, 211, 223, 235, 237, 296n33

Ateneo Musical Mexicano, 39, 42

aural, the, 15, 39, 59, 210

aurality, 4, 9–10, 12, 15, 56, 59, 143, 249; archive of, 210; *archivo muerto* and, 209; *Disco pirata* (Blume) and, 187; *Islas Resonantes* and, 247; modern, 17; Prieto Acevedo and, 96

authenticity, 36, 74, 164, 179; of the archive, 5, 154; Carrillo Pianos and, 208; cultural, 31; *Disco pirata* (Blume) and, 180; in film, 186; identitarian, 102; inauthenticity, 305n65; indigenous communities and, 77; Mexican Rarities and, 189; national, 177; Sonido 13 (Carrillo) and, 194; sound and, 186, 279; of sound objects, 21; Valdovinos and, 147, 151

authority, 232, 258; archival, 27, 274, 277, 280, 282; archival labor and, 276; of the archive, 154–55, 164, 173, 188, 254; archives and, 7, 171, 189, 255, 273; of cultural elites, 20; infinite archive and, 266

Avándaro Festival, 104, 110

avant-garde, 99–100; music, 165–68; music movements, 235, 296n31; nationalism and, 91

Avar, Peter, 67

Bahamonde, Emilia, 240, 244–46, 250

Baker, Geoffrey, 14

Balboa, Laura, 247

Bancquart, Alain, 223, 308n61

Baqueiro Foster, Gerónimo, 30, 38, 42; *Huapangos*, 30

Barskova, Polina, 26, 87, 114, 193, 209, 226, 264

Bastian, Adolf, 123–24, 136, 178

Baudrillard, Jean, 70, 86

Bell Laboratories, 48–49

belonging, 6, 14, 143, 239, 266, 284; cosmopolitan, 98, 163

Benjamin, Walter, 32, 295n24

Berlin Palace, 117–18, 143, 298n1

Berliner Phonogramm-Archiv, 21, 24, 118, 121*f*, 138*f*

Bernardelli, Félix, 126, 298n20

Bieletto-Bueno, Natalia, 14, 66, 259, 291n10

Billier, Sylvaine, 221, 308n61, 308n63

Bioy Casares, Adolfo, 55

Blume, Félix, 180, 295n17, 304n47, 305n65; *Coro informal*, 179; *Coro polifónico*, 179–80; *Los gritos de México*, 179, 181; *Memoria del hierro*, 108–9, 297n57; *Mientras escucho*, 114, 115*f*; *Polifonía ambulante*, 179, 182, 184–85. *See also Disco pirata*

Boas, Franz, 123, 126–27, 139

body without organs, 107, 116

body politic, 88, 104, 106–7; Mexican, 99, 105–6, 110

Bohlman, Andrea, 108–9, 113

Bolsonaro, Jair, 272–73

borders, 6–7, 12, 60, 247, 266; of the archive, 64

Borges, Jorge Luis, 83, 228, 230, 261, 269; "La biblioteca de Babel," 83, 228, 261, 269

Brazil, 61, 126, 244, 272–73

Bryson, Norman, 255

Bulla, 247–48, 249*f*

Burke, John, 223, 308n63

Buschmann, Federico, 197–98, 199*f*, 306n23. *See also* Carrillo Pianos

Camacho, Lidia, 62–63, 67. *See also* Fonoteca Nacional

Campos, Rubén M., 38

Campos Fonseca, Susan, 77, 240

canon, 114; of great men, 18; Mexican music, 24, 90, 98, 275, 280; Western art music, 235

capital, 143, 303n20; cultural, 2, 11–12, 19, 38, 59, 143, 172, 179; emotional, 27, 284

capitalism, 250–51

Cárdenas, Alexandra, 237–38, 240–41, 243, 267

Carmona, Gloria, 53

Carrillo, Dolores, 197, 203, 206, 209

Carrillo, Julián, 17, 26, 42, 101–2, 193–208, 210–19, 221–22, 225–26; archive, 206, 209, 231–32; *Balbuceos*, 212, 213*f*, 217, 221; cromometrofonía, 304n29; *Estudios*, 212, 214, 215*f*; *Leyes de metamorfosis musicales*, 307n43; Mexican Rarities and, 165, 167, 170, 173–74, 176; microtonal pianos, 196–97, 200, 211–12,

342   Index

214, 225, 306n18; *El Sonido 13* magazine, 306n14. See also Centro Julián Carrillo; equal temperament; microtonality; microtonal scales; Sonido 13

Carrillo Archive, 209, 231–32

Carrillo Pianos, 193–95, 197–200, 204–8, 210–12, 214, 218, 222, 225, 306n24, 307nn26–27; as archives, 223–24, 226; at Expo 58, 26, 200–202; Fuentes and, 219–21, 223, 278, 308n57; Joste and, 221; Lavista and, 206; as open-source archives, 26, 277; Waller and, 216–17, 221, 223, 278

Castañeda, Daniel, 31, 38, 41–42, 290n66. See also Chávez, Carlos; *Instrumental precortesiano*

Castillo, Arturo, 17, 165–69, 171–74, 176, 177f, 303n16. See also Mexican Rarities

Celestino Celestino, Bolívar, 146–47

Centro Cultural Border, 93–94, 240

Centro Experimental Oído Salvaje, 245–46

Centro de Investigación y Estudios Musicales (CIEM), 216, 219

Centro Julián Carrillo, 206, 208, 219, 306n20, 306n23

Centro Multimedia, 236–38, 241

Centro Nacional de las Artes (CENART), 236–37

chants, 120, 126–27, 132–34, 151, 158, 300n52; mitote, 128–30, 133, 146–48, 153; Náayeri (Cora), 24, 128–34, 139–40, 143, 146, 154, 158, 299n38; *pachitas*, 153; reduction to speech, 132, 141; repatriation of, 150, 158; street vendors', 179; texts, 147; Wixárika (Huichol), 24, 126, 130–34, 143, 154

chaos, 93, 105, 110, 114, 116, 195, 225; archive and, 7–8, 24, 83–84, 88, 116, 224; nation-state and, 95

Chapultepec Castle, 206, 208

Chávez, Carlos, 23, 29–30, 38–42, 48–49, 104, 110, 290n62; *Los cuatro soles*, 30; *El fuego nuevo*, 98, 296n27; *Hacía una nueva música/Toward a New Music*, 23, 34–37, 48, 50–55, 277, 288n16, 290n69; *Instrumental precortesiano* (Castañeda and Mendoza) and, 289n44; as intellectual, 288n16; *La paloma azul*, 30; *Sones de mariachi*, 30; *Xochipili-Macuilxochitl* (*Xochipilli*), 30–32, 53, 287n7, 287n9. See also *A Program of Mexican Music*

chilangos, 25, 181, 184, 187

El Chopo, 18, 165–66, 171, 303n14

circulation, 159; of archival knowledge, 121; of archives, 256; of avant-garde and experimental music, 166; collector's logic of, 176; democratization of, 15; informal economy strategies of, 25; of information, 23, 253, 266; of instruments, 207, 216; of knowledge, 4, 6, 11–12, 20, 23, 84, 92, 264, 272, 276–77; of materials, 171–72, 189; of *noriginales*, 179, 277; open-access archives and, 26; in pirate music economies, 182; of Preuss Collection, 122, 149, 156, 158; of sonic products/objects/materials, 73, 159, 170; of sound, 59

civilization, 10, 136, 286n23; Aztec, 123, 126; German identity and, 141; Mexica, 125; modern national, 178; Western, 124

class, 92, 238; oppression, 240; relations, 243; struggles, 11

El Colegio Nacional, 48, 290n69

Colombia, 119, 238, 241–42, 244, 277, 311n62; Manizales, 241, 310n37

colonialism, 117, 124, 158; interior microcolonialism, 77

Columbia Records, 29–32, 37, 48, 55

Cometa 1973/Cromometrofonía No. 1 (Vargas Leal and Espejo), 165, 174, 176, 304n29

communication, 6, 49, 66, 101; interpersonal, 288n24; performative listening and, 314n15

complexity, 3, 49–50, 227–28, 260–61

composition, 216–17, 219, 237–38; practices, 211; programs, 59; sound, 240; students, 41

Concierto para fotógrafos (Cosmos and Márquez), 106–7

Conservatorio Nacional de Música (CNM, National Conservatory of Music), 34, 39–43, 92

consciousness, 9, 85–87

constellation, 33; archive as, 35; Mexican, 95; rhizomatic, 84. See also archival constellations

consumption, 12, 81, 173, 276; cultural, 77–78; goods, 30–31, 51; of local musics, 51; of sonic products, 73

control, 4, 7, 11–12, 37, 150, 232; science as instrument of, 53

Conway, Kellyanne, 271–72

Index 343

Cook, Terry, 1–2
Cora language, 127–29, 139, 145–46
Cortés, Malitzin, 237, 240
cosmopolitanism, 91, 103
Cosmos, Ángel, 106–7, 170
Critical Constellations of the Audio-Machine in Mexico (ccamm), 23, 88–90, 92–99, 101, 105–13, 116, 277, 297n67
ctm Festival, 23, 88, 90, 94
cultural practices, 3, 16, 77, 133, 147, 158, 163
culture, 9, 60, 62, 167, 261; aural, 9; bourgeois, 296n31; democratization of, 10, 74; diy, 266; European colonial, 55; European elite, 225; expressive, 6, 37, 139, 141; free, 250–51; indigenous, 98; instruments and, 193; material, 111, 124–26, 149, 212; memory production of, 1; Mexican, 30, 90, 112, 243, 254; ministries, 242; modern, 280; music, 38; Náayeri, 133–34; national, 41; of open sharing, 238; popular, 181, 185; of sonic awareness, 64; sound, 59, 64, 95; vernacular, 11; visual, 64, 185; war, 270; Wixárika, 125, 133–34, 292n22; written, 11
Cvetkovich, Ann, 53–54, 265
cyber archive, 85–86
cyborgs, 85, 101, 103, 115, 162, 189, 239, 303n2

D'Harcourt, Raoul and Marguerite, 45
data, 45, 154, 168, 265, 272; affective, 264; archives and, 86–87, 192, 209, 264, 279, 284; big, 81–82; empirical, 122; internet, 81; sonic, 279; storage, 53; truthful, 273
databases, 2, 25, 243
Deleuze, Gilles, 24, 86–88, 107, 116, 232
Denning, Michael, 50, 289n53
Department of Fine Arts, 40, 51
Derrida, Jacques, 5, 173, 285n6
design, 111; album cover, 30; of archive(s), 8, 24, 122, 152, 159, 254, 265, 281; of Carrillo Pianos, 195–96, 199, 207, 211, 221, 223–25, 278; of ccamm (Prieto Acevedo), 99, 110; of *Disco pirata* cd jacket, 185; graphic, 237; Humboldt Forum and, 155; of instruments, 192–93; prints, 45; of technology, 235; urban, 11; of *Variación de voltaje* (Prieto Acevedo), 93. *See also* sound design
desire, 158–59, 211, 282; archival, 154; cosmopolitan, 101, 203; objects of, 77, 100, 140, 275

De Vega, Mario, 105, 267, 295n17, 303n16
Díaz, Ascensión, 128–29
Díaz, Porfirio, 301n82; Edison and, 58, 79, 291n4
dictatorship, 12, 272
difference, 102, 122, 127, 163, 189; gender, 236; performative listening and, 288n24
Diguet, Léon, 125–27, 298n21
Disco pirata (Blume), 25–26, 179–80, 182–89, 277, 279–80, 284
Domínguez Ruiz, Ana Lidia, 17, 259–61, 266, 276, 314n15

Ecuador, 242, 244
Edison, Thomas Alva, 132; Díaz and, 58, 79, 291n4
electronic music, 94, 105, 167, 216, 234; Mexican, 18, 88, 90, 92
elite, 66, 193; artistic, 99; audiophiles, 171, 173; cultural, 11, 20; economic, 227; European, 225; experimental music, 77; intellectual, 11, 16–19, 38, 48, 99, 181; Latin American, 15, 38; political, 10, 99; urban, 14; urban intellectual, 4, 10–11, 13, 15
empire, 141; audible, 143; building, 4, 119, 121, 148, 155–56. *See also* German Empire
Enríquez, Leocadio, 128–29
Enríquez, Manuel, 103
entextualization, 36, 45, 50, 254, 277
Epifonías, 250, 252
epistemic placeholders, 3, 6–7, 163
equal temperament (et), 211, 222; microtonality, 102, 222–23
Escuela de Música, Teatro y Danza, 40–41
Escuela Nacional de Antropología e Historia (enah), 144–45, 249, 260
essentialism, 3, 23, 75, 110
Estévez Trujillo, Mayra, 246
Estrada, Julio, 103
estrangement, 8, 87–88, 96, 280; in *Balbuceos* (Carrillo), 214; Carrillo Pianos and, 210, 212, 216–17, 225; *Critical Constellations of the Audio-Machine in Mexico* and, 110, 116; decolonial, 276; epistemological, 224; *lo inaudito* and, 24; instruments and, 195, 210; listening as, 278; montage and, 111; Prieto Acevedo and, 102; Sonido 13 (Carrillo) and, 225

344   Index

Estridentismo, 38, 100, 296n31
ethnography, 26, 181, 277
Ethnologisches Museum Berlin, 24, 117–18, 134, 137, 140, 148, 150, 301n96
ethnology, 124, 127, 136, 145
European Union, 120, 149
experimental music, 17–18, 25, 195, 229, 234–35, 244–45, 247; concerts, 238; elite, 77; hubs, 310n37; Latin American, 248; Mexican Rarities and, 166–68, 173–74; in peripheral countries, 94; RDA and, 256; scene, 240, 245
Expo 58 (Exposition Universelle et Internationale de Bruxelles), 26, 200–203, 205–6, 212, 216
Expo Vinylo Oaxaca (EVO), 168, 174
Ex Teresa Arte Actual, 240, 254, 256–57, 297n67
extractivism, 132, 142–43

falsehood, 27, 271
fantasy, 74, 141; canonic, 88, 90; dystopian futurist, 163; of indigenous culture, 98; modernist, 45; national, 102, 164; nationalist, 163; past and, 119; sonic, 95
fanzines, 92–93, 111
fascism, 280, 296n31
Feminoise Latinoamérica, 240
Fernández, Nona, 106, 296n51
fetishization, 21, 160; of archival objects, 159; of the archive, 5, 16; of musical work, 18; of sound objects, 16, 21, 79
Figueroa, Libertad, 237–39
Fish, Stanley, 274–75
Flammer, Ernst Helmuth, 221, 223, 308n61
Fokker, Adriaan, 202, 204; organ, 217–18
folklore, 59; European, 47; musical, 42; regional, 288n25
Fonoteca Nacional (National Sound Archive), 23, 57–60, 62–65, 81–84, 88, 172, 174, 253, 276, 291n10, 312n87; Baruj "Beno" Lieberman, Enrique Ramírez de Arellano, and Eduardo Llerenas Collection, 64, 292n23; budget cuts, 273; Caminatas Sonoras, 57, 79–80, 294n59; *De Puntitas*, Collection, 64, 292n23; *Disco pirata* (Blume) at, 25, 184; Encuentros de Música y Danza Indígena Collection, 64, 292n23; Estudios Churubusco Collection, 64, 292n23; *El Foro de la Mujer* Collection, 64, 292n23; García and, 250; Henrietta Yurchenco Collection, 64, 291n2, 292n23; patrimony and, 21, 77–78, 84; Paz and, 290n1; PoéticaSonoraMX and, 254; *Polifonía ambulante* (Blume) at, 179, 185; Raúl Hellmer Collection, 64, 98, 292n23; Red de Estudios sobre el Sonido y la Escucha and, 260; sound map project (México Suena Así), 23, 65–66, 73–74, 78–79, 293n29; soundscape project, 23, 57, 66–71, 73–75, 77–79, 182, 293n32; Thomas Stanford Collection, 63–64, 290–91n2, 292n23
Foucault, Michel, 5
Franco, Jean, 10, 12
Fratti, Mabe, 239, 267
Fuentes, Arturo, 219–21, 223, 278, 308n57; *Agua*, 219, 308n54. *See also* Carrillo Pianos
futurism, 295n13; indio-futurism, 96, 98–99, 110; Italian, 296n31 (*see also* Estridentismo)
futurity, 26, 54–55, 193, 223, 226

Gagnon, Vincent-Olivier, 223, 308n63
Galindo, Blas, 30, 104
galvanos, 135–37, 300n61
Garay, Víctor, 17, 165, 168, 171
García, Jorge David, 17, 249–53, 259, 263–65, 276, 278
García, Leslie (Microhm), 240–41
García Leyva, Cinthya, 254, 256
gender, 236, 238–40, 242, 247–48, 263–64; abuse, 283; dissidence, 244; nonnormative, 252; violence, 243
Generx Experimentación Latinoamérica (GEXLAT), 243–45
Gerber Bicecci, Verónica, 105–7, 296n51
German, Daniel, 273, 275
German Empire, 124, 148, 298n14
Germany, 122, 137, 139, 149–50; CCAMM in, 113; colonial legacy of, 118; Germany/Mexico Cultural Year, 94; *Los gritos de México* (Blume) in, 181; Heilbronn, 221; Kaiserreich, 122, 124, 127, 136, 141, 298n14, 301n80; Kulturgutschutzgesetz, 149, 156; patrimony in, 153; pianos built in, 196, 200, 211, 222f, 306n18; unification of, 124
globalization, 15, 75, 181, 189, 292n22

Index 345

Göbel, Barbara, 143, 148–50, 152
Gonneville, Michel, 223, 308n63
González Aktories, Susana, 254–56, 312n83
González Cortés, César, 186
Google, 65, 74, 81
Grimmel, Werner, 221, 308n61
Grupo de los Nueve, 39, 42
Guattari, Félix, 24, 86–88, 107, 116, 232
Gutiérrez Rafael, Antonio, 146–47
Guzik, Ariel, 105, 107

Hába, Alois, 202, 204–5, 306n18
Haraway, Donna, 239
Herder, Johann Gottfried, 124, 298n15
heritage, 20, 22, 63, 117, 118; cultural, 21, 122, 149, 153, 156, 207, 291n10; Filmoteca de la UNAM and, 62; German, 148–49, 152, 157–58; Mexican, 78, 208; Preuss Collection as, 137; sound and, 21, 58, 60; world, 61, 134, 148–49, 156
Herman, Nick, 113, 297n67
Hernández, Carlos, 252–53, 263–65, 278, 312n75
Hernández, Rolando, 176–77
Herrera y Ogazón, Alba, 38, 91
Híbridas y Quimeras, 239–40, 241f, 310n33
history, 5, 14, 50, 118, 124–25, 139, 294n60; archive and, 262; of the archive(s), 1–3; art, 18; colonial, 117; of conceptual and performance art in Mexico, 106; of electronic music and sound art in Mexico, 88, 90; of Europeans in Gran Nayar, 127; German, 149; human, 10, 81; as humanist endeavor, 141; instability of, 113; Latin American, 11–12; material and, 6; memory and, 282; Mexican, 95, 99, 107, 112, 178, 207–8; of Mexican music, 90; of music, 95; outside of, 142; teleological understanding of, 90–91, 295n12; of Western music, 195
Hörbst, Viola, 146, 300n69, 301n90
Hornbostel, Erich von, 44, 135–37, 139–40, 300n61
humanities, 4–6, 9–10, 40, 139, 141, 251, 275
Humboldt Forum, 24, 117–19, 119f–21f, 134, 135f, 148, 155–56, 160, 298n1. *See also* Ethnologisches Museum Berlin
Humboldt, Wilhelm von, 124, 298n15

hungry listening, 32, 45, 47, 54–55, 134, 141, 156, 178
hyperreal, the, 86–87

Ibero-Amerikanisches Institut zu Berlin, 143, 148, 150
identity, 12, 22, 153, 158–59, 217; archive(s) and, 1, 6–7; collective, 283; cultural, 14; Fonoteca Nacional and, 64, 77, 84; German, 122, 141; local, 69, 79, 184; memes and, 3; Mexican, 92, 98, 105; Mexican Rarities and, 167–68; multitude and, 104; national, 75, 91, 93, 98, 103, 163; patrimony and, 47; politics of, 106; sonic, 80, 180; of sound, 44, 49; sound culture and, 95; state institutions and, 60
ideology, 91, 99, 224, 271; extractivist, 51; nationalist, 90; neoliberal, 273
immobility, 223, 281, 284
imperial archive, 25, 121, 280
improvisation, 68, 102, 217, 234–35, 239, 247, 258
lo inaudito, 7–9, 22, 26–27, 83, 96, 233, 267, 278; affective modes of listening and, 283; *Critical Constellations of the Audio-Machine in Mexico* and, 116; estrangement and, 24; Mexican Rarities and, 17, 179f; Preuss archive and, 152, 160
indianismo, 38, 97
indigenismo, 38, 97, 295n13
indigeneity, 98–99
indigenous communities, 24, 98, 140, 156, 178; Mexican, 124; radio initiatives and, 247
indigenous cultures, 98, 177–78; Hispanic, 45
information, 34, 175, 228; access to, 12, 59, 83, 150, 265; accurate, 273; activism, 36; acoustic, 192, 212, 219, 221, 223; affective, 209; archives and, 2–3, 6, 8, 28, 119–20, 122, 152–53, 155, 164, 187, 205, 232–33, 263, 266, 273–76, 278, 280; *archivos muertos* and, 208–9; aural, 266; Carrillo archive and, 231–32; circulation of, 23, 253, 266; digital, 158; digitized, 168; GEXLAT and, 244; imperial formations and, 121; instruments and, 43, 195, 210–11, 223–24, 226 (*see also* Carrillo Pianos); Mexican Rarities and, 170–72; networks, 6; open-source archives and, 194; somatic, 216; sonic, 212; sound

recording and, 131; transmission of, 252, 264–65; transparency of, 150
Institut de Recherche et Coordination Acoustique/Musique (IRCAM), 216, 219
institutional archives, 79, 164, 173, 176, 178, 188–89
institutional Aural City, 20, 60, 78, 84, 90, 92, 105, 233
Instituto Nacional de Antropología e Historia (INAH), 62, 170, 176–78; Fonoteca del, 62, 176–77
Instrumental precortesiano (Castañeda and Mendoza), 23, 34–37, 41–49, 51–55, 91, 277, 288n16, 289nn44–45
instruments, 6, 36, 49–50, 54, 129, 223; absence of, 132; acoustic, 52; acoustic information in, 221; affectivity and, 192, 224; as archive(s), 26, 28, 190, 193–94, 210, 212, 216; Carrillo as inventor of, 196; communal, 238; as cultural heritage, 21; electric, 34, 50–52; indigenous, 34, 41, 98, 118; materiality of, 193, 195, 211, 221; Mexican, 41; microtonal, 174; musical theory and, 195, 305n6; as open-source archives, 194–95, 280; origin of sound and, 44; percussion, 31, 43–44, 289n45; pre-Hispanic, 31–32, 43–45, 47, 287n6; traditional, 41, 210. *See also* Carrillo Pianos; *xaweri*
intellectuals, 42, 54, 63, 100, 102, 247, 260; organic, 19; postrevoluntionary, 38
Interface, 103
internet, 3, 12, 81, 165, 172, 245; archival projects/archives, 25–26, 252, 266; habits, 65; hacking, 251; Mexican Rarities and, 168, 170, 172; networks, 6, 229; radio broadcasting, 247. *See also Disco pirata* (Blume)
invisible archives, 56, 192, 230, 234, 259, 261, 276
Islas Resonantes, 247–48, 260, 268
Iturbide, Graciela, 104–5, 110

James, Robin, 19–20
Janequin, Clément, 179–80
Jáuregui, Jesús, 140, 145
Jiménez Mabarak, Carlos, 98, 104
Johnes, Martin, 4, 6
Joste, Martine, 221–22, 308n61, 308n63
Juliastuti, Nuraini, 36, 288n19

Kane, Brian, 20, 157
Khoury, Nadim, 25, 163, 176
kinetic action, 210, 212, 214, 217, 220–21
Kirshenblatt-Gimblett, Barbara, 208
kitsch, 185, 187–89, 305n59
Königliches Museum für Völkerkunde (KMV), 123–24, 127, 134, 141–42, 178, 298n17
Kulturvölker, 124, 141, 301n80
Kunstraum Kreuzberg/Bethanien, 23, 88, 94, 97f, 108, 109f, 111

labor, 16, 27–28, 56, 143, 250, 267; aesthetic, 247; algoraves and, 238; artistic, 234, 251; aural, 60, 233, 265; of the Aural City, 13, 17–18, 20, 60, 174, 254, 261, 264, 275, 278, 282; collective, 283; cultural, 16, 246; curatorial, 95–96, 108, 110, 114–16, 280 (*see also* Prieto Acevedo, Carlos); division of, 41, 168, 236; editorial, 259; epistemic, 4, 14; Fonoteca Nacional and, 23; force, 12, 303n1; intellectual, 28, 53, 234, 251, 261; of the Lettered City, 13; libidinal, 116; listening as, 14–15; modernist, 55; museographic, 50; Náayeri and Wixáritari, 302n119; performative, 24, 122; performative listening and, 278; political, 53–54, 246; relations, 87; researchers', 232–33; virtual, 161; women's, 242. *See also* archival labor
Laboratorio Arte Alameda, 18, 254, 256–57
Laprida, Alma, 244
Lara Velázquez, Rossana, 18, 256–59, 265–67, 277, 291n7, 312n87
Latin America, 4, 12, 181, 272, 283; archival work in, 8; the aural in, 15; cultural projects in, 20. *See also* Lettered City
Lavista, Mario, 103–5, 110, 206
Le Guin, Ursula K., 227–28, 230–31, 262; *The Dispossessed*, 227–28, 231, 269
letrados, 11, 14
Lettered City, 4, 10–17, 38, 48, 100, 287n51; Aural City and, 19–20, 37, 54, 60, 84, 105, 261, 276; Carrillo and, 208; knowledge circulation in, 84; listening practices and, 34; Mexican, 32, 53, 92, 194; music and, 23; power relations of, 78
Lilly, John C., 63, 292n22
Lispector, Clarice, 230, 234

Index 347

Valdovinos, Margarita, 120, 122–23, 127–28, 139, 160, 280, 299n34; chants and, 131–33, 146–47, 299n38, 300n52; *La expedición al Nayarit* (Preuss) and, 143, 145–48; Hörbst and, 146, 301n90; Preuss Collection and, 24, 122, 146, 152–53, 159, 275, 277; repatriation of Preuss's recordings and, 148–49, 152, 156–57; Rodríguez Reséndiz and, 267; *Walzenaufnahmen der Cora und Huichol aus Mexico 1905–1907/Grabaciones en cilindros de cera de los coras y los huicholes de México* and, 148, 150–52

Vasconcelos, José, 38–39, 97, 100, 104, 295n26

Vazquez, Alexandra, 277, 302n112

Villa-Flores, Javier, 88

Villaseñor, Hernani, 236–37

Villegas, Juan Pablo, 17, 165–69, 171, 173–76, 295n17, 303n16

violence, 82, 121, 131, 162–63, 178, 272; archive and, 5; gender, 243; racial, 99, 243; sounds of, 70, 79

virtual archives, 82–83, 233, 243, 267

Waller, Juan Felipe, 216–21, 223, 278; *Lhorong, 31°N 96°E*, 217–18, 307n52; notation system of, 307n49. *See also* Carrillo Pianos

walls, 7, 161; of archives, 7, 266; of the Aural City, 185; of the Berlin Palace, 118; of Buenos Aires, 262; of institutions, 233, 266, 276; of the lettered city, 10

Walzenaufnahmen der Cora und Huichol aus Mexico 1905–1907/Grabaciones en cilindros de cera de los coras y los huicholes de México, 144, 148, 153

wax cylinders, 4, 118, 121, 151, 302n119; BPA and, 136–37, 148–50, 156, 158; Edison, 58, 291n4; Lumholtz and, 126; Preuss and, 127, 129–32, 134–35, 140, 142–43, 146, 156–58; Preuss's voice and, 154–55. *See also* galvanos; Preuss Collection

Westerkamp, Hildegard, 79

Wiedmann, Albrecht, 151

Wixárika (Huichol), 298n9; chants, 24, 126, 130–34, 143, 154; communities, 24, 123, 130, 142, 148, 151, 153; culture, 125, 133–34, 292n22; language, 130; music, 126, 133; mythology, 138 *neixas*, 130–31, 133; objects, 153; people (Wixáritari), 120, 125–27, 130, 132, 149–50, 155, 158, 299n22, 302n119; territories, 119, 130; texts, 130, 134, 136

World Soundscape Project, 71, 79

Wyschnegradsky, Ivan, 202–4, 211, 308n63

xaweri, 126, 298n21

Young, Nigel, 25, 163, 179